PHYSICS OF GLASSES

PHYSICS OF GLASSES

Structure and Dynamics

Cargèse, Corsica May 1999

EDITORS
Philippe Jund
Rémi Jullien
Université Montpellier 2
France

American Institute of Physics

AIP CONFERENCE
PROCEEDINGS 489

Melville, New York

Editors:

Philippe Jund
Rémi Jullien
Université Montpellier 2
Laboratoire des Verres
Place Eugène Bataillon - CC069
34095 Montpellier, Cedex 5
FRANCE

E-mail: jund@ldv.univ-montp2.fr
         remi@ldv.univ-montp2.fr

Authorization to photocopy items for internal or personal use, beyond the free copying permitted under the 1978 U.S. Copyright Law (see statement below), is granted by the American Institute of Physics for users registered with the Copyright Clearance Center (CCC) Transactional Reporting Service, provided that the base fee of $15.00 per copy is paid directly to CCC, 222 Rosewood Drive, Danvers, MA 01923. For those organizations that have been granted a photocopy license by CCC, a separate system of payment has been arranged. The fee code for users of the Transactional Reporting Service is: 1-56396-903-3/99/$15.00.

© 1999 American Institute of Physics

Individual readers of this volume and nonprofit libraries, acting for them, are permitted to make fair use of the material in it, such as copying an article for use in teaching or research. Permission is granted to quote from this volume in scientific work with the customary acknowledgment of the source. To reprint a figure, table, or other excerpt requires the consent of one of the original authors and notification to AIP. Republication or systematic or multiple reproduction of any material in this volume is permitted only under license from AIP. Address inquiries to Office of Rights and Permissions, Suite 1NO1, 2 Huntington Quadrangle, Melville, N.Y. 11747-4502; phone: 516-576-2268; fax: 516-576-2499; e-mail: rights@aip.org.

L.C. Catalog Card No. 99-067179
ISBN 1-56396-903-3
ISSN 0094-243X
DOE CONF- 990520

Printed in the United States of America

Contents

Preface vii
Photographs. viii

I EXPERIMENTAL BACKGROUND

Neutron and X-ray Scattering from Glasses 3
    U. Buchenau
Relaxations and Vibrations in Glasses: Experiments 24
    J. Pelous, C. Levelut, and F. Terki
Medium-Range Structure in Amorphous and Crystalline GeSe$_2$ 38
    P. H. Gaskell

II MODE COUPLING THEORIES

Mode Coupling Theories 47
    J.-L. Barrat
Aging in Glassy Systems: Traps and Mode-Coupling Theory 63
    J.-P. Bouchaud
Numerical Tests of Mode Coupling Theory 68
    W. Kob, T. Gleim, and K. Binder

III GEOMETRICAL APPROACHES

Phenomenological Analysis of Supercooled Liquids 83
    D. Kivelson and G. Tarjus
Geometric Frustration and the Glass Transition 105
    J.-F. Sadoc
Two-Level Systems/Tunnelling Modes in Glass 119
    N. Rivier
Frustration in Model Glass Systems: Numerical Investigations 129
    R. Jullien, P. Jund, D. Caprion, and J.-F. Sadoc

IV NUMERICAL SIMULATIONS

**Multimillion Atom Molecular Dynamics Simulations of Glasses
and Ceramic Materials** 149
    P. Vashishta, R. K. Kalia, and A. Nakano
Scattering of Plane-Wave Atomic Vibrations in Disordered Structures 171
    S. N. Taraskin and S. R. Elliott
Molecular Dynamics in Amorphous Solids and Liquids. 191
    H. R. Schober

The Liquid-Glass Transition in the $Ni_{0.5}Zr_{0.5}$ System:
Molecular Dynamics Studies... 203
    H. Teichler

V CONTRIBUTED TALKS

Dynamical Transition in Orientationally Disordered Crystals 217
    F. Affouard and M. Descamps
Simulation of a Silica Glass from Combined Classical and *ab initio*
Molecular Dynamics.. 222
    M. Benoit, S. Ispas, P. Jund, and R. Jullien
A Study of Conformational Jumping of Terrylene in *p*-terphenyl
by Molecular Modelling.. 227
    P. Bordat and R. Brown
Molecular Dynamics Simulations of Crack Propagation Mode in Silica 231
    L. van Brutzel and E. Bouchaud
Empirical Potential for Si-B-N Ceramics 237
    M. Gastreich, J. Gale, and C. M. Marian
Schematic Mode-Coupling Approach to Experimental Data................... 243
    V. Krakoviack and C. Alba-Simionesco
Structural and Elastic Disorder in Lennard Jones Glass..................... 248
    T. Kustanovich and Z. Olami
Solvable Models of Glass Transition 253
    M. Micoulaut
Crystallization in Hard Sphere Systems: A Structural Analysis............... 259
    P. Richard, A. Gervois, L. Oger, and J.-P. Troadec
EPR, NMR and Molecular Dynamics in Fluoride Glasses................... 264
    G. Silly, J. Y. Buzaré, B. Bureau, C. Legein, and B. Boulard
Slow Dynamics of Supercooled Colloidal Fluids:
Spatial Heterogeneities and Nonequilibrium Density Fluctuations............ 268
    M. Tokuyama, Y. Enomoto, and I. Oppenheim

Author Index... 275

Preface

The school "Physics of Glasses: Structure and Dynamics" was held during the period May 10-22, 1999 at the Institut d'Etudes Scientifiques de Cargèse (IESC) in Cargèse, Corsica, France.

Approximately 60 participants from 11 countries gathered to discuss the main aspects of the physics of liquids, amorphous solids and glasses. Concerning both structural and dynamical aspects of glasses, many theoretical approaches are either complementary or still in conflict: geometrical frustration approaches, mode coupling theory, extension of the two-level theory (soft modes approach), theoretical models for slow relaxation, etc...The aim of the school was to give to young (as well as not so young) physicists a picture of the state of the art on both theories and simulation results in close connection with experiments. Therefore fifteen specialists in the different fields gave a 4-hour invited lecture summarized here in this volume and grouped in sections with relatively loose boundaries. A specific section dedicated to the contributions of some of the participants has been included to show some of the applications of the methods discussed in the lectures.

At this date, the school was sponsored and supported by the "Formation Permanente du Centre National de la Recherche Scientifique (CNRS)", by the "Ministère de l'Education Nationale, de la Recherche et de la Technologie (MENRT)" and by St Gobain.

We would like to thank the Director of the IESC, Elisabeth Dubois-Violette, as well as her staff for their hospitality, their help and their dedication in the day-to-day organisation of the school. We are grateful also to Gilbert Bounaud and the staff of the Formation Permanente du CNRS in Montpellier for their help in the administrative and financial aspects of the school. Our thanks go also to the persons (M. Gastreich, L. Oger, M. Micoulaut and J.F. Sadoc) who gave us the permission to include some of their photographs and drawings in this volume.

Philippe Jund
Rémi Jullien

I EXPERIMENTAL BACKGROUND

Neutron and X-ray scattering from glasses

U. Buchenau

*Institut für Festkörperforschung, Forschungszentrum Jülich
Postfach 1913, D-52425 Jülich, Federal Republic of Germany.*

I FUNDAMENTALS

There is an extensive quantum mechanical treatment of the scattering from atoms moving in a crystal [1]. This treatment can be partially carried over to glasses [2–5]. To keep the treatment simple, we will formulate the scattering laws in terms of classical physics. The equations for the one-phonon scattering from vibrations can be transformed into their quantum mechanical counterparts replacing the prefactor $k_B T$ by $\hbar\omega/(exp(\hbar\omega/k_B T) - 1)$. Here k_B is the Boltzmann factor, T is the temperature and $\hbar\omega$ is the energy transfer of the scattering process, taken to be positive in energy gain of the scattered radiation.

A Scattering process and cross sections

Consider a scattering process from a single atom interacting with an incoming plane wave. Let the wavelength of the incoming radiation be λ_i. The wavevector $\mathbf{k}_i$ of the incoming radiation has the direction of propagation of the wave and the absolute value

$$k_i = \frac{2\pi}{\lambda_i}. \tag{1}$$

The scattered radiation is characterized by a sperical wave centered at the atom.

The scattering measurement selects a small part of the scattered wave at the solid angle Ω over a small solid angle region $d\Omega$. The amount of scattered radiation is characterized by the differential cross section

$$\frac{d\sigma}{d\Omega} = \frac{\text{stream of particles scattered into } d\Omega \text{ at } \Omega}{\text{stream per unit area of incoming particles x } d\Omega}, \tag{2}$$

which has the dimension of an area. Integrating the differential cross section over the solid angle gives the total cross section of the atom

$$\sigma = \int \frac{d\sigma}{d\Omega} d\Omega. \tag{3}$$

If the scattering is from a moving atom, the wavelength λ_f of the scattered radiation will be in general different from the one of the incoming radiation. Thus the wavevector $\mathbf{k}_f$ of the scattered wave will have an absolute value

$$k_f = \frac{2\pi}{\lambda_f} \qquad (4)$$

which can differ from the one of the incoming radiation. The direction of $\mathbf{k}_f$ is defined by the solid angle of the measurement. One defines the momentum vector $\mathbf{Q}$ of the scattering process by

$$\mathbf{Q} = \mathbf{k}_i - \mathbf{k}_f. \qquad (5)$$

The reason for this definition becomes clear when one calculates the phase difference for the scattering from two different atoms.

If the first atom is at the origin and the second at a position vector $\mathbf{r}_j$, the phase difference is

$$\frac{\mathbf{k}_i \mathbf{r}_j}{k_i} \frac{2\pi}{\lambda_i} - \frac{\mathbf{k}_f \mathbf{r}_j}{k_f} \frac{2\pi}{\lambda_f} = (\mathbf{k}_i - \mathbf{k}_f)\mathbf{r}_j = \mathbf{Q}\mathbf{r}_j. \qquad (6)$$

If the wavelength of the radiation is changed, there is an energy transfer $\hbar\omega$ in the scattering process. One calls such a process an *inelastic* scattering process. In the presence of inelastic scattering, the scattering process must be characterized by a double differential cross section $d\sigma/d\Omega d\omega$ in order to specify hoch much is scattered per solid angle and frequency interval.

B Neutron and X-ray scattering

For the case of neutron scattering, one has to distinguish between coherent and incoherent scattering. It depends on the nuclei of the atoms in the sample how much one has of the one or the other. If a nucleus has spin zero and no isotopes, it scatters purely coherently, i.e. the scattering from different nuclei interferes with a phase factor depending only on their relative positions. For a nucleus with nonzero spin, however, the scattering amplitude depends on the relative orientation of nuclear spin and neutron spin. Then one has an average scattering amplitude providing coherent scattering, and in addition a part which is nonzero at each single nucleus, but averages to zero in a sum over all nuclei. Since the scattering is proportional to the square of the scattering amplitude, this fluctuating part contributes to the scattering as well, but without interference between different atoms. This latter part of the scattering is the incoherent scattering which reflects the single particle behaviour of the atoms. Since the scattering amplitudes can sometimes adopt negative values, one has even some nuclei with predominantly incoherent scattering, like hydrogen and vanadium.

For an ensemble of atoms of the same element, one describes the coherent scattering by the average scattering length b and the incoherent scattering by the incoherent scattering cross section σ. b and σ depend on the element, and result from the averaging over spin states and isotopes. If one knows the chemical composition of a sample, and the isotopic distribution for each of the elements in the sample, one can take b and σ from tables.

For X-rays, we consider only the dipole radiation by the forced oscillations of the electrons in the electric field of the radiation (no Compton effect and no electronic excitation). Then the scattered radiation is purely coherent. Taking the electronic density around the atoms as sperical, one has to replace b by the atomic form factor, resulting from the interference of the scattered X-rays from different parts of the electron cloud around the atom. Unlike b in the neutron case, the atomic form factor decreases slowly with increasing scattering angle. At the scattering angle zero, one has still a positive interference from all Z electrons of the atom; typically, at the first sharp diffraction peak of a glass, the form factor has diminished to about 80 % of its zero angle value. Otherwise, one can carry over the formulae for the coherent scattering of neutrons to the X-ray case, just replacing b by the atomic form factors.

C Coherent and incoherent scattering

Let us consider a glass with N atoms. The coherent part of the double differential cross section is given by [1]

$$\frac{d^2\sigma_{coh}}{d\omega d\Omega} = \frac{k_f}{k_i} \frac{1}{2\pi} \int_{-\infty}^{\infty} dt e^{-i\omega t} \sum_{j,j'=1}^{N} b_j b_{j'} <e^{-i\mathbf{Q}\mathbf{r}_j(0)} e^{i\mathbf{Q}\mathbf{r}_{j'}(t)}>. \quad (7)$$

Here Ω is the solid angle of the scattered beam. k_f and k_i are the absolute values of the wave vectors of the scattered radiation $\mathbf{k}_f$ and the incoming radiation $\mathbf{k}_i$. The scattering vector $\mathbf{Q}$ is given by the difference of the wavevectors, $\mathbf{Q} = \mathbf{k}_i - \mathbf{k}_f$. $\mathbf{r}_j$ is the position vector of atom j and b_j is the average scattering length for this atom as defined above. In our classical treatment, the brackets indicate an average over the beginning time $t = 0$, taken over infinite time.

Similarly, the double differential cross section of the incoherent scattering is given by

$$\frac{d^2\sigma_{inc}}{d\omega d\Omega} = \frac{k_f}{k_i} \frac{1}{2\pi} \int_{-\infty}^{\infty} dt e^{-i\omega t} \sum_{j=1}^{N} \frac{\sigma_j}{4\pi} <e^{-i\mathbf{Q}\mathbf{r}_j(0)} e^{i\mathbf{Q}\mathbf{r}_j(t)}>, \quad (8)$$

where now σ_j is the incoherent scattering cross section for the atom j. The incoherent scattering is usually easier to understand and to interpret than the coherent one.

¿From a theoretical point of view, it is preferable to express the scattering by a normalized $S(\mathbf{Q},\omega)$ that can be integrated over all frequencies. For that integration, one needs to remove the disturbing factor k_f/k_i which still depends on k_i for a given frequency. One defines

$$S_{coh}(\mathbf{Q},\omega) = \frac{k_i}{k_f} \frac{1}{N\overline{b^2}} \frac{d^2\sigma_{coh}}{d\omega d\Omega}, \quad (9)$$

where $\overline{b^2}$ is the average mean square scattering length of all atoms in the sample.

For the incoherent case one defines

$$S_{inc}(\mathbf{Q},\omega) = \frac{k_i}{k_f} \frac{4\pi}{N\overline{\sigma}} \frac{d^2\sigma_{inc}}{d\omega d\Omega}. \tag{10}$$

Here $\overline{\sigma}$ is the average of the incoherent cross sections of the N atoms.

If there is no preferred direction in the glass or the liquid, as is very often the case, the scattering functions become isotropic

$$S(\mathbf{Q},\omega) = S(Q,\omega), \tag{11}$$

for both coherent and incoherent scattering.

D Intermediate scattering function and sum rules

Let us treat the intermediate scattering function for the simple case of a monoatomic substance; afterwards, we will switch back to the general case. The fundamental concept, the sum rules and the gaussian approximation can be extended to glasses with more than one element, as we will see in the discussion of the examples. Also, we follow the usual way of not inserting any scattering length or cross section in this function. Thus the following holds for both coherent and incoherent scattering.

One defines the intermediate scattering function

$$F(\mathbf{Q},t) = \frac{1}{N} \sum_{j,k=1}^{N} < e^{-i\mathbf{Q}\mathbf{r}_j(0)} e^{i\mathbf{Q}\mathbf{r}_k(t)} >. \tag{12}$$

The scattering function $S(\mathbf{Q},\omega)$ is given by the Fourier transform of the intermediate scattering function

$$S(\mathbf{Q},\omega) = \frac{1}{2\pi} \int_{-\infty}^{\infty} e^{i\omega t} F(\mathbf{Q},t) dt, \tag{13}$$

with the back transform

$$F(\mathbf{Q},t) = \int_{-\infty}^{\infty} e^{-i\omega t} S(\mathbf{Q},\omega) d\omega. \tag{14}$$

Taking the intermediate scattering function at $t=0$, one obtains the integral $S(\mathbf{Q})$ of $S(\mathbf{Q},\omega)$ over all frequencies

$$S(\mathbf{Q}) = \frac{1}{N} \sum_{j,k=1}^{N} < e^{-i\mathbf{Q}\mathbf{r}_j} e^{i\mathbf{Q}\mathbf{r}_k} >. \tag{15}$$

This is now a much easier time average, where one takes the correlations between different atoms at the *same* time. To see the fundamental importance of this result, let us consider the consequences for the coherent and the incoherent scattering. For the coherent case, one has a snapshot of the instantaneous spatial atomic correlation, averaged over long times, so one gets information about the structure by measuring the integral of the scattering over all possible energy transfers. Naturally, one does not really measure $S(\mathbf{Q})$, but rather the integral over the double differential cross section at a given scattering angle, but it turns out to be possible to correct for that (the Plazek correction). For the incoherent case, the correlation between different atoms is not visible, so $S(\mathbf{Q}) = 1$, a very simple and convenient sum rule.

Differentiating the intermediate scattering function twice with respect to t at $t = 0$ yields

$$\frac{\partial^2}{\partial t^2} F(\mathbf{Q}, t)|_{t=0} = -\int_{-\infty}^{\infty} \omega^2 S(\mathbf{Q}, \omega) d\omega. \tag{16}$$

To calculate the second derivative of the intermediate scattering function, we recall the definition

$$<e^{-i\mathbf{Q}\mathbf{r}_j(0)} e^{i\mathbf{Q}\mathbf{r}_k(t)}> = \frac{1}{T_m} \int_0^{T_m} e^{-i\mathbf{Q}\mathbf{r}_j(t')} e^{i\mathbf{Q}\mathbf{r}_k(t'+t)} dt', \tag{17}$$

where T_m, the measuring time, is long compared to the microscopic times. Differentiating this integral with respect to t is equivalent to differentiate the second factor with respect to t', so one has an integral of the form uv'', which can be converted into an integral of the form $u'v'$

$$\frac{\partial^2}{\partial t^2} <e^{-i\mathbf{Q}\mathbf{r}_j(0)} e^{i\mathbf{Q}\mathbf{r}_k(t)}> = -\frac{1}{T_m} \int_0^{T_m} \frac{\partial}{\partial t'} e^{-i\mathbf{Q}\mathbf{r}_j(t')} \frac{\partial}{\partial t'} e^{i\mathbf{Q}\mathbf{r}_k(t'+t)} dt'. \tag{18}$$

The second moment sum rule follows in the form

$$\int_{-\infty}^{\infty} \omega^2 S(\mathbf{Q}, \omega) d\omega = \frac{1}{N} \sum_{j,k=1}^{N} <(\mathbf{Q}\dot{\mathbf{r}}_j)(\mathbf{Q}\dot{\mathbf{r}}_k) e^{-i\mathbf{Q}\mathbf{r}_j} e^{i\mathbf{Q}\mathbf{r}_k}>, \tag{19}$$

where the time average is only over quantities taken at the same time.

The textbook argument [9] continues: "Positions and velocities at a given instant are statistically independent". This is taken to imply a vanishing of the terms with $j \neq k$. Since the average kinetic energy per particle is $3k_B T/2$, one obtains the second moment sum rule [10]

$$\int_{-\infty}^{\infty} \omega^2 S(\mathbf{Q}, \omega) d\omega = \frac{1}{N} \sum_{j=1}^{N} <(\mathbf{Q}\dot{\mathbf{r}}_j)^2> = \frac{k_B T Q^2}{M}, \tag{20}$$

where M is the atomic mass.

The second moment sum rule holds again for both coherent and incoherent scattering.

E Bloch identity and gaussian approximation

Let us denote by $r_{jQ}(t)$ the component of the atomic position vector parallel to the momentum vector $\mathbf{Q}$. If $r_{jQ}(t) - r_{jQ}(0)$ has a gaussian distribution for different choices of the zero point in time, one can make use of the Bloch identity

$$< e^{-i\mathbf{Q}\mathbf{r}_j(0)} e^{i\mathbf{Q}\mathbf{r}_j(t)} > = e^{-(Q^2 <(r_{jQ}(t)-r_{jQ}(0))^2>)/2}. \tag{21}$$

A gaussian distribution of the time-dependent displacement of an atom is not infrequent; according to the central limit theorem one expects a gaussian distribution whenever a quantity results from many statistically independent processes. Examples for a gaussian distribution of the atomic displacement are the classical diffusion or the atomic motion in a crystal, the latter because it results from the motion in a harmonic potential.

One can use the Bloch identity to formulate an approximation to the incoherent scattering law, the gaussian approximation for the intermediate incoherent scattering function

$$F_{inc}(\mathbf{Q}, t) = e^{-Q^2 \gamma(t)}, \tag{22}$$

where $\gamma(t)$ is an average time-dependent mean square displacement for all the atoms in the sample. The approximation is often a useful simplification. It has been used, for instance, to study the gradual crossover from vibrations to the viscous flow in undercooled liquids.

F Nongaussianity and heterogeneity

One can imagine several reasons for a failure of the gaussian approximation [6]. An obvious one is the deviation of the distribution of the time-dependent atomic displacement from a gaussian. To take a simple example: Imagine an atom jumping between two potential minima. Then the displacement distribution function has two maxima, obviously different from a gaussian.

But even if all atoms have a gaussian distribution of their displacements in all three spatial directions, as for instance in a crystal, the gaussian approximation will very often fail, because it requires the same width of the distribution for each atom in each direction. Let us discuss this effect for an ensemble of N atoms with different time-dependent mean square displacements $\gamma_j(t)$ for the direction of the scattering vector $\mathbf{Q}$. Expanding the incoherent scattering function, one gets

$$\frac{1}{N} \sum_{j=1}^{N} e^{-\gamma_j Q^2} = 1 - \overline{\gamma} Q^2 + \frac{1}{2}\overline{\gamma^2} Q^4 + O(Q^6) = e^{-\overline{\gamma} Q^2}\left[1 + \frac{1}{2}A_0 Q^4 + O(Q^6)\right], \tag{23}$$

where the dimensionless nongaussianity coefficient A_0 describes the deviation from the gaussian approximation to first order, and is given by

$$A_0 = (\overline{\gamma^2} - \overline{\gamma}^2)/\overline{\gamma}^2. \qquad (24)$$

We have omitted the time here, but it is clear that $A_0 = A_0(t)$.

It is interesting to consider the connection between nongaussianity and dynamical heterogeneity. Obviously, one can have nongaussianity already in a homogeneous medium, either because the motional distribution of the atoms is not gaussian or because of anisotropies on the microscopic level. On the other hand, if there is dynamical heterogeneity, one has perforce nongaussianity. So heterogeneity implies nongaussianity, but the reverse is not true.

A second important point is the comparison to simulation data on model glasses and liquids. There, one defines [7] a second order nongaussianity coefficient $\alpha_2(t)$

$$\alpha_2(t) = \frac{3\overline{r^4(t)}}{5\overline{r^2(t)}^2} - 1. \qquad (25)$$

This coefficient is connected to the $A_0(t)$ defined from the scattering, but it is not exactly the same. The difference is that the measurement looks only in one direction, while the simulation definition does not care about the direction of the displacement. Thus an ensemble of anisotropic oscillators, all with the same anisotropy, but randomly oriented in space, would have a zero $\alpha_2(t)$ and a nonzero $A_0(t)$. Apart from this detail, the two definitions agree with each other. If we just consider the displacements in one direction, and again assume N atoms with a different gaussian distribution of $r_j(t)$, one has

$$< r_j^4(t) > = \frac{5}{3} < r_j^2(t) >^2 = \frac{5}{3}\gamma_j^2. \qquad (26)$$

Averaging over all atoms, one sees the agreement of the two definitions.

II PHONON EXPANSION

A Equilibrium positions and displacements

Let us assume time-independent equilibrium positions of the N atoms of the glass (in other words, the atomic diffusion should be slow enough to be negligible). Then the time-dependent atomic position $\mathbf{r}_j(t)$ of atom j is the sum of equilibrium position $\mathbf{R}_j$ and displacement $\mathbf{u}_j(t)$

$$\mathbf{r}_j(t) = \mathbf{R}_j + \mathbf{u}_j(t). \qquad (27)$$

The phonon expansion is not only applicable to glasses, but also to undercooled liquids, where one has a difference of many orders of magnitude between the time scale of the α-relaxation, connected with the viscosity, and vibrational time scales. One should keep in mind, however, that the elastic scattering of the glass becomes a quasielastic scattering with a small, but nonzero quasielastic width in the undercooled liquid.

One makes use of the decomposition of the atomic coordinate into equilibrium position and displacement to rewrite the correlation function

$$< e^{-i\mathbf{Q}\mathbf{r}_j(0)} e^{i\mathbf{Q}\mathbf{r}_{j'}(t)} > = e^{i\mathbf{Q}(\mathbf{R}_{j'}-\mathbf{R}_j)} < e^{-i\mathbf{Q}\mathbf{u}_j(0)} e^{i\mathbf{Q}\mathbf{u}_{j'}(t)} > . \tag{28}$$

For small displacements, one can expand the exponential function. The first order terms disappear, because their time average is zero. The second order terms give the one-phonon scattering, in many cases a good approximation for the inelastic scattering. The multiphonon scattering from the higher terms can often be considered as a small correction. An elegant way to take them into account is to introduce an effective frequency-dependent Deby-Waller factor giving the correct Q^4-term of the scattering. We will come back to this point later.

In the phonon expansion the correlation function is given by

$$< e^{-i\mathbf{Q}\mathbf{r}_j(0)} e^{i\mathbf{Q}\mathbf{r}_{j'}(t)} > = e^{i\mathbf{Q}(\mathbf{R}_{j'}-\mathbf{R}_j)} e^{-W_j} e^{-W_{j'}} [1+ < (\mathbf{Q}\mathbf{u}_j(0))(\mathbf{Q}\mathbf{u}_{j'}(t)) > + \mathcal{O}(Q^4)]. \tag{29}$$

The Debye-Waller exponents W_j are given by

$$W_j = \frac{1}{2} < (\mathbf{Q}\mathbf{u}_j)^2 > \tag{30}$$

provided one has a gaussian distribution of the atomic displacements.

B Elastic and integrated scattering

The elastic scattering is calculated from the time-independent first term of the phonon expansion. It is easy to convince oneself that a constant term in the intermediate scattering function provides a δ-function in $S(Q,\omega)$. Consider

$$\int_{-a}^{a} dt e^{i\omega t} = 2a \frac{\sin \omega a}{\omega a} \tag{31}$$

and

$$\int_{-\infty}^{\infty} dx \frac{\sin x}{x} = \pi, \tag{32}$$

then you see that one of the representations of the δ-function is

$$\delta(\omega) = \frac{1}{2\pi} \int_{-\infty}^{\infty} dt e^{i\omega t}. \tag{33}$$

With this relation, one gets

$$S(Q,0) = \delta(\omega)\frac{1}{N\overline{b^2}}\sum_{j,j'=1}^{N} b_j e^{-W_j} b_{j'} e^{-W_{j'}} e^{-i\mathbf{Q}\mathbf{r}_j} e^{i\mathbf{Q}\mathbf{r}_{j'}} = \delta(\omega)\frac{1}{N\overline{b^2}} |\sum_{j=1}^{N} b_j e^{-W_j} e^{i\mathbf{Q}\mathbf{r}_j}|^2 .$$
(34)

The integral over all frequencies corresponds to a snapshot of the atomic positions at time 0

$$S(Q) \equiv \int_{-\infty}^{\infty} d\omega S_{coh}(Q,\omega) = \frac{1}{N\overline{b^2}} <|\sum_{j=1}^{N} b_j e^{i\mathbf{Q}\mathbf{r}_j(0)}|^2>.$$
(35)

$S(Q)$ oscillates around 1 according to the pair correlations between the atomic positions at the time zero. At high Q, the oscillations die out and $S(Q)$ approaches 1.

Analogous to $S(Q)$, one can define the quantity $S_{el}(Q)$ according to

$$S_{el}(Q) \equiv \frac{1}{N\overline{b^2}} |\sum_{j=1}^{N} b_j e^{-W_j} e^{i\mathbf{Q}\mathbf{r}_j}|^2 \qquad S_{coh}(Q,0) = S_{el}(Q)\delta(\omega),$$
(36)

in which the pair correlations of the equilibrium positions play the same role as the instantaneous positions in $S(Q)$. $S_{el}(Q)$ does not extrapolate to 1 at high Q, but because of the additional Debye-Wallerfactors to an average Debye-Wallerfactor e^{-2W}, which goes to zero at large Q. At smaller Q the two functions $S_{el}(Q)$ and $S(Q)$ are very similar, because the instantaneous atomic positions differ only by the small displacement from the equilibrium positions, and the effect of the displacement tends to average out in the time average.

At $Q = 0$ both functions are a δ-function in $\mathbf{Q}$, as can be shown by the following consideration: At small Q one can reckon with the homogenous scattering density $\overline{b}N/V$, where $\overline{b}$ is the average atomic scattering length. Let the sample be a cube with the volume $V = 8a^3$ and let us assume a cartesian coordinate system with the axes parallel to the edges of the cube and its origin in the center of the cube. Then

$$\sum_{j=1}^{N} b_j e^{i\mathbf{Q}\mathbf{r}_j} \approx \int_{-a}^{a}\int_{-a}^{a}\int_{-a}^{a} dxdydz \frac{N\overline{b}}{V} e^{iQ_x x} e^{iQ_y y} e^{iQ_z z}$$

$$= \frac{8N\overline{b}}{V} \frac{\sin Q_x a}{Q_x} \frac{\sin Q_y a}{Q_y} \frac{\sin Q_z a}{Q_z}.$$
(37)

Since

$$\int_{-\infty}^{\infty} \frac{\sin^2 x}{x^2} = \pi$$
(38)

one has the small Q limit

$$S(\mathbf{Q}) \approx S_{el}(\mathbf{Q}) \approx \frac{8\pi^3 N\overline{b}^2}{V\overline{b^2}} \delta(\mathbf{Q}).$$
(39)

For small but finite Q the scattering interference averages to zero in a perfectly homogeneous medium. Thus one registers only the weak thermal density fluctuations, which are proportional to the absolute temperature and to the isothermal compressibility κ_T.

$$\lim_{Q \to 0} S(Q) = \frac{N\overline{b}^2}{V\overline{b^2}} k_B T \kappa_T. \tag{40}$$

For glasses below the glass transition temperature T_g, $S(Q)$ in this region is usually of the order 0.1 or smaller. Then follows a slow rise to the first sharp diffraction peak at about 1.5 bis 2 Å^{-1}, corresponding to atomic distances of 3 to 5 Å. This first sharp diffraction peak of the glass corresponds to the first Bragg reflections of the crystal. Very often, it shows a pronounced maximum in $S(Q)$ with values between 1.5 and 2 and halfwidths (FWHM) of about 0.4 Å^{-1}. In the Q-region of this peak, the Debye-Wallerfactors do not yet differ strongly from 1, so these atom correlations correspond closely to those of the equilibrium positions reflected in the elastic scattering. In other words, the first sharp diffraction peak of a glass is to a large extent elastic scattering. One can use this and the following peaks of $S(Q)$ to analyse the eigenvectors in the inelastic scattering, as we will see below.

The elastic incoherent scattering is given by

$$S_{inc}(Q, 0) = \delta(\omega) \frac{1}{N\overline{\sigma}} \sum_{j=1}^{N} \sigma_j e^{-2W_j} \tag{41}$$

and the integral over all frequencies gives the sum rule

$$\int_{-\infty}^{\infty} d\omega S_{inc}(Q, \omega) = 1. \tag{42}$$

III INELASTIC SCATTERING FROM SOUND WAVES

A The elastic medium

There is one point where the dynamics of glasses is simpler than the dynamics of crystals: the low frequency sound waves. A glass is an isotropic elastic medium, described only by three material parameters, namely the density ρ, the bulk modulus B (the inverse of the compressibility) and the shear modulus G. Instead of B and G, one could also take the two elastic constants c_{11} and c_{44}; they are related by

$$B = c_{11} - \frac{4}{3} c_{44} \qquad G = c_{44}. \tag{43}$$

Similarly, one could take the two sound velocities v_l for the longitudinal sound waves (those vibrating *parallel* to the direction of propagation) and v_t for the transverse sound waves(those vibrating *perpendicular* to the direction of propagation). They are given by

$$v_l^2 = \frac{c_{11}}{\rho} \quad v_t^2 = \frac{c_{44}}{\rho}. \tag{44}$$

Rewriting the isothermal compressibility in terms of the low frequency sound velocities v_l and v_t, one obtains

$$\lim_{Q \to 0} S(Q) = \frac{\overline{b}^2}{\overline{b^2}} \frac{k_B T}{M(v_l^2 - 4v_t^2/3)}. \tag{45}$$

As an example, we consider the case of selenium in the glass phase, close to the glass transition temperature T_g of 305 K. There, v_l is 1850 m/s and v_t is 930 m/s. With an M of 78.9 atomic units, one calculates a $S(Q)$ of 0.015 at 305 K, a rather low value.

B Debye model

Debye's famous consideration yields a density $V/(2\pi)^3$ in wave vector space for these sound waves. For a given wavevector direction, one has one longitudinal branch and two transverse branches.

The total density of sound waves (normalized to 1) is

$$g(\omega) = \frac{3\omega^2}{\omega_D^3}, \tag{46}$$

with the Debye frequency ω_D given by

$$\omega_D^3 = \frac{18\pi^2 \rho}{\overline{M}(1/v_l^3 + 2/v_t^3)}. \tag{47}$$

There is general agreement that the Debye model is a valid description of the sound waves in glasses at low frequencies, at least up to about 100 GHz. What happens above that frequency is again controversial. Neutron and X-ray experiments are able to supply information at these higher frequencies. Therefore it is important to know the dynamic structure factor of sound waves in scattering experiments. This was first derived in detail by Carpenter and Pelizzarri [3].

C Scattering from a single sound wave

The sound wave at the time t is described by

$$\mathbf{u}_j(t) = \mathbf{u}_0 \cos(\mathbf{q}\mathbf{R}_j - \omega_s t - \Phi_0). \tag{48}$$

Here $\mathbf{q}$ is the wave vector and ω_s is the frequency of the sound wave. The amplitude $\mathbf{u}_0$ and the phase Φ_0 depend on the choice of the time zero; this choice is averaged in the brackets

$$< (\mathbf{Q}\mathbf{u}_j(0))(\mathbf{Q}\mathbf{u}_{j'}(t)) >=< (\mathbf{Q}\mathbf{u}_0)^2 \cos(\mathbf{q}\mathbf{R}_j - \Phi_0) \cos(\mathbf{q}\mathbf{R}_{j'} - \omega_s t - \Phi_0) >. \quad (49)$$

Taking into account the statistical independence of amplitude and phase, the average over the choice of the zero point in time gives

$$< \mathbf{u}_0^2 \cos^2 \Phi_0 >=< \mathbf{u}_0^2 \sin^2 \Phi_0 >= \frac{k_B T}{N \overline{M} \omega_s^2} \quad < \mathbf{u}_0^2 \sin \Phi_0 \cos \Phi_0 >= 0, \quad (50)$$

where again $\overline{M}$ is the average atomic mass.

With these relations one finds

$$< (\mathbf{Q}\mathbf{u}_j(0))(\mathbf{Q}\mathbf{u}_{j'}(t)) >=$$
$$\frac{k_B T Q^2 \cos^2 \Theta}{N \overline{M} \omega_s^2} \left[\cos \mathbf{q}\mathbf{R}_j \cos(\mathbf{q}\mathbf{R}_{j'} - \omega_s t) + \sin \mathbf{q}\mathbf{R}_j \sin(\mathbf{q}\mathbf{R}_{j'} - \omega_s t) \right]. \quad (51)$$

where Θ is the angle between $\mathbf{Q}$ and the amplitude of the sound wave.

Using the addition relations for the cosine and decomposing $\cos \omega_s t$ and $\sin \omega_s t$ in complex parts, one gets

$$< (\mathbf{Q}\mathbf{u}_j(0))(\mathbf{Q}\mathbf{u}_{j'}(t)) >= \frac{k_B T Q^2 \cos^2 \Theta}{2N \overline{M} \omega_s^2} \left(e^{i\mathbf{q}(\mathbf{R}_j - \mathbf{R}_{j'})} e^{i\omega_s t} + e^{-i\mathbf{q}(\mathbf{R}_j - \mathbf{R}_{j'})} e^{-i\omega_s t} \right). \quad (52)$$

Inserting this result into the scattering formulae, one gets the one phonon scattering from the sound wave

$$S_{coh}^{1,s,e}(Q,\omega) = \frac{1}{N\overline{b^2}} | \sum_{j=1}^N b_j e^{-W_j} e^{i(\mathbf{Q}\pm\mathbf{q})\mathbf{R}_j} |^2 \frac{k_B T Q^2 \cos^2 \Theta}{2N \overline{M} \omega_s^2} \delta(\omega \pm \omega_s). \quad (53)$$

Here eq. (33) was again used to obtain the two δ-functions in ω, as in the case of the elastic scattering.

The expression can be simplified using eq. (36) for the elastic scattering. One obtains

$$S_{coh}^{1,s,e}(Q,\omega) = S_{el}(\mathbf{Q} \pm \mathbf{q}) \frac{k_B T Q^2 \cos^2 \Theta}{2N \overline{M} \omega_s^2} \delta(\omega \pm \omega_s). \quad (54)$$

This is a simple and convenient form for the coherent scattering from a single sound wave.

The incoherent scattering for a single sound wave reads

$$S_{inc}^{1,s,e}(Q,\omega) = e^{-2W} \frac{k_B T Q^2 \cos^2 \Theta}{2N \overline{M} \omega_s^2} \delta(\omega \pm \omega_s), \quad (55)$$

where W is the average Debye-Waller exponent for all atoms.

D Incoherent scattering from all sound waves

It is easy to derive the incoherent sound wave scattering within the Debye model.

The average over $\cos^2 \Theta$, i.e. over the angle between $\mathbf{Q}$ and the direction of vibration of the sound waves, supplies a factor $1/3$. With the Debye density of states, eq. (46), the total incoherent one phonon scattering from the sound waves reads

$$S_{inc}^{1,s}(Q,\omega) = Q^2 e^{-2W} \frac{3k_B T}{2\overline{M}\omega_D^3} = Q^2 e^{-2W} \frac{k_B T(1/v_l^3 + 2/v_t^3)}{12\pi^2 \rho}. \tag{56}$$

This provides a frequency-independent spectrum from zero up to the Debye frequency. One has to keep in mind that the Debye model is an oversimplification, in particular in glasses. However, at low frequencies it is a solid basis for at least one part of the vibrational spectrum. As long as one has reasonably well defined sound waves, there is a frequency-independent contribution in the incoherent spectrum which can be calculated from the density and the sound velocities.

E Brillouin scattering

The coherent one phonon scattering from the sound waves is similarly well defined, but a bit more complicated. One has to distinguish two kinds of scattering, namely the Brillouin scattering at small Q from the longitudinal sound wave at $Q = q$, and the so-called umklapp scattering from both longitudinal and transverse sound waves at higher Q. Let us begin with the Brillouin scattering at small Q.

In the case of Brillouin scattering, the scattering vector equals the phonon wave vector, $\mathbf{Q} = \pm\mathbf{q}$, where the positive sign stands for the creation and the negative sign for the annihilation of the phonon. In eq. (54) one then has to take $S_{el}(\mathbf{Q})$ at $\mathbf{Q} = 0$. As seen from eq. (39), this is a δ-function in $\mathbf{Q}$. Integrating eq. (54) over all possible phonon wave vectors with the density of states $V/(2\pi)^3$ in wave vector space, and taking into account that for $\mathbf{Q} = \mathbf{q}$ only the longitudinal sound wave has a displacement parallel to $\mathbf{Q}$, one gets the Brillouin scattering

$$S_{Brillouin}(Q,\omega) = \frac{k_B T Q^2}{2\overline{M}\omega^2} \frac{\overline{b}^2}{\overline{b^2}} \delta(\omega \pm v_l Q). \tag{57}$$

For a monoatomic glass, the prefactor from the scattering lengths is 1, because $\overline{b^2} = \overline{b}^2$. If we multiply the intensity of the Brillouin lines with ω^2, we see immediately that the Brillouin spectrum exhausts the second moment sum rule, eq. (20). That implies there should be no other inelastic scattering at this Q, unless there is some reduction of the intensity of the Brillouin line. If this holds for an elastic medium composed of only one element, it should hold also for an elastic medium composed of many elements, because the properties of an elastic medium depend only on its density and on its sound velocities. We will come back to this point when discussing the experimental results.

F Coherent umklapp scattering

Next one proceeds to the umklapp scattering from longitudinal and transverse phonons at higher Q. Let us begin with the umklapp scattering from the longitudinal sound waves. Their density of states $g_l(\omega)$ is given by

$$g_l(\omega) = \frac{M\omega^2}{6\pi^2 \rho v_l^3}. \tag{58}$$

At fixed ω one has a fixed absolute value ω/v_l of the wave vector $\mathbf{q}$, and one has to average the scattering in eq. (54) over all possible directions of this wave vector. Let θ be the angle between $\mathbf{Q}$ and $\mathbf{q}$. Then $\cos\Theta = \cos\theta$, and the coherent one phonon scattering from the longitudinal sound waves at the frequency ω and higher Q reads

$$S_{coh}^{1,long}(Q,\omega) = S_{lq}(Q,\omega) \frac{k_B T Q^2}{12\pi^2 \rho v_l^3}. \tag{59}$$

where $S_{lq}(Q,\omega)$ is given by the following average over $S_{el}(Q)$

$$S_{lq}(Q,\omega) = \frac{3}{2} \int_{-1}^{1} d\mu\, \mu^2 S_{el}((Q^2 + q^2 - 2Qq\mu)^{1/2}) \tag{60}$$

with $q = \omega/v_l$.

Dealing with the transverse sound waves, one can choose one of the two directions of vibrations perpendicular to $\mathbf{Q}$ in the elastically isotropic glass. Let us denote this first kind of transverse sound wave by the index $t1$ and the other one by $t2$. Then the sound waves indexed by $t1$ give no contribution to the inelastic umklapp scattering; for them $\cos\Theta = 0$. For the sound waves indexed by $t2$ one has $\cos\Theta = \sin\theta$, where again θ denotes the angle between $\mathbf{Q}$ and $\mathbf{q}$. The density of the sound waves indexed by $t2$ is

$$g_{t2}(\omega) = \frac{M\omega^2}{6\pi^2 \rho v_t^3}. \tag{61}$$

The integration of eq. (54) over all sound waves $t2$ yields

$$S_{coh}^{1,trans}(Q,\omega) = S_{tq}(Q,\omega) \frac{k_B T Q^2}{6\pi^2 \rho v_t^3}. \tag{62}$$

where now $S_{tq}(Q,\omega)$ is given by a slightly different average over $S_{el}(Q)$

$$S_{tq}(Q,\omega) = \frac{3}{4} \int_{-1}^{1} d\mu (1-\mu^2) S_{el}((Q^2 + q^2 - 2Qq\mu)^{1/2}) \tag{63}$$

with $q = \omega/v_t$.

In the limit of frequency zero the absolute value q of the phonon wave vector goes to zero as well. Then $S_{lq}(Q,\omega) = S_{tq}(Q,\omega) = S_{el}(Q)$ and the total coherent one phonon scattering from the sound waves is

$$S^{1,s}_{coh}(Q,\omega) = Q^2 S_{el}(Q)\frac{3k_B T}{2\overline{M}\omega_D^3} = Q^2 S_{el}(Q)\frac{k_B T(1/v_l^3 + 2/v_t^3)}{12\pi^2 \rho}. \tag{64}$$

This corresponds to the incoherent one phonon scattering from the sound waves, with the average Debye-Waller factor replaced by $S_{el}(Q)$.

IV OTHER VIBRATIONAL AND RELAXATIONAL MODES

A The eigenmode concept

The next step is to extend the concept of eigenmodes, valid for crystals and molecules, to glasses. The extension requires an unproved assumption:

We assume that the displacement is a sum of independent eigenmodes

$$\mathbf{u}_j(t) = \sum_{k=1}^{3N} \frac{A_k(t)}{M_j^{1/2}} \mathbf{e}_{jk}, \tag{65}$$

where M_j is the mass of atom j, $A_k(t)$ is the time-dependent amplitude of the eigenmode k and $\mathbf{e}_{jk}$ is the component of the eigenvector of mode k at the atom j. The eigenvector is normalized

$$\sum_{j=1}^{N} \mathbf{e}_{jk}^2 = 1. \tag{66}$$

The formula is valid in the harmonic approximation, which holds in the case of a single minimum of the potential energy with negligible anharmonic potential terms (in this case the minimum is at $\mathbf{r}_j = \mathbf{R}_j$ for all j). We assume its validity for the case of a complicated energy landscape with many minima, allowing for both quasiharmonic and relaxational eigenmodes. The former should appear as δ-peaks at a finite frequency in the scattering spectrum, the latter should give rise to a quasielastic feature under the elastic line.

The sound waves treated in the preceding section belong to the quasiharmonic modes (in fact, their quasiharmonic character is indeed obvious, because the sound velocities depend on temperature). In terms of the real eigenvector picture described above, one should describe the sound waves as standing waves. However, this is no serious problem, but just a different choice of the basis.

A more serious problem is the assumption of independence. Though sound waves at low frequencies are well defined in glasses, they scatter much more strongly than sound waves in crystals [11,12]. One of the possible explanations of this behaviour is the soft potential model [13], which postulates localized modes which have a bilinear interaction with both the longitudinal and the transverse sound waves. We will come back to this question in the comparison to experiment. Evidence for the existence of localized modes in glasses has been also found in simulations on model glasses [14].

For the time being, let us assume the independence of the eigenmodes. Following eq. (65), one can write the atomic displacement of atom j as a sum over eigenmode amplitudes. The amplitudes of different eigenmodes should be statistically independent, so the mixed terms disappear in the time average. Then the mean square displacement of atom j is

$$<\mathbf{u}_j^2> = \sum_{k=1}^{3N} \frac{\mathbf{e}_{jk}^2}{M_j} <A_k^2> \qquad (67)$$

and the correlation function reads

$$<(\mathbf{Qu}_j(0))(\mathbf{Qu}_{j'}(t))> = \sum_{k=1}^{3N} \frac{(\mathbf{Qe}_{jk})(\mathbf{Qe}_{j'k})}{M_j^{1/2} M_{j'}^{1/2}} <A_k(0)A_k(t)> \qquad (68)$$

This reduces the scattering functions to sums over single mode correlations. As a consequence, one can consider the inelastic scattering as a sum over independent contributions, each of them coming from one of the eigenmodes of the glass. In the following section, we calculate the mode structure factor (i.e. the Q dependence of the scattering from the eigenmode) in terms of the eigenvector of the mode.

B Mode structure factor

For a single mode with index k the inelastic coherent one phonon scattering is

$$S_{coh}^{1,k}(Q,\omega) = \frac{1}{N\overline{b^2}} |\sum_{j=1}^{N} b_j e^{-W_j} e^{-i\mathbf{Q}\mathbf{r}_j} \frac{\mathbf{Qe}_{kj}}{M_j^{1/2}}|^2 \frac{1}{2\pi} \int_{-\infty}^{\infty} dt e^{-i\omega t} <A_k(0)A_k(t)> . \qquad (69)$$

The interference of this inelastic scattering depends both on the phase factors from the atomic position and on the displacement pattern of the mode, characterized by the eigenvector. Thus the interference maxima can differ strongly from those of the elastic scattering. Later, we will see examples for such differences.

In analogy to $S(Q)$ and to $S_{el}(Q)$, one can define a mode structure factor $S_k(Q)$

$$S_k(Q) = \frac{3\overline{M}}{Q^2 \overline{b^2}} |\sum_{j=1}^{N} b_j e^{-W_j} e^{-i\mathbf{Q}\mathbf{r}_j} \frac{\mathbf{Qe}_{kj}}{M_j^{1/2}}|^2, \qquad (70)$$

where $\overline{M}$ is the average atomic mass. Note the factor 3, which normalizes $S_k(Q)$ again to the average Debye-Wallerfactor. For long wavelength excitations as, for instance, the sound waves it has been shown above that this $S_k(Q)$ agrees essentially with $S_{el}(Q)$. For short-wavelength excitations, however, this inelastic dynamic structure factor can differ strongly from the elastic one. This is demonstrated by the following simple example of two vibrating atoms.

C Two-atom example

Consider two atoms at a distance d, the first at $z = d/2$ and the second at $z = -d/2$ in the z-direction of a cartesian coordinate system. Let θ denote the angle between the scattering vector $\mathbf{Q}$ and the z-direction. Assume an ensemble of such biatomic molecules with all possible orientations to the scattering vector $\mathbf{Q}$. The directional average over the elastic scattering (neglecting Debye-Wallerfactors) gives

$$S_{el}(Q) = \frac{1}{4} \int_0^\pi d\theta \sin\theta \, | e^{\frac{iQd\cos\theta}{2}} + e^{\frac{-iQd\cos\theta}{2}} |^2 = 1 + \frac{\sin(Qd)}{Qd}. \tag{71}$$

The elastic diffraction maximum lies at $Qd \approx 5\pi/2$.

Turning to the inelastic scattering, let us consider a mode, in which the first atom has the amplitude $\mathbf{u}_1$ and the second an amplitude $\mathbf{u}_2$. The inelastic structure factor is proportional to

$$| (\mathbf{Qu}_1)e^{\frac{iQd\cos\theta}{2}} + (\mathbf{Qu}_2)e^{\frac{-iQd\cos\theta}{2}} |^2 =$$
$$4(\mathbf{Qu}_i)^2 \cos^2\frac{Qd\cos\theta}{2} + 4(\mathbf{Qu}_a)^2 \sin^2\frac{Qd\cos\theta}{2} \tag{72}$$

where

$$\mathbf{u}_i = \frac{\mathbf{u}_1 + \mathbf{u}_2}{2} \qquad \mathbf{u}_a = \frac{\mathbf{u}_1 - \mathbf{u}_2}{2} \tag{73}$$

are the in-phase and antiphase components of the mode motion of the two atoms, respectively.

One sees that the two components decouple. Consequently, their inelastic scattering intensities can be calculated separately and summed afterwards.

Similarly, one can decouple the contributions parallel and perpendicular to z, at least after the directional average, because the motion parallel to z is proportional to $\cos\theta$ and the motion perpendicular to z is proportional to $\sin\theta$.

It follows that one can restrict oneself to the four pure cases, the in-phase motions parallel and perpendicular to the connection line between the two atoms and the corresponding antiphase motions. The most important case is the in-phase motion along the connecting line. The directional average contains an additional factor $\cos^2\theta$ as compared to the elastic scattering. One gets the mode structure factor

$$S_k(Q) = 1 + \frac{3\sin(Qd)}{Qd} + \frac{6\cos(Qd)}{Q^2 d^2} - \frac{6\sin(Qd)}{Q^3 d^3}. \tag{74}$$

In the neighbourhood of the diffraction maximum at $Qd \approx 5\pi/2$ the last two terms of this equation are unimportant. The main interference is contained in $\sin(Qd)/Qd$, which reproduces the elastic diffraction maximum in the inelastic scattering, but a factor of 3 stronger than in the elastic case. We see that the inelastic interference can even be dramatically more pronounced than the elastic one.

The same holds for the antiphase motion along the connection line of the two atoms. One then gets eq. (74) with a change of sign of all trigonometric functions. That means that maxima of the structure factor transform into minima, and viceversa.

For the motion perpendicular to the connection line, the term $\sin(Qd)/Qd$ disappears, leaving only the weaker oscillation of the additional terms. Since all interference terms stem from sums over pairs of atoms, one concludes that the main part of the interference in the coherent inelastic scattering is due to correlated motions of atom pairs along their line of connection.

V   EIGENMODES AND SPECTRUM

A   Spectral shapes

For a vibrational mode with frequency ω

$$< A_k(0)A_k(t) > = < A_k^2 > \cos \omega t \tag{75}$$

and for a relaxational mode with the relaxation time τ

$$< A_k(0)A_k(t) > = < A_k^2 > e^{-t/\tau}. \tag{76}$$

In the vibrational case, one has to insert eq. (75) into eq. (69). Decomposing

$$\cos \omega t = \frac{e^{i\omega t} + e^{-i\omega t}}{2} \tag{77}$$

and considering eq. (33), one finds that the substitution of the two exponential functions in the scattering function provides two δ-functions at $\pm \omega$, again the same argumentation as in the case of the δ-function in the elastic scattering.

Thus the coherent one phonon scattering from the vibrational mode reads

$$S_{coh}^{1,k}(Q,\omega) = Q^2 S_k(Q) \frac{k_B T}{6N\overline{M}\omega_k^2} \delta(\omega \pm \omega_k). \tag{78}$$

The mean square amplitude of a relaxation depends on the distance and on the asymmetry of the two potential minima of the relaxational jump. For the simplest case of a symmetric double well potential (in a glass probably rather the exception than the rule) the jump width is temperature-independent, given by the distance between the two minima. But the jump rate depends markedly on temperature. In the simple case of thermally activated jumps over a potential barrier V the relaxation time τ follows the Arrhenius equation

$$\tau = \tau_0 e^{\frac{V}{k_B T}}. \tag{79}$$

Here τ_0 is a microscopic time of the order of 10^{-13} s.

Such a stochastic jump motion gives rise to a Lorentzian in the spectrum, centered around the frequency zero with the halfwidth $1/\tau$. According to the Arrhenius equation, one expects this halfwidth to rise strongly with rising temperature. At low temperatures, the relaxation should be practically invisible, because the Lorentzian is much narrower than the resolution, but then one should find a Lorentzian which broadens rapidly with increasing temperature.

In the relaxational case eq. (76) yields the coherent one phonon scattering

$$S_{coh}^{1,k}(Q,\omega) = Q^2 S_k(Q) \frac{<A_k^2>}{3N\overline{M}} \frac{1}{\pi\tau} \frac{1}{1/\tau^2 + \omega^2}, \tag{80}$$

because the Fourier transform of the exponential function is a Lorentzian. The validity of the one phonon approximation requires atomic jump widths considerably smaller than $1/Q$. If this holds, the Q-dependence is identical to the one of a vibration with the same displacement pattern. The spectrum differs in any case. The vibration gives rise to two sharp maxima at both sides of the elastic line, the relaxation supplies a Lorentzian under the elastic line.

The incoherent one phonon scattering of the mode k is given by

$$S_{inc}^{1,k}(Q,\omega) = \frac{1}{N\overline{\sigma}} \sum_{j=1}^{N} \sigma_j e^{-2W_j} \frac{(\mathbf{Q}\mathbf{e}_{kj})^2}{M_j} \frac{1}{2\pi} \int_{-\infty}^{\infty} dt e^{-i\omega t} <A_k(0)A_k(t)>. \tag{81}$$

Here again the correlation function of the mode amplitude supplies two δ-functions at $\pm\omega$ for a vibration at the frequency ω and a Lorentzian under the elastic line for a relaxation.

B Effective vibrational density of states

The equations derived above allow to describe the coherent and incoherent inelastic scattering from glasses in terms of the one phonon approximation. One can use these equations to determine an effective vibrational density of states from measured neutron or X-ray data (the word *effective* admits our lack of knowledge to which extent we deal with true vibrations) In the case of coherent scattering, this determination often requires additional assumptions on the eigenvectors of the modes.

One can try to see how much is true vibration and how much is relaxation by looking at the temperature dependence of the resulting effective density of states. For a truly harmonic solid, the vibrational density of states is temperature-independent. In the presence of relaxations, one expects a strong rise of the effective density of states near to the elastic line. That is in fact a finding in most glasses, in particular as one approaches the glass transition temperature T_g. But a straightforward interpretation with a clear distinction between quasiharmonic vibrational softening and relaxational jumps has turned out to be so difficult, that one cannot help thinking that the whole ansatz is defective. Nevertheless, it is instructive to represent the scattering results in terms of a temperature dependent effective density of states, preferably in the form $g(\omega)/\omega^2$, where one has approximately the measured spectrum divided by temperature.

For a quantitative treatment, in particular at higher temperatures, one has to take the multiphonon scattering into account. This contribution can be calculated in the gaussian approximation, assuming not only a gaussian distribution of the atomic displacements, but also the same distribution width for each atom in each spatial direction. Again, this is a crude oversimplification, but it allows to calculate a reasonably accurate multiphonon correction.

With the gaussian approximation, the intermediate incoherent scattering function $F_{inc}(Q,t)$ is characterized by the time-dependent mean square displacement $\gamma(t)$ according to eq. (22). The function $\gamma(t)$ is given by the effective vibrational density of states

$$\gamma(t) = \frac{k_B T}{\overline{M}} \int_0^{\omega_{max}} d\omega \frac{g(\omega)}{\omega^2}(1 - \cos\omega t). \tag{82}$$

The incoherent scattering is obtained from the Fourier transform of the intermediate scattering function in time

$$S_{inc}(Q,\omega) = \frac{1}{\pi} \int_0^\infty dt \cos\omega t e^{-\gamma(t)Q^2}. \tag{83}$$

In this approximation, one does not only get the one phonon scattering, but the entire inelastic scattering including the multiphonon terms.

As pointed out above, one can take the multiphonon terms into account by introducing a frequency dependent effective Debye-Waller factors for the inelastic scattering in the one phonon approximation. One can use the method both for coherent and for incoherent scattering. One proceeds as follows: first one determines an effective density of states in the one phonon approximation (in the case of coherent scattering one needs assumptions on the eigenvectors to choose an appropriate mode structure factor). Then one calculates numerically the complete incoherent scattering in the Q- and ω domain of the measurement. To that calculated scattering, one fits the effective frequency dependent Debye-Waller factors one is looking for. With these, one again applies the one phonon approximation to determine again the effective density of states. The procedure is simple and converges even at higher temperatures after a few iterations.

REFERENCES

1. W. Marshall and S. W. Lovesey, *Theory of Thermal Neutron Scattering*, Oxford, Clarendon Press 1971, p. 65 ff.
2. J. D. Axe, in *Physics of Structurally Disordered Solids*, ed. by S. S. Mitra, New York, Plenum Press 1974, p. 507
3. J. M. Carpenter and C. A. Pelizzari, Phys. Rev **B12**, 2391 (1975)
4. K. Fröbose and J. Jäckle, J. Phys. **F7**, 2331 (1977); J. Jäckle and K. Fröbose, J. Phys. **F9**, 967 (1979)

5. L. v. Heimendahl, J. Phys. **F9**, 161 (1979)
6. R. Zorn, Phys. Rev. B **55**, 6249 (1997)
7. A. Rahman, Phys. Rev. **136**, A405 (1964)
8. W. Kob and H. C. Andersen, Phys. Rev. E **51**, 4626 (1995)
9. J.-P. Hansen and I. R. McDonald, *Theory of simple liquids*, 2nd ed. (Academic Press, New York 1986). Second moment sum rule: ch. 7.4, eq. (7.4.40), p. 220; Rayleigh-Brillouin scattering: ch. 8.5, p. 275 ff.
10. de Gennes, P. G. 1959 Physica **25**, 825
11. R. C. Zeller and R. O. Pohl, Phys. Rev. B **4**, 2029 (1971)
12. Phillips, W. A. (ed.) 1981, *Amorphous Solids: Low temperature properties*, (Springer, Berlin)
13. Parshin, D. A. 1994, Phys. Solid State **36**, 991
14. H. R. Schober and B. B. Laird, Phys. Rev. B **44**, 6746 (1991); B. B. Laird and H. R. Schober, Phys. Rev. Lett. **66**, 636 (1991); H. R. Schober, C. Oligschleger and B. B. Laird, J. Non-Crystalline Solids **156-158**, 965 (1993)
15. Ramos, M. A., and Buchenau, U. 1997, Phys. Rev. B **55**, 5749
16. Sokolov, A. P., Buchenau, U., Richter, D., Masciovecchio, C., Sette, F., Mermet, A., Fioretto, D., Ruocco, G., Willner L., and Frick B. 1999 (submitted)

Relaxations and vibrations in glasses: experiments

J. Pelous, C. Levelut and F. Terki

Laboratoire des Verres, Université Montpellier II, 34000 Montpellier, France

Abstract. Light scattering is a powerful tool to analyze relaxational processes and vibrational excitations in glasses. This paper presents an overview about relaxational and vibrational processes in amorphous materials, followed by a more specific discussion about the aspects related to the analysis of Brillouin light scattering and low frequency Raman scattering.

BACKGROUND ABOUT RELAXATIONS AND VIBRATIONS IN GLASSES

$\Phi(t)$ and $\chi(\omega)$

In general, two different quantities can be used in order to extract information from experimental data for comparison with theoretical predictions or modelization: the correlation or relaxation function, $\Phi(t)$, and the spectral susceptibility, $\chi(\omega)$. The first one, $\Phi(t)$, is deduced for example from nuclear magnetic resonance, photon correlation or neutron spin echo experiments. The second one, $\chi(\omega)$, is obtained from mechanical spectroscopy, dielectric measurements, as well as neutron, X-ray or light scattering –including Brillouin, Raman or infrared– techniques. In fact the two quantities $\Phi(t)$ and $\chi(\omega)$ are not independent; the Fourier transform $\Phi(\omega)$ of the relaxation function is connected to the spectral susceptibility by: $\Phi(\omega) \sim \frac{\chi(\omega)}{\omega} \sim S(q,\omega)$, where $S(q,\omega)$ is the dynamic structure factor and q the wavevector of the excitation. Experimental determination of those quantities and their variation with temperature will be illustrated in the following.

α relaxation

Several kind of temperature dependent relaxations are observed in glasses. For temperatures located above the glass transition temperature, T_g, the so-called α or structural relaxation corresponds to an increase of the shear viscosity and shear modulus during the cooling. The behavior of the glass-forming material changes from a liquid behavior, to a viscoelastic one, where the movements are cooperative. This α relaxation is a signature of the glass transition. The historical experimental approach, still in use, is to study the shear modulus G dependence versus temperature and frequency, using a torsional pendulum. The comparison between the shear response and the stress excitation (or its inverse) give the real part $G'(\omega)$ and imaginary part $G''(\omega)$ of the shear modulus and related quantities such as $\tan\Phi = \frac{G''(\omega)}{G'(\omega)}$. The typical investigated frequency range is $10^{-4} - 1$ Hz. Two examples are shown in figure 1 for a polymeric glass and an electrolyte glass. The α relaxation is characterized by a step in G' above the glass transition temperature, this step is related to an elastic response of the material. A peak in the imaginary part G''', is also observed at the same temperature. Similar curves can be obtained from capacity measurements. Using dielectric spectroscopy, the real and imaginary part of the electric permittivity, $\varepsilon'(\omega)$ and $\varepsilon''(\omega)$, can be investigated in a more extended frequency range, typically form 10 Hz to 10 GHz.

Using $\chi''(\omega)$ obtained for several temperatures, a "master" curve can be obtained by rescaling the curves along both axis. An example of such plot is presented in figure 2. Such curves, usually plotted using a logarithmic-linear scale, are not symmetric. Several analytical functions can be used to modelize the asymmetry. As pointed in the introduction, because $\Phi(t)$ and $\Phi(\omega)$ are related through a Fourier transform relation, any asymmetric fitting expression used for $\Phi(\omega)$ should correspond to a specific behaviour for $\Phi(t)$. The most usual description of $\Phi(t)$ is the stretched exponential

$$\exp(-t/\tau)^\beta, \tag{1}$$

where the exponent β, also called the β_{KWW} (Kohlrausch, Williams, Watt) parameter, can be related to other exponents used for the direct fit of $\chi(\omega)$. An example of fit using a Cole Davidson function :

$$\chi(\omega) = \frac{1}{1 + (i\omega\tau)^{\beta_{CD}}}, \tag{2}$$

where τ is a characteristic time for the relaxation and the exponent β_{CD} is related to the width of the distribution function of times, is also given in figure 2. It can be noted that $\beta_{CD} = 1$ correspond to the Debye relaxation for which the curves $\chi(\omega)$ are symmetric. The theoretical explanation of the experimental values obtained from β_{KWW} or $\beta_{CD} < 1$ is a challenge for most contributions in the field of the glass transition. Another important characteristic quantity, the relaxational time $\tau(T)$ can be deduced from the fit using formulas (1) or (2). In the shear experiments mentioned above, the relaxation time τ_s can be related to the viscosity η_S using the classical relation $\eta_S = G'_\infty \tau_S$, where G'_∞ is the shear modulus at a frequency higher than the characteristic frequency of the relaxation. Near the glass transition temperature T_g, the temperature dependence of G'_∞ is negligible compared to the exponential variation of τ_S. Thus, the shear viscosity and the relaxation time follow a similar temperature variation, well described in most glasses by a Vogel Fulcher

Tamman law, $\exp \frac{E_a}{T-T_0}$, where E_a is an activation energy and T_0 a characteristic temperature.

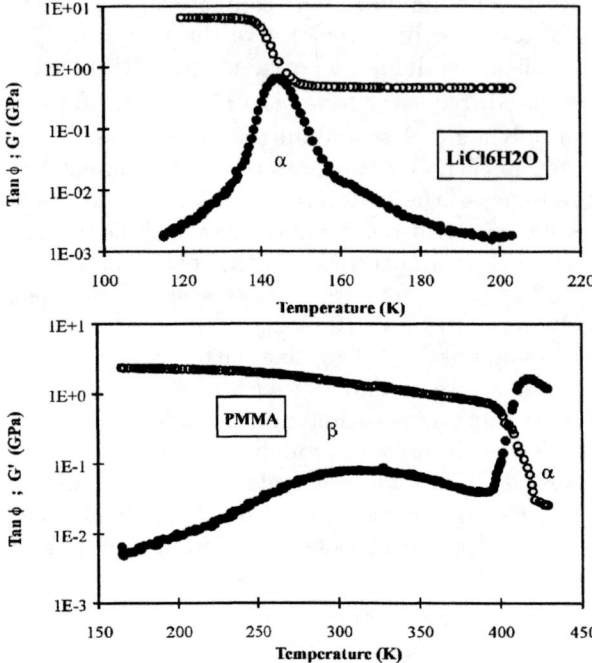

FIGURE 1. Real part of the shear modulus $G'(\omega)$ and shear loss $\tan \Phi = \frac{G''(\omega)}{G'(\omega)}$, where G'' is the imaginary part of the shear modulus, plotted as a function of temperature, for two very different glasses: a polymeric one, PMMA (T_g=293K) and an electrolyte, LiCl(H$_2$O)$_6$ (T_g=142K). The step observed in the real part of the modulus and the strong peak in the imaginary part are due to the α relaxation. A broader peak at lower temperature, related to the β relaxation, can also be observed in PMMA. Reproduced from reference [1].

One important result is the fact that the temperature behavior of different determinations of the relaxation times τ_α associated to different experimental probes is the same: E_a and T_0 are identical within the accuracy of the experiments. Even if some discrepancies in the absolute τ_α values are observed, a judicious choice of the scales in a logarithmic plot $\log \tau_\alpha$ versus the inverse of the temperature $1/T$ illustrates these conclusions (see figure 3).

β relaxation

In most glass forming materials, in particular in polymers, near and below T_g a second peak is observed for the imaginary part of the susceptibility $\chi''(\omega)$ (figure 1b); this feature is called the β relaxational process. Extensively studied from dielectric measurements by Johari and Goldstein, it is also referred to as Johari and Goldstein relaxation. Moreover, in order to distinguish this process from another relaxational process introduced in the recent theory of the glass transition (cf. contributions in this book related to the Mode Coupling theory) it is also called the slow β relaxation. The relaxation time τ_β follows an Arrhenius behaviour in contrast to τ_α described above (see figure 3). At high temperatures, the characteristic times

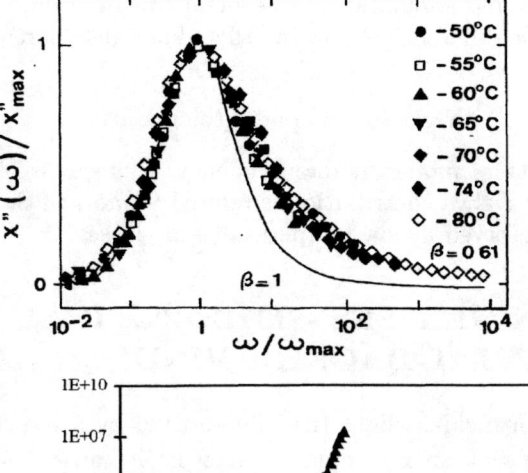

FIGURE 2. Master curve, $\chi''(\omega)/\chi''_{max}$ as a function of ω/ω_{max}, obtained by rescaling susceptibility curves obtained for several temperatures in meta-fluoro aniline [2]. Symbols corresponds to experimental curves. The solid lines correspond to fits using Cole and Davidson distribution function (see text), with a Cole and Davidson exponent β_{CD} equal to 1 and 0.61, respectively.

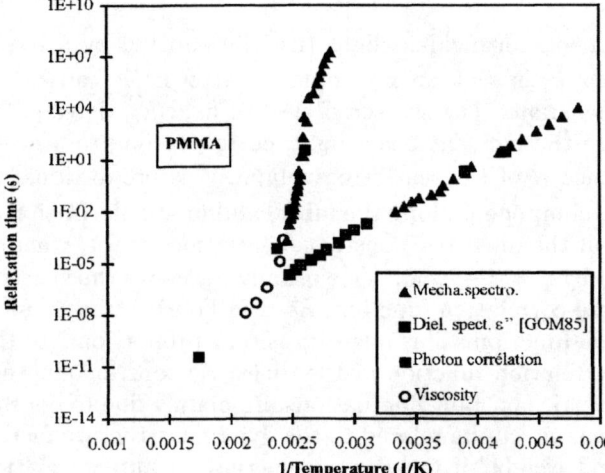

FIGURE 3. Relaxation times obtained for PMMA in a wide temperature and frequency range using several experimental techniques. The relaxation times are plotted in an Arrhenius plot, $\log \tau$ versus $1/T$. Reproduced from reference [1].

of the α and β processes, τ_α and τ_β, collapse and a bifurcation between the two processes appears in the supercooled liquid.

Vibrations

The disorder is responsible for differences between the vibrational excitations in crystals and in amorphous materials and glasses. Due to the lack of periodicity, the wavevector is no longer a good quantum number to describe these excitations. The description in terms of dispersion curves, ω versus q, valid for crystals, is in principle inappropriate. In fact, the concept of phonons is still often used, associated with a reduction of the lifetime (or the mean free path) of the phonons, due to the disorder. The description of the vibrations must include the concept of spatial localization. On the other hand, the density of modes $g(\omega)$ in disordered systems remains as a quantity available from experiments and simulations. The main contribution of $g(\omega)$ arises from short range order. The differences between ordered and disordered media appears in two ways:

1. a broadening of the features in $g(\omega)$ for glasses compared to crystals

2. an enhanced value for the density of modes at low frequency; this excess of modes compared to the expected Debye contribution is related to most of the anomalous physical properties observed at low temperatures in glasses.

LIGHT SCATTERING IN GLASSES - THEORETICAL AND EXPERIMENTAL BACKGROUND

In a light scattering experiment, an incident light (usually emitted by a laser) of pulsation ω_i and wavevector $\vec{k}_i$ is sent on a sample. Scattering is caused by fluctuations of the dielectric constant. The scattered electric field $\vec{E}_s$, detected at a given angle θ with respect to the incident beam light, corresponding to a given wavevector $\vec{k}_f$, and at a distance R of the scattering volume V is proportional to the component $\delta\bar{\bar{\epsilon}}_{if}(\vec{q},t)$ of the component along the initial and final polarization of the spatial Fourier transform of the dielectric constant fluctuation tensor, where $\vec{q}$ is the scattering vector defined by $\vec{q} = \vec{k}_i - \vec{k}_f$ [3]. One usually measures the spectral density of the electric field time-correlation function, i.e. the Fourier transform of the electric field autocorrelation function. This quantity is thus proportional to the spectral density of the autocorrelation function of the dielectric constant fluctuations. Assuming that the dielectric constant fluctuations are mainly due to density fluctuations, the scattered intensity can be related to the dynamic structure factor, $S(\vec{q},\omega)$, defined as the spectral weight of the density fluctuation autocorrelation function, by the relation: $I_{if}(\vec{q},\omega) = \frac{E_0^2 k_f^4}{R^2} \left(\frac{\partial \bar{\bar{\epsilon}}_{if}}{\partial \rho}\right)^2 S(\vec{q},\omega)$.

Different quantities can be measured by selecting the polarization of the incident and scattered light. If the scattering plane is defined as the $(\vec{k}_i, \vec{k}_s)$ plane, and if the indexes V and H are used for light polarized perpendicularly to and parallel to the scattering plane, respectively, the quantities are: I_{VV}, I_{VH}, I_{HV} and I_{HH}, where the first index is related to the incident light and the second to the scattered light. It must be noted that if there are no molecular rotations, $\delta\bar{\bar{\epsilon}}$ is a scalar and $I_{VH} = I_{HV} = 0$.

Brillouin

Brillouin scattering corresponds to light scattered with a frequency change, with respect to the incident light, in the GHz range. It results from the scattering of light by coherent propagative motions, or acoustic phonons, arising from thermal excitations. The frequency range investigated is 10^9–10^{11} GHz, corresponding to time scales in the 10^{-11}–10^{-9} s range. The wavevector are in the 10^{-3}–10^{-2} nm^{-1} range, corresponding to length scale in the 10^2–10^3 nm range, in the mesoscopic range.

The analysis of such small frequency shifts requires the dispersive power of Fabry-Pérot interferometers. A very common type of spectrometer is made of two plane Fabry-Pérot interferometers with very slightly different thicknesses, in multipass use in order to increase the contrast [4]. Another kind of spectrometer uses a very high resolution of a spherical Fabry-Pérot interferometer, following a two of four-pass plane Fabry-Pérot used as a filter whose maximum transmission is adjusted to the frequency of the Brillouin line to study [5]. An example of spectrum obtained using such high-resolution spectrometer is shown in figure 4.

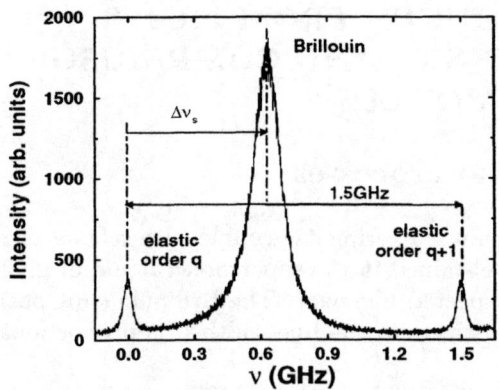

FIGURE 4. Example of Brillouin spectra obtained in silica using the high-resolution experiment described in reference [5]. The Brillouin shift is the sum of the apparent shift $\Delta\nu_s$ determined with the spherical interferometer and an integer number q, to determine separately, of orders of this spherical interferometer: $\Delta\nu_s + q \times 1.5$ GHz.

The position of the Brillouin line $\delta\nu$ is related to the sound velocity V from which elastic modulus can be deduced. The line width Γ is related to the sound attenuation and to the dynamic viscosity. Rather than Γ, one usually determines the inverse mean free path, or spatial attenuation of the phonons, $\ell^{-1} = \alpha = \frac{4\pi\Gamma}{V}$, and the internal friction coefficient $Q^{-1} = \frac{\Gamma}{\delta\nu}$. The intensity of the Brillouin lines are related to elasto-optic constants.

Raman

In disordered systems, Raman scattering corresponds to inelastic scattering due to the interaction with optical modes and incoherent contribution from high frequency acoustic modes. The mechanism of scattering can be direct or through dipole-induced dipole interaction. Frequency shifts are in the 10^{11}–10^{14} Hz range, the 10^{11}–10^{12} Hz range is called the low frequency Raman range. Martin and Brenig [6] explained the Raman intensity by disorder-induced first order scattering: because of the disorder, the selection rules that govern the Raman scattering in crystal are relaxed and all the modes contributes, leading to broad bands instead of discrete series of lines. Shuker and Gammon [7] assumed that all the modes originating from one band b contribute with the same weight, $C_b(\omega)$. Thus the Raman intensity can be written as a function of a sum over the band of the density of modes in the band b, $g_b(\omega)$, multiplied by the coupling constant for this band. A simplified expression using the total vibrational density of modes $g(\omega)$ and an effective coupling constant $C(\omega) = \sum_b C_b g_b(\omega)/g(\omega)$, representing the efficiency of the interaction between light and phonons is also often used. Some authors, on the contrary, attributed the depolarized (I_{VH}) spectra mainly to second-order dipole-induced-dipole scattering [8,9]. The correspondence, at least qualitative, between the Raman spectrum and the vibrational density of modes is illustrated below in figure 7.

MAIN RESULTS DEDUCED FROM LIGHT SCATTERING EXPERIMENTS AND COMPARISON WITH MODELS

Relaxational processes

Acoustical phonons detected by Brillouin scattering can couple with relaxational processes. In figure 5, Brillouin spectra obtained in the supercooled liquid of LiCl-6H$_2$O mixture give an example of the expected behavior. The Brillouin shift position $\delta\nu$ is proportional to the sound velocity and the line-width Γ is proportional to the hypersonic attenuation.

As a first step of the analysis, $\Gamma(T)$ (figure 5b) and $\delta\nu(T)$ (figure 5c) can be deduced from the spectra (figure 5a) using an hydrodynamic description of the sound waves [10]. It is interesting to compare theses results with those obtained by mechanical spectroscopy in the same kind of glasses and shown in figure 1a. The change, of 9 orders of magnitude, of the frequency probed does not change the behavior observed for $\chi(T)$ (G', G'' at low frequencies and C'_{11}, C''_{11} at high frequencies). This is a universal behavior observed in very different glasses as oxide glass (B$_2$O$_3$ [11]) or polymeric ones [12].

More sophisticated analyses of the Brillouin profile (in fact $S(q,\omega)$) have been proposed in order to extract further information. For example, such analysis has been applied by our group in polyurethanes [13,14]. The spectra were fitted to a generalized linear expression including a non-exponential relaxation taken into account by a Cole and Davidson function:

$$S(q,\omega) = \frac{I_0}{\omega}\Im\left[\left(\frac{M^\star q^2}{\rho} - \omega^2\right)^{-1}\right] \text{ with } M^\star/\rho = V_\infty - \frac{V_\infty - V_0}{(1+i\omega\tau)^{\beta_{CD}}}, \tag{3}$$

where $M^\star$ is a complex elastic modulus, ρ the density of the sample, V_∞ and V_0 the high and low frequency limits of the sound velocity. The Brillouin velocity in the

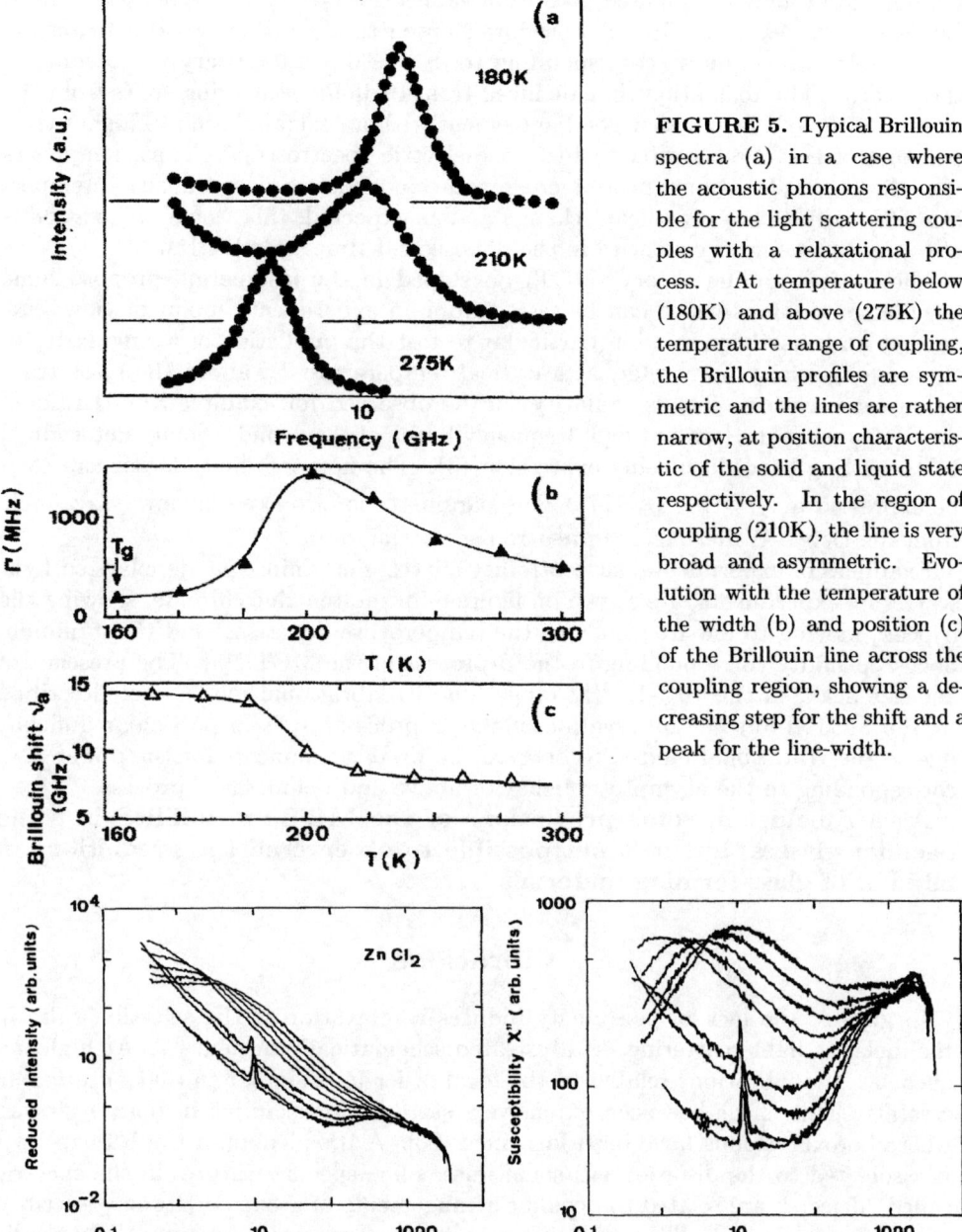

FIGURE 5. Typical Brillouin spectra (a) in a case where the acoustic phonons responsible for the light scattering couples with a relaxational process. At temperature below (180K) and above (275K) the temperature range of coupling, the Brillouin profiles are symmetric and the lines are rather narrow, at position characteristic of the solid and liquid state respectively. In the region of coupling (210K), the line is very broad and asymmetric. Evolution with the temperature of the width (b) and position (c) of the Brillouin line across the coupling region, showing a decreasing step for the shift and a peak for the line-width.

FIGURE 6. Reduced intensity and susceptibility as a function of frequency, in log-log scale, in molten zinc chloride, measured at 650, 600, 500, 450, 400, 350 and 300° C. Reproduced from [18].

temperature range investigated goes from values close to V_0 at low temperature to values close to V_∞. The fitting procedure showed that a rather broad distribution for the relaxation time τ, corresponding to $\beta_{CD} \simeq 0.2$ is necessary to account for the spectra. The relaxation time deduced from Brillouin scattering splits from the α relaxation, and is in rather good agreement with an extrapolation to higher temperature of the β relaxation probed by dielectric spectroscopy. Thus, it appears that the β_{slow} relaxational process crosses the calorimetric glass transition temperature T_g without any significant change and as expected, this "local" relaxation is also present at high frequencies in the nanosecond time scale [14,15].

The mode coupling theory (MCT) developed in the last period propose some peculiar predictions which can be tested from an experimental point of view. Extensive work has been done in particular to test the prediction of a singularity in the non-ergodicity parameter, at a critical temperature T_c, above the glass transition temperature. This singularity can be observed for example from Brillouin data [16], using the low and high frequency limits of the sound velocity determined by the fitting procedure using expression (3). The non-ergodicity factor can then be expressed as $f = 1 - \frac{V_0^2}{V_\infty^2}$ [17]. This parameter is also experimentally available from the Debye-Waller factor in neutron scattering data.

Examples of experimental susceptibility curves, determined by depolarized light scattering experiments, are shown on figure 6 for molten zinc chloride, showing the α peak, moving to low frequency as the temperature decreases, and the minimum of susceptibility corresponding to the β process of the MCT [19]. The presence of another peak, in the 10^{11}-10^{12}Hz range, due to vibrational modes, not described by the MCT, blurs the observation of this β process. It is in particular difficult, due to the vibrational modes, to observe the proper exponents for the power laws corresponding to the asymptotic behavior above and below the β process.

As a conclusion, some predictions of the MCT are fulfilled in some peculiar glasses, but it is not possible to observe all the predictions in all kind of glass-forming materials.

Vibrations

In glasses, the lack of periodicity induces a relaxation of the selection rules in the inelastic light scattering, as illustrated schematically in figure 7. At high frequencies, the vibrations related to the local order look similar to that observed in crystals. This effect has been extensively studied, for example in borate glasses, where boroxol groups have been identified [20]. A broadening of the Raman lines is associated to the disorder and sometimes, some specific features in the spectra, called "defects" are related to peculiar arrangements of groups of atoms, described in terms of rings [21]. The main difference between crystals and amorphous media appears at low frequency, where two specific features are observed in glasses for $S(q,\omega)$ measured in Raman spectra:

1. a quasielastic contribution, called the Rayleigh wing or light scattering excess

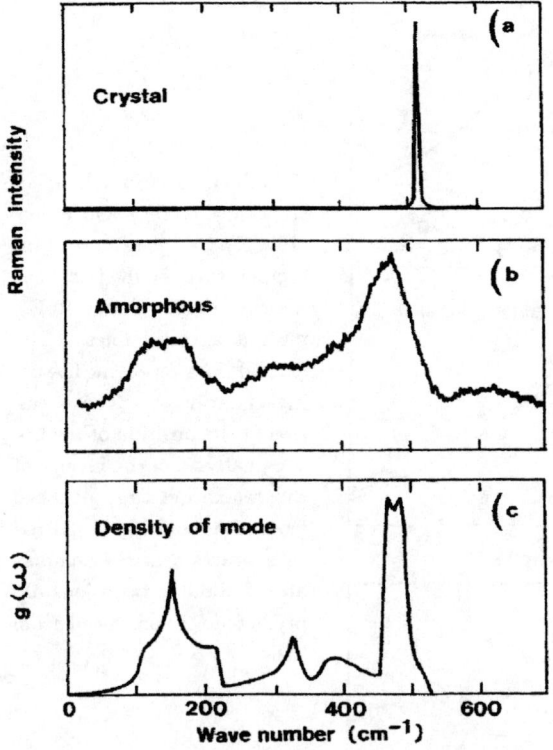

FIGURE 7. Raman intensity measured in crystalline (a) and amorphous (b) silicon. Vibrational density of modes in the amorphous material (c).

2. an inelastic contribution, the Boson peak, whose intensity follows the Bose factor variation with the temperature, as for phonons.

An example of these two contributions, is presented in figure 8, below and above the glass transition temperature, for an industrial oxide glass. Many conflicting models have been proposed to explain the origin of the experimental data. Usually, one assumes that the vibrations responsible for the boson peak are damped by the relaxational processes who are also responsible for the quasielastic lines [23–25]. More recently, fluctuations of the density of phonons have been proposed to account for the excess wing [26]. In this approach, near the glass transition temperature, the Raman intensity can be described by:

$$\frac{I_{QLS}}{n(\omega)+1} \simeq \frac{\omega\tau}{1+\omega^2\tau^2} \int \frac{\delta^2 I_{vib(\Omega)}}{\Omega^3 d\Omega}, \qquad (4)$$

where τ is the relaxation time of the phonons, I_{vib} is the Raman contribution at low temperature, where the quasielastic contribution I_{QLS} is negligible, $\delta^2 \simeq \gamma^2 T$, where γ is the Grüneisen parameter, related to the anharmonicity, and T is the temperature. Fits using equation (4) are presented in figure 8. The model predicts

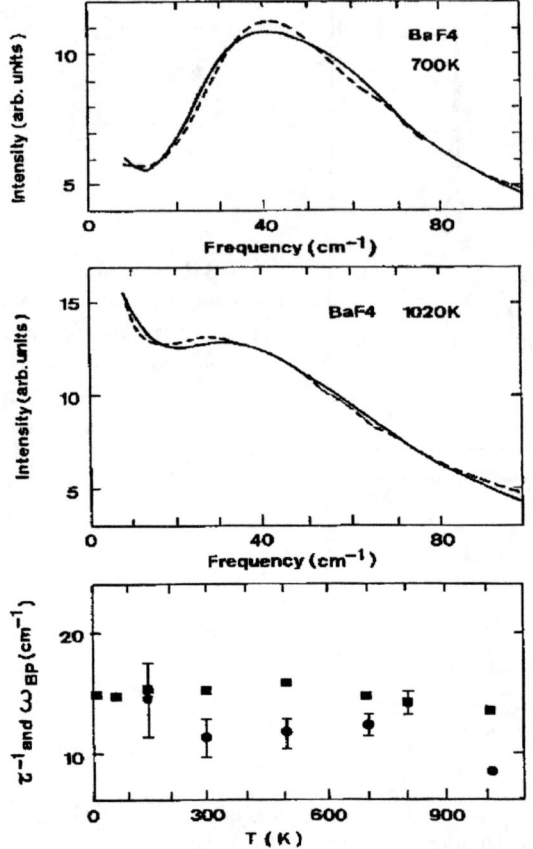

FIGURE 8. Raman spectra below (700K) and above (1020K) the glass transition temperature (800K) in a commercial oxide glass, BaF$_4$, plotted as solid lines. The dashed lines represent fits using equation 4. The variation of the position of the boson peak and of the inverse of the relaxation time deduced from the model are plotted as a function of the temperature. Similar variations are predicted. Reproduced from [22].

that the relaxation time τ is proportional to the Boson peak maximum. This prediction is fulfilled in most glasses at room temperature [26]. In figure 8, this prediction has been tested for an oxide glass as a function of temperature. Due to the approximations of the model and to the uncertainties in the determination of the fit parameters, the discussion on this field is still open.

Acoustical phonons

In crystals, ultrasonic and hypersonic properties can be explained by anharmonicity. The sound attenuation is due to interaction of the acoustic phonons with thermal vibrations also described by phonons. In the "Akhieser regime", the attenuation is given by $\Gamma = \frac{\omega^2}{\rho v^3} C T \gamma^2 \tau$, where γ is the Grüneisen constant, ρ the density, v the sound velocity, C the specific heat of the material, τ the lifetime of the thermal phonons and T the temperature [27]. At low temperature, another regime, called "Landau-Rumer regime" is observed where the attenuation is proportional to a power n of the temperature, $\Gamma \propto T^n$, with $n > 4$ [28].

The disorder induces a completely different behavior at low temperatures. For example, the addition of less than one percent of OH impurities introduces a peak in the hypersonic attenuation around 20K [29]. The disorder induced by neutron irradiation in quartz induces an enhancement of the acoustical attenuation at low temperature [30]. In fact, all the physical properties of disordered systems and glasses exhibit very different behaviors than that of crystals at low temperature. This universal behavior has been explained by assuming the existence of tunneling entities acting as two-level-systems (TLS). Thus, the acoustical properties in the 1kHz–100GHz frequency range are dominated, at low enough temperatures (typically below about 20K) by the interaction with TLS [31]. At higher temperatures, the same entities can be thermally activated and the ultrasonic properties in glasses have been well described by assuming double-well potentials with a large distribution of barriers.

An extension of the TLS model, the "soft potential model", has also been proposed to explain the observed excess in the density of modes $g(\omega)$ at low temperature [32–34]. The additional harmonic modes –soft modes– interact with phonons and can explain the $\Gamma \propto T$ law observed for the acoustical attenuation at hypersonic frequencies in many glasses, up to 100K [36]. Moreover, this phenomenological model can explain also the proportionality between the intensity of the quasielastic contribution in the Raman spectra and the acoustical attenuation [35].

In most glasses, for temperatures above 100K, the activated relaxational processes alone cannot explain the temperature dependence of the friction coefficient (or attenuation). At hypersonic frequencies, an extended plateau is observed up to T_g, in oxide glasses as well as in polymeric glasses. Moreover, in this temperature range, the acoustical attenuation in glasses appears very close to that observed in the corresponding crystals [30]. So the most efficient process is probably anharmonicity or, in other words, interaction of the phonons with other vibrational modes (including resonant or soft modes). This revisited Akhieser regime in disordered media has been calculated recently in amorphous silicon [38] and appears as the dominant contribution, at least up to 100GHz.

For higher frequencies, a large debate concerns the concept of localization of phonons in glasses. Recent results from X-ray inelastic scattering have been discussed in terms of localization in the meaning of Ioffe-Regel [39], or as an evidence of propagative excitations in a large q range, assuming that the dispersion curve description is appropriate [40]. Up to now, the discussion is not closed. Nevertheless, at hypersonic frequencies, it is possible to test the sensitivity of the acoustical attenuation to the spatial fluctuations of the elastic constants or to density heterogeneities. The scattering of phonons by disorder is often described in terms of Rayleigh-like process. An analytical formalism to describe this process has been developed in silica aerogels, which are highly porous materials, extensively studied as an example of material with a fractal structure. Changing the initial conditions of preparation, it is possible to prepare less porous materials (xerogels or densified aerogels). For the latter materials the distribution of pores is narrow and the pore size decreases with increasing density. The results for the acoustical attenuation (in fact for the friction coefficient Q^{-1}) is plotted in figure 9.

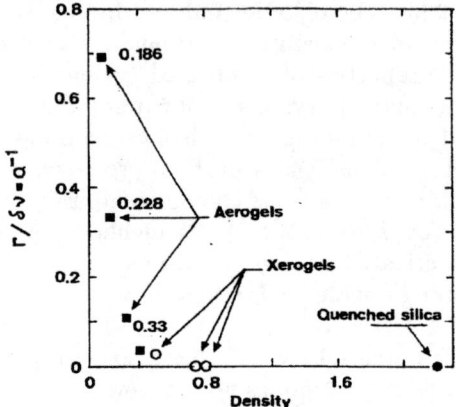

FIGURE 9. Friction coefficient for dense and porous silica. Several materials, such as aerogels and xerogels, with very different density and pore sizes, are compared. The porosity yields a significant increase of the friction coefficient, only for the less dense materials.

It appears that the temperature independent contribution to the acoustical attenuation at hypersonic frequencies, due to Rayleigh-like scattering, is significant only when the characteristic size of the density or elastic fluctuations is not negligible compared to the acoustic wavelength (about 200 nm).

ACKNOWLEDGEMENTS

Thanks to the different authors who gave us the permission to use their data as illustration of some parts of this paper.

REFERENCES

1. Faivre, A., *Ph D thesis*, INSA-Lyon (France), 1997.
2. Alba-Simionesco, C., Fan, J., Angell, C. A., *J. Chem. Phys.* **110**, 5262 (1999).
3. Berne, B. J. and Pecora, R., *Dynamic Light Scattering*, New York: Wiley, 1976, p24.
4. Sandercock, J. R. *J. Phys. E* **9**, 566 (1976).
5. Vacher, R. *Thesis*, Montpellier 1972; Vacher, R., Vialla, R. and Cavaillé, D., to be published.
6. Martin, A. and Brenig, W., *Phys. Stat. Sol. B* **64**, 163 (1974).
7. Shuker, R. and Gammon, W., *Phys. Rev. B* **25**, 222 (1970).
8. Stephen, M. J., *Phys. Rev.* **187**, 279 (1969).
9. Li, G., Du, W.M., Chen, X.K., Cummins, H. Z., and Tao, N. J., *Phys. Rev. A* **45**, 3867 (1992).
10. Mountain, R. D., *J. Res. Nat. Bur. Stand.* **70A**, 207 (1966).
11. Hassan, A. K., Torell, L. M. and Börjesson, L., *J. Physique IV* **C2**, 265 (1992).
12. Patterson, G. D., *J. Macr. Sci.-Phys.* **B13**, 647 (1977).

13. Levelut, C., Scheyer, Y., Boissier, M., Pelous, J., Durand, D., Emery, J., *J. Phys.: Cond. Matt.* **8**, 941-957 (1996).
14. Scheyer, Y., Levelut, C., Pelous, J., and Durand, D., *Phys. Rev. B.* **57**, 11212 (1998).
15. Levelut,C., Scheyer, Y., Pelous, J., and Durand, D., to be published.
16. Dreyfus, C., Lebon, M.-J., Cummins, H. Z., Toulouse, J., Bonello, B., and Pick, R. M., *Phys. Rev. Lett.* **69**, 3666 (1992).
17. Fuchs, M., Gotze, W., and Latz, A., *Chem. Phys.* **149**, 185 (1990).
18. Lebon, M.-J., *Ph D thesis*, University Paris 7 (France), 1995.
19. Lebon, M.-J., Dreyfus, C., Pick, R. M., *Phys. Rev. E*, 4537 (1995).
20. Lörosch J. , Couzi M. , Pelous J., Vacher R., and Levasseur A., *J. Non-Cryst. Sol.* **69**, 1 (1984).
21. Galeener, F. L., Barrio, R. A., Martinez, E. and Elliot, R. J., *Phys. Rev. Lett* **53**, 2429 (1998); Barrio, R. A., Galeener, F. L., Martinez, E., Elliott, R.J., *Phys. Rev. B* **48**, 15672 (1993).
22. Cavaillé, D., *PhD thesis*, University Montpellier II (France), 1998.
23. Winterling, G., *Phys. Rev. B* **12**, 2432 (1975).
24. J. Jäckle, "Low-frequency Raman scattering in glasses", in *Amorphous Solids: Low Temperature Properties*, ed. W. A. Phillips, Berlin: Springer, 1981, p171.
25. Gochiyaev V. Z. , Malinovsky V. K. , Novikov V. N. and Sokolov A. P. *Phil. Mag. B* **63** 777, 1991.
26. Novikov, V. N., *Phys. Rev. B* **58**, 8367 (1998); Novikov, V. N., *Phys. Rev. B***55** R14685 (1997).
27. Akhieser, A., *J. Phys. (USSR)* **1**, 277 (1939).
28. Landau, L. , and G. Rumer, G., *Phys. Z. Sovvjetunion* **11**, 18 (1937).
29. Berret, J.-F., Pelous, J., Vacher, R., *J. Phys. Lett.* **44**, L433 (1983).
30. Bonnet, J.P., Vacher, R., Pelous, J., and Laermans, C. *Phys. Rev. B* **45**, 557 (1992).
31. Hunklinger, S., and Arnold, W., "Ultrasonic Properties of Glasses at Low Temperatures", in *Physical Acoustics, Principles and methods, vol XII*, eds. Mason and Thurston, New York: Academic Press, 1976, p155; Vacher, R., Sussner, H., and Hunklinger, S., *Phys. Rev. B* **21**, 5850 (1980).
32. Karpov, V. G. Klinger, M. I., Ignat'ev, F. N., *Sov. Phys. JETP* **57**, 439 (1983).
33. Klinger, M. I., *Phys. Lett. A* **170**, 222 (1992).
34. Parshin, D. A., *Phys. Solid State* **36**, 991 (1994).
35. Gurevich, V. L., Parshin, D. A., Pelous, J., and Schober, H. R., *Phys. Rev. B* **48** 16318 (1993).
36. Prat, J. L., *PhD thesis*, University Montpellier II (France), 1996
37. Terki, F., Levelut, C., Prat, J.-L., Boissier, M., and Pelous, J., *J. Phys. Cond. Matt.* **9** 3955 (1997).
38. Fabian, J., and Allen, P.B., *Phys. Rev. Lett.* **82**, 1478 (1999).
39. Foret, M., Courtens, E., Vacher, R., and Suck, J.-B., *Phys. Rev. Lett.* **77**, 3831 (1996); Foret, M., Hehlen, B., Taillades, G., Courtens, E., Vacher, R., Casalta, H., Dorner, B., *Phys. Rev. Lett.* **81**, 2100 (1998).
40. Masciovecchio, C., Ruocco,G., Sette, F., Krisch, M., Verbeni, R., Bergmann, U., and Soltwisch, M., *Phys. Rev. Lett.* **76**, 3356 (1996).

Medium-range structure in amorphous and crystalline GeSe$_2$

Philip H. Gaskell

Cavendish Laboratory, University of Cambridge,

Madingley Road, Cambridge CB3 0HE, U.K.

Abstract. The known structure of a crystalline phase is useful in investigating the unknown structure of the compositionally-equivalent glass. One essential clue is given by reciprocal-space features at low Q (scattering vector) in X-ray or neutron scattering data, which are clearly related to the medium-range structure of the glass. Interpretation of these features as "quasi-Bragg" scattering allows direct comparison between the structures of the glass and equivalent crystalline phases.

An application of this method will be illustrated in the case of amorphous GeSe$_2$. For this material the experimental data is particularly rich as partial structure factors have been measured. Correspondence between low-Q features in the neutron diffraction data for the glassy and crystalline phases is qualitatively good and extends to the partial functions too. Thus, essential features of the medium-range structure of this glass appear to be interpretable, rather easily.

LOW-Q STRUCTURE IN GLASSES

In many glasses and melts, one or more prominent features appear in neutron and X-ray scattering data at low values of the scattering vector, **Q**. Here $|\mathbf{Q}| = Q = 4\pi\sin\theta/\lambda$, where 2θ is the scattering angle and λ, the wavelength of the probing radiation. In glasses like SiO$_2$, or GeSe$_2$, one sharp prominent peak is seen at values of Q between 10 - 15 nm^{-1}. Since this is the first feature in the diffraction pattern and the peak is very narrow, it has become known as the "First sharp diffraction peak", FSDP. In more complex, multicomponent glasses, such as the silicates, germanates and borates containing large cations like Rb and Cs, the single peak is replaced by a series of two or three distinguishable features. For this reason the term "Low-Q structure" (LQS) seems preferable.

The origins of the various types of LQS have been the subject of vigorous debate over a number of years. It is quite clear that the features are related to medium-range structure in the glass or melt – that is, inter-atomic correlations over the range 0.5-1.5 nm are involved. In earlier work, Gaskell and Wallis (1) argued that in vitreous silica – a prototypical network glass - the LQS is related to periodic variations in atomic density with a characteristic spacing of about 0.42 nm. Moreover, the atomic density fluctuations have a similar origin to the {111} Bragg peak in high cristobalite, the crystalline phase of silica that can be considered the thermodynamic neighbour of amorphous SiO_2. Briefly, the LQS structure seen in this glass reflects similar medium-range order in the amorphous glassy (and liquid, perhaps) and crystalline phases of the material. The term "quasi-Bragg planes" was coined to describe these atomic density fluctuations in the glass – although it is necessary to stress that this does not imply a micro-crystalline structure. Merely that the medium-range inter-atomic correlations in the glassy and crystalline phases have the same essential character.

APPLICATION TO AMORPHOUS $GeSe_2$

The comparison between the LQS in glasses and their crystalline equivalents has been extended by a detailed examination of the neutron and x-ray data for several borates, silicates and chalcogenide glasses. The correspondence in the Q-space values of features in crystalline and amorphous phases is good and, with more qualification, in their relative intensities in both X-ray and neutron scattering data; see, for example reference (2). The most rigorous test is afforded by measurements of partial structure factors, obtained usually by neutron scattering measurements on isotopically-substituted glasses and melts. One of the more interesting materials is $GeSe_2$. Since isotopes of both elements are available, the three partial functions for the diatomic glass can be resolved. The partial structure factors of the melt were determined some time ago by Penfold and Salmon (3) and more recently similar measurements have been published for glassy $GeSe_2$ by Penfold, Salmon and Fischer (4). The two sets of data are in good agreement although the features in the data for the glass are sharper due to reduced thermal vibrations. In particular, the first sharp diffraction peak seen in the total structure factor at about 11 nm^{-1} is also seen in the data for $S_{GeGe}(Q)$. This, in particular, has proved to be difficult to reproduce by molecular dynamics modelling (5), including *ab initio* MD (6). The result has been to polarise the community to a degree: distinguishing those who retain faith in experiment and those who cling to computation.

In this paper, I point out, simply, that the low-Q features in both the total and partial structure factors of glassy $GeSe_2$ are paralleled by similar features in the crystalline phase of $GeSe_2$. Thus, arguments for similarities in the medium range structures of the two phases become strengthened.

The first peak in the total structure factor of GeSe$_2$ is observed at $Q = 11$ nm^{-1}, almost exactly the same value as the prominent (002) peak of the crystalline phase. This leads to the supposition that the LQS in the glass is related to the pronounced layering of the lattice projected normal to this direction (Figure 1). Experimental partial structure factor data, $S_{\alpha\beta}(Q)$, for glassy GeSe$_2$ is shown in Figure 2 from the recent work of Salmon et al (4). Note the fact that the first peak in the total structure factor at 11 nm^{-1} appears strongly in the Ge-Ge partial, as a very weak feature in the Ge-Se partial, and is almost completely absent in $S_{SeSe}(Q)$.

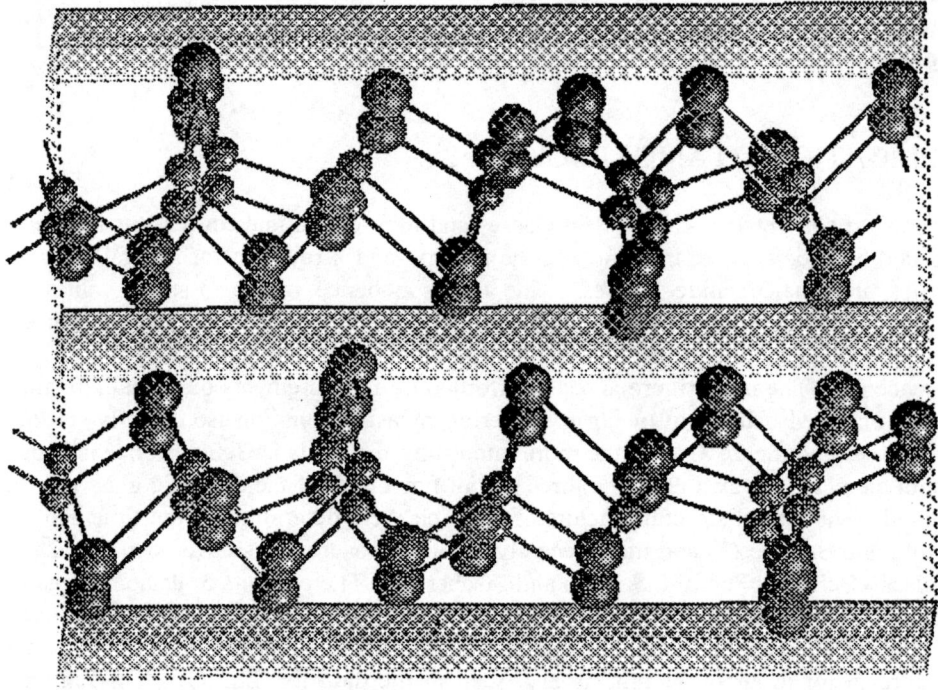

FIGURE 1. The structure of crystalline GeSe$_2$ (7) projected normal to (002). The (002) planes are positioned between the GeSe$_2$ layers for clarity.

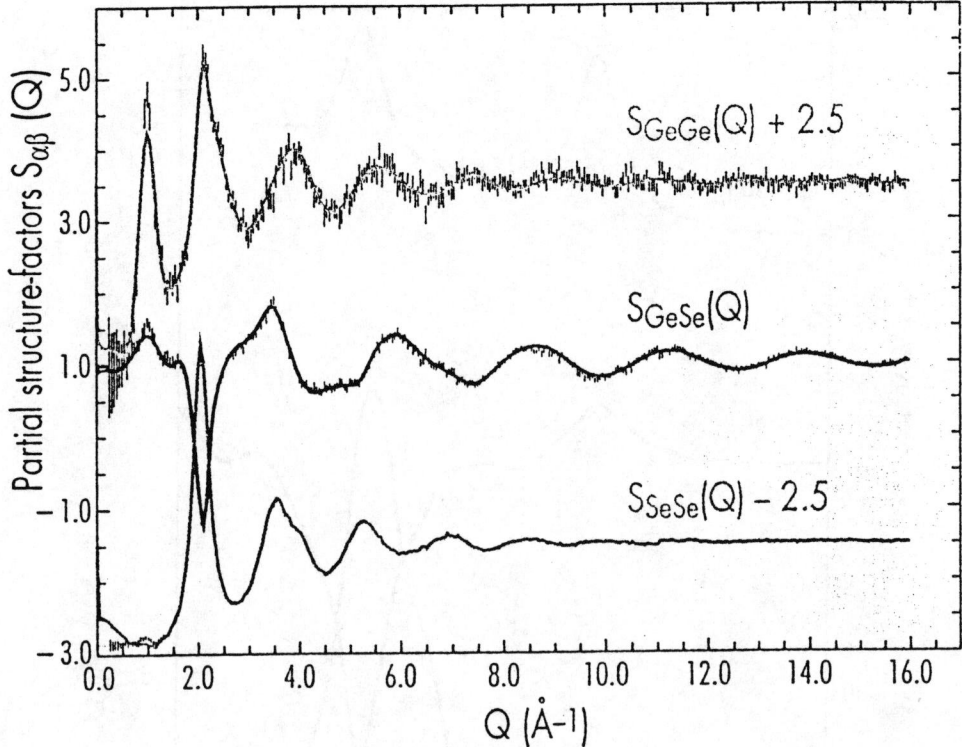

FIGURE 2. Measured partial structure factors S(Q) for glassy GeSe$_2$ from the recent work of Petri, Salmon and Fischer (4). These data are similar to previous results for molten GeSe$_2$ (3).

First-order simulation of the diffraction data for the glass can be obtained by calculating the diffracted intensity from the unit cell of the crystal (7). The calculations were carried out on a standard crystal diffraction package (Molecular Simulations, Cerius). In order to give an approximate indication of effects due to disorder, the calculations were performed for a 3x3x3 nm crystallite. The partials were obtained by calculating the scattered intensity for the complete structure, then for the Ge and Se sub-lattices. The last two give directly quantities related to $S_{GeGe}(Q)$ and $S_{SeSe}(Q)$. The Ge-Se partial can be simulated by subtracting the contributions of the scattered intensities from the Ge and Se sub-lattices from the total scattered intensity. Note however that these values are *not* identical to the partial structure factors. Results are shown in Figure 3.

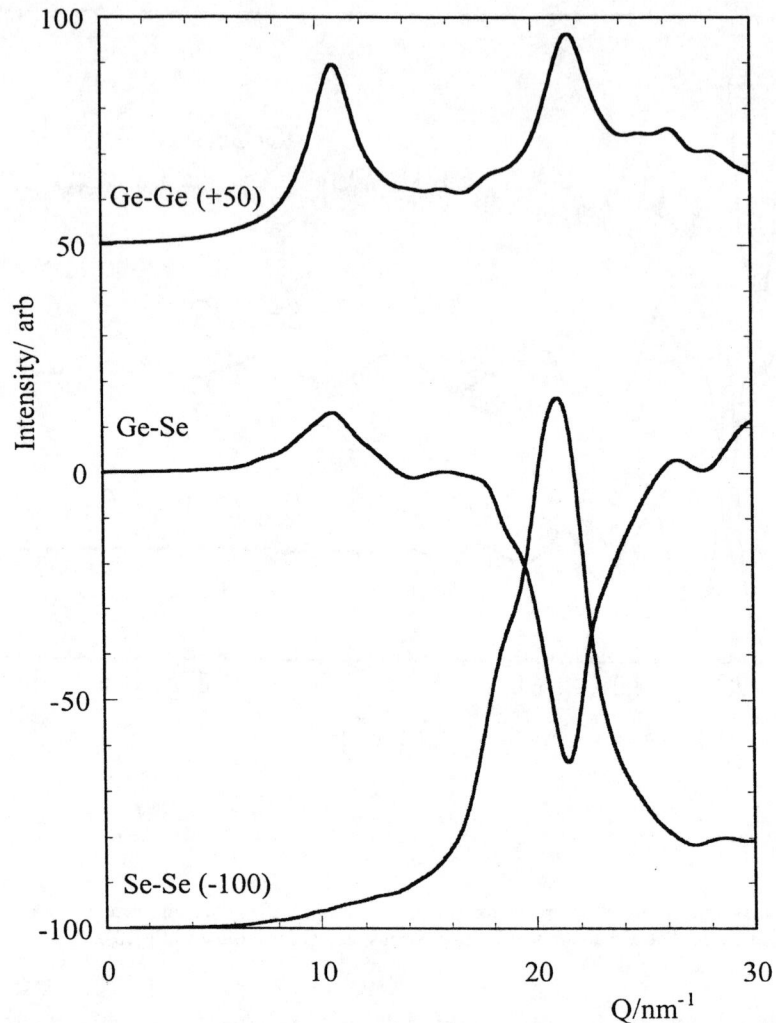

FIGURE 3. Intensity of neutron scattering from the Ge-Ge, Se-Se and Ge-Se correlations simulated using the structure of crystalline GeSe$_2$ (7). Note that the ordinate is common to all three functions but that the scale is arbitrary. These data do not correspond directly to the results shown in Figure 2 as the former have not yet been correctly normalised. This does not affect the Q scale, of course.

The similarity between the simulated results and experimental data is obvious – even though the ordinates of Figure 3 are on arbitrary scales. In particular, the appearance of the sharp first peak at 11nm^{-1} in the simulated data for Ge-Ge scattering and its absence in Se-Se scattering is encouraging. The small shoulder at 11 nm^{-1} in the Ge-Se contribution should also be noted, together with the negative-going feature at the position of the second peak in the total structure factor.

The detailed shape of the partial structure factors at low Q becomes obvious from Figure 1. Planes with a spacing of $d_{002} = 2\pi/Q_1$, where $Q_1 = 11$ nm^{-1}, corresponding to position of the first peak are shown in Figure 1 For clarity, these have been positioned between the GeSe$_2$ layers, but it is obvious from the figure that they could have been positioned to pass through the centre of the layers - and approximately through the Ge atom sites. Se atoms on the other hand lie above and below the Ge-containing Bragg planes. Any "quasi-Bragg" planes in the glass with medium-range structure similar to that of the crystal will give rise to a diffracted peak from the Ge sub-structure at 11 nm^{-1} and at approximately twice this value from the Se substructure, but with no contribution at 11 nm^{-1}. The correspondence between the weak feature at 11 nm^{-1} in the Ge-Se scattering and the sharp first peak at the same value of Q seems, on this argument, a fortuitous coincidence. The spacing between the Se atoms in one GeSe$_2$ layer and those in neighbouring layers is not obviously related to Ge-Se nearest neighbour distances.

SUMMARY

The calculations described here are trivial compared to the previous MD simulations. Nonetheless they give an interesting representation of the experimental data – even though the results are, at best, only a semi-quantitative representation of the intensity distributions. Further calculations to provide a more direct representation of the partial structure factors are in progress. The indications are, however, that models which adequately represent the structure of glassy and liquid GeSe$_2$ will need to reproduce the essential medium-range structure-forming elements seen in the crystalline phase – the most obvious being the pronounced layering.

REFERENCES

1. Gaskell, P.H. and Wallis, D.J., *Phys. Rev. Lett.* **76**, 66-69 (1996)
2. Gaskell, P.H. (1997), in *2nd Int. Conf. on borate glasses, crystals and melts* (eds A.C. Wright and S.A. Feller) Sheffield: Soc. Glass Tech., 71-9
3. Penfold, I.T. and Salmon, P.S., Phys. Rev. Lett., **67**, 97-100 (1991)
4. Petri, I., Salmon, P.S. and Fischer, H.E., *Structure of the binary network glass GeSe$_2$* Annual report of the ILL 52-53 (1998)
5. Vashishta, P., Kalia, R.K., and Ebbsjö, I. Phys. Rev. **B39** 6034-47 (1989)
6. Massobrio, C., Pasquarello, A. and Car, R., Phys. Rev Lett. **80** 2342-45 (1998)
7. Dittmar, von G., and Schäfer, H., Acta Crystall., **B32** 2726 (1976)

II MODE COUPLING THEORIES

Mode Coupling Theories

Jean-Louis Barrat*

*Département de Physique des Matériaux
Université Claude Bernard and CNRS
43 Boulevard du 11 Novembre
69622 Villeurbanne Cedex, France

INTRODUCTION: LANGEVIN EQUATION AND BEYOND

The Langevin equation was initially formulated as a semi-empirical tool to describe the motion of a large (slow) brownian particle interacting with the small (fast) molecules of a solvent. In this context, it reads

$$M\dot{\mathbf{V}} = -\zeta M \mathbf{V} + \mathbf{R}(t) \qquad (1)$$

where $\mathbf{V}$ is the brownian particle velocity, $-\zeta M\mathbf{V}$ is the "systematic" contribution to the force that acts on this particle (i.e. the average force the solvent exerts on a particle with constant velocity $\mathbf{V}$), and $\mathbf{R}$ is the so called "random force" that corresponds to the many collisions experienced by the slow particle. The essential assumption in the Langevin approach is that this random force does not have any memory, i.e. that its correlation function is

$$<R_\alpha(t)R_\beta(t')> = 2\pi R_0 \delta_{\alpha\beta}\delta(t-t') \qquad (2)$$

In order for the equilibrium distribution of $\mathbf{V}$ to satisfy thermodynamic equilibrium in the long time limit, it can be shown that the strength R_0 of the fluctuating force has to satisfy the so called fluctuation dissipation relation [1]

$$\pi R_0 = k_B T \zeta M. \qquad (3)$$

This Langevin approach (and the related Fokker-Planck equation) have been applied to describe a number of physical situations in which a (supposedly) slow degree of freedom is coupled to many "fast" degrees of freedom. Such applications are described in detail in the books by van Kampen or Risken [2,3].

Now, the separation on time scales implicit in the Langevin approach is obviously not present if one wants to study the motion of a molecule comparable in size to other solvent molecules. A possible approach, still phenomenological at this stage, is to write a "Generalized Langevin Equation" (GLE) in the form

$$M\dot{\mathbf{V}} = -\int_0^t M\zeta(t-s)M\mathbf{V}(s)ds + \mathbf{R}(t) \tag{4}$$

thus allowing for a memory in the "systematic" part of the force. A non-obvious consequence of this is that consistency with thermodynamic equilibrium then requires a generalized fluctuation dissipation relation of the form

$$<\mathbf{R}_\alpha(t)\mathbf{R}_\beta(t')> = k_B T M \zeta(t-t')\delta_{\alpha\beta} \tag{5}$$

which implies that the "random force" R and the "memory function" ζ have correlations on the same time scale. In this approach, the general strategy would then be to first isolate some slow variable ($\mathbf{V}$ in that case), then to write (using the approach of the next section) a generalized Langevin equation for this variable, and to formulate simple approximations for the random force (e.g. $<R(t)R(t')>\sim \exp(-(t-t')/\tau)$). The friction on the "slow" variable is then $\int_0^\infty <R(t)R(0)>dt$, entirely determined by the "fast" variables.

Although it is more sophisticated that the original Langevin approach, this reasoning (which will be described in the following as the "Markov approximation") is not fully satisfactory. In particular, it has proven unable to explain such features as the so called "long time tails" in correlation functions ($<\mathbf{V}(t)\mathbf{V}(0)>\sim t^{-\alpha}$, or the critical anomalies in the transport coefficients, the divergences in transport coefficients in two dimensions, or the slowing down in supercooled liquids. All these prominent features in the dynamics of liquids have been treated within "mode-coupling" theories, which go beyond the simple Markov approximation by accounting for long range time correlations in the "random forces".

THE MORI-ZWANZIG FORMALISM

The Mori-Zwanzig formalism [4] is essentially a mathematical device that allows to write *formally exact* GLE equations starting from the microscopic equations of motion. If one considers a classical system of points particles characterized by positions and velocities, and defines an observable A as being some function of these microscopic variables. $A(t) = A(R_i(t), p_i(t))$ is then a function of time whose evolution is governed by the Liouville operator L of the system. The general idea behind the formalism is to interpret the correlation function $c_{AA}(t) = <A(t)A(0)>$ as a projection of $A(t)$ on $A(0)$. More precisely, for any observable B the projection on $A(0)$ is defined as

$$PB = <BA(0)>/<A(0)A(0)> \tag{6}$$

(for simplicity I consider only real observables here. Complex conjugation must be introduced in appropriate places for complex quantities) The orthogonal projector Q is defined as $Q = 1 - P$. The general idea, whose details can be found in many textbooks [5,4] is then to treat separately the components PA ("slow" component) and QA (fast component). The result is a formally exact equation for the evolution of A

$$\dot{A}(t) - i\Omega A(t) + \int_0^t M(t-s)A(s)ds = R(t) \tag{7}$$

In this GLE, the oscillation frequency Ω, memory function M and random force R have now the precise definitions

$$\Omega = <\dot{A}(0)A(0)> / <A(0)A(0)> \tag{8}$$

$$R(t) = \exp(iQLQt)Q\dot{A}(0) \tag{9}$$

$$M(t) = <R(t)R(0)> / <A(0)A(0)> \tag{10}$$

As a consequence, the correlation function c_{AA} obeys

$$\dot{c}_{AA}(t) - i\Omega c_{AA}(t) + \int_0^t M(t-s)c_{AA}(s)ds = 0 \tag{11}$$

All this can be generalized without difficulties to the case where A stands for a set $A_1...A_p$ of observables, in which case both Ω and M become pxp matrices.

The Markov approximation

If one is able, based on some physical intuition, to identify in the set A all the "slow" variables in the problems (I.e. A would correspond to variables evolving on time scales well separated from the microscopic ones), one can expect that $M(t)$ has a short memory (it corresponds to the correlation function of R, which is uncorrelated to the slow variables). The GLE becomes then a simple Langevin equation

$$\dot{A}(t) - i\Omega A(t) + \Gamma A(t) = R(t) \tag{12}$$

with $\Gamma = \int_0^\infty M(s)ds$. It is in fact possible to improve systematically on this Markovian approximation. The general scheme consists in using the Laplace transformed equation

$$C_{AA}(z) = \frac{1}{z - i\Omega_0 + M(z)} \tag{13}$$

Extending the Mori Zwanzig scheme to the random force itself, one eventually gets a continued fraction expansion for $c(z)$

$$C_{AA}(z) = \cfrac{1}{z - i\Omega_0 + \cfrac{\Delta_1}{z - i\Omega_2 + \cfrac{\Delta_2}{z - \ldots\ldots}}} \qquad (14)$$

The coefficients in this expansion can in fact be obtained from the knowledge of the short time expansion of $c_{AA}(t)$, which in turn involves only static correlation functions, like $\dot{c}_{AA}(0) = <\dot{A}(0)A(0)>$. There are, however, severe limitations to this approach. First of all, high order derivatives of $c_{AA}(t)$ at $t=0$ are difficult to compute exactly. More fundamental is the fact that such an approach, truncated at finite order, can only yield exponential relaxation (the inverse Laplace transform of a rational fraction being a discrete sum of exponentials).

An example: shear viscosity of a fluid

Let us consider a classical fluid made of point particles. We take as the slow variable A the Fourier components of the transverse momentum density for small wave-vectors, i.e.

$$A_q = j_q^x = \sum_i p_{ix} \exp(iqz_i) \qquad (15)$$

for a wave-vector $\mathbf{q}$ parallel to Oz. Because of translational invariance different wave-vectors are completely decoupled and we can as well consider a single A_q (rather than the entire set). Also, the transverse momentum density decouples from all other variables such as energy density or number density. Nevertheless, it is a slow variable because it is the density of a conserved variable.

In that case the time derivative of A_q is by definition related to the off diagonal stress tensor, i.e.

$$\dot{A}_q = q\sigma_{xz}(\mathbf{q}) \qquad (16)$$

The Mori Zwanzig equations applied to this slow variable gives for the Laplace transform of the transverse momentum density autocorrelation function

$$C_{AA}(q, z) = \frac{1}{z + M(q, z)} \qquad (17)$$

(in that case the frequency Ω vanishes because the observable is odd under time reversal.). The memory function is formally given by

$$M(q,z) = (Nk_BT)^{-1} <\sigma_{xz}(-\mathbf{q})Q \exp(iQLQt)Q\sigma_{xz}(\mathbf{q})> \qquad (18)$$

In the small wave-vector limit ($q \to 0$), $M(q,z)$ can be replaced by $q^2 G(z)$, where $G(z)$ is the Laplace transform of the stress stress autocorrelation function (in this limit it can be shown that all the projectors Q can safely be ignored). If all the slow variables have indeed been included in the set A, a Markovian approximation can be made at this stage, yielding

$$C_{AA}(q,z) = \frac{1:}{z + \nu q^2} \tag{19}$$

where ν is the usual (kinematic) shear viscosity, which is related to the integral of the stress stress autocorrelation function through the usual Kubo formula [5,4]. This form is typical of the correlation of a quantity (here transverse momentum) whose evolution is governed by a diffusion equation.

In practice, the limiting behaviour described by equation is observed as soon as one reaches the hydrodynamic limit, such that $1/q$ is larger than any internal length scale and $1/\nu q^2$ is larger than any internal relaxation time of the fluid. This will of course always be achieved in principle for small enough q, but this limit is not necessarily relevant. Example of fluids with large internal length scales or internal relaxation times will be described in the following.

WRITING MODE COUPLING EQUATIONS

The Markovian approximation described above is valid only in the case where all the slow variables have been included in the original set A. In many cases, the difference between slow and fast variables is not so clear-cut, so that the "random" force will actually depend on relatively slow variables. Mode coupling equations or theories have been formulated in order to account for this dependence in an approximate manner.

As a first example, let us consider the case where A_q is the Fourier component of the density of a conserved quantity. Its time evolution is then given by

$$\dot{A}_{\mathbf{q}} = -i\mathbf{q}.\mathbf{j}_{\mathbf{q}} \tag{20}$$

where $\mathbf{j}_{\mathbf{q}}$ is a current. The memory function appearing in equation is then

$$M_{\mathbf{q}}(t) = q^2 <\mathbf{j}_{-\mathbf{q}} Q \exp(iQLQt) Q\mathbf{j}_{\mathbf{q}}> / <A_{-\mathbf{q}} A_{\mathbf{q}}> \tag{21}$$

Unfortunately, although $Q\mathbf{j}_{\mathbf{q}}$ is, by construction, orthogonal (uncorrelated) to $A_{\mathbf{q}}$, it is not orthogonal to variables such as the bilinear variables $B_{\mathbf{qq'}} = A_{q-q'} A'_q$, which can be expected to be rather slow. The basic idea will then be to separate in Qj_q a part that is really fast, and a part that is correlated to the $B_{\mathbf{qq'}}$. After some algebra, this leads to a decomposition of the memory function in the form

$$<\mathbf{j}_{-\mathbf{q}} Q \exp(iQLQt) Q\mathbf{j}_{\mathbf{q}}> / <A_{-\mathbf{q}} A_{\mathbf{q}}> = K^*(t) + K_{MC}(t) \tag{22}$$

where K^* is a rapidly decaying function and

$$K_{MC}(q,t) = \sum_{qq'} <B_{\mathbf{qq'}} \exp(iQLQt) B_{\mathbf{qq'}}> U_{\mathbf{qq'}} U_{\mathbf{qq'}} \tag{23}$$

the "vertices" $U_{\mathbf{qq'}}$ are time independent quantities that can be calculated (usually) from the statics of the system. One then proceeds by applying a crude and uncontrolled *factorization approximation* to the correlation function of the $B_{\mathbf{qq'}}$

$$< B_{\mathbf{qq'}} \exp(iQLQt) B_{\mathbf{qq'}} > \sim \delta_{q'q''} C_{AA}(q',t) C_{AA}(q+q',t) \qquad (24)$$

This obviously allows to obtain a closed set of coupled equations for the correlation functions $C_{AA}(q',t)$. If the Markovian approximation is made for the part $K^*e(t)$ one has

$$\dot{C}_{AA}(q,t) = (i\Omega \Lambda^* q^2) C_{AA}(q,t)$$
$$-q^2 \sum_q {}' U_{qq'}^2 \int_0^t C_{AA}(q',t-s) C_{AA}(q+q',t-s) C_{AA}(q,s) ds \qquad (25)$$

These equations have been obtained by considering only the coupling to bilinear variables (modes). Obviously one would expect a contribution from higher order terms like AAA. It turns out; however; that coupling to bilinear modes seems to yield at least qualitatively correct results, in the sense that it captures the most singular (i.e. slowly decaying) part in the time dependence of the memory function.

Equation 25 are nonlinear coupled integrodifferential equations. In order to see how they can yield a singular dependence, let us first ignore the "mode-coupling" term and assume $\Omega = 0$. One obtains the diffusive behaviour $C_{AA}(q,t) \sim \exp \Lambda^* q^2 t)$. The mode coupling term is then easily shown to behave as $K_{MC}(q,t) \sim \Lambda^* t)^{-d/2}$ for small wavelength. Hence even at this very simple level we see the failure of the Markov approximation, and also the peculiar role of space dimension $d = 2$ for which the mode coupling contribution to transport coefficients becomes divergent.

A seemingly different approach to the writing of mode coupling equations was taken by Kawasaki [6]. His method essentially consists in three steps. First, identify the set of interesting slow variables A_i. Second, add to the set all possible products of the A_i, and formally write the corresponding Mori-Zwanzig equations. As now all the potentially slow variables have been included in the set one can use a Markov approximation, so that the evolution equation for A_i reads (repeated indexes are summed over)

$$\dot{A}_i = i\Omega_{ij} A_j - \Lambda_{ij} A_j + R_i + V_{ijk} A_j A_k. + ... \qquad (26)$$

In principle, one should simultaneously write the equations for all the products. Instead, Kawasaki limits himself to 26, *truncated at second order in A* (this is equivalent to the coupling to bilinear variables only in the other method) and transforms it into an equation for the time correlation function using a diagrammatic method. To illustrate the method, consider the simple equation

$$\dot{A} = i\Omega A - \Lambda A + R + vAA. \qquad (27)$$

If $C(t)$ is the correlation function of A for a nonzero value of the nonlinear coupling v, and C_0 its value for $v = 0$, then one has the Dyson equation

$$G(t) = G_0(t) + G * \Sigma * G_0 \tag{28}$$

where $\Sigma(t)$ is the self energy (equivalent to the memory function) formed by all irreducible diagrams involving links G_0 and vertices v [7]. The mode coupling approximation then consists in treating Σ within the "self consistent one loop approximation", i.e. using only the first diagram $\Sigma(t) = 2v^2 G_0(t)^2$ and replacing $G_0(t)$ by $G(t)$, so that a self consistent equation for $G(t)$ is eventually obtained.

Although the approximations used in Kawasaki's approach are apparently quite different from those presented in the first approach (which is usually preferred by people working in liquid state theory), the final results turn out to be identical. One advantage of the Kawasaki approach is that the "self consistent one loop" approximation turns out to be exact when dealing with disordered systems in high spatial dimension, as discussed in [8] or in the lectures by J.P. Bouchaud.

MODE COUPLING THEORIES AND CRITICAL PHENOMENA

Although this is not directly related to the main topic of this school, I would like to very briefly describe one of the most significant success of mode coupling theories, namely their application to critical phenomena. For details, I refer the reader to e.g [9]. Let us consider an incompressible, symmetric binary mixture undergoing a phase separation at a temperature T_c. From a static point of view, the growth of the concentration fluctuations $c_\mathbf{q}$ close to T_c can be described by the Ornstein Zernike form of the concentration-concentration structure factor [10]

$$S(q) = <c_\mathbf{q} c_{-\mathbf{q}}> = \frac{S(0)}{1 + q^2 \xi^2} \tag{29}$$

where the correlation length ξ diverges close to T_c as $(T - T_c)^{-\nu}$, and $S(0) \sim \xi^2$. The classical theory of the dynamics of fluctuations near the critical point assumes that $c_\mathbf{q}(t)$ obeys the Langevin type equation [9]

$$\dot{c}_\mathbf{q}(t) = -\lambda \nabla^2 \frac{\partial F}{\partial c_\mathbf{q}} + R_\mathbf{q}(t) \tag{30}$$

For small fluctuations, the derivative of the free energy with respect to $c_\mathbf{q}$ is just $1/S(q)$. Hence in this theory one eventually finds that the collective diffusion constant D vanishes as ξ^{-2} near to the critical point ("critical slowing down" [11]). Other transport coefficients are, in this classical theory, not affected by the proximity of the critical point.

Some objections can immediately apparent. In particular, based on physical intuition, one would expect that the large concentration fluctuations that form near T_c behave in some sense as "droplets" of size ξ, so that hydrodynamics would predict a diffusion constant of order $1/\xi$. One should also be aware that large concentration fluctuations give rise to fluctuating interfaces, and the associated fluctuating stresses (surface tension) can be expected to give a contribution to the viscosity. All these effects, which are not captured by the classical van Hove theory, are accounted for by a "mode coupling" approach to the problem. As "slow" variables, one takes the concentration c_q and the transverse part of the momentum density, jq (the longitudinal part turns out to be irrelevant for an incompressible system). It is easily shown that within the Mori formalism, the 2x2 memory function matrix is diagonal, and that the part corresponding to the concentration fluctuations is

$$M_{cc}(q,t) = q^2 < \mathbf{g_q} Q \exp(iQLQt) Q \mathbf{g_q} > /S(q) \tag{31}$$

where by definition $\dot{c}_\mathbf{q} = i\mathbf{q}.\mathbf{g_q}$. The mode coupling contribution to M_{cc} is obtained by projecting on bilinear modes of the form $c_{-q'+q} j'_q$. Following the approximation scheme described above, one eventually ends up with a "mode coupling" contribution to $M_{cc}(q,t)$ of the form

$$M_{cc}^{MC}(q,t) = \frac{k_B T q^2}{NmS(q)} \sum_q{}' (1 - (\mathbf{q}.\mathbf{q'})^2/q^2 q'^2) S(\mathbf{q}-\mathbf{q'}) \exp(-q'^2 \nu t) \exp(-(\mathbf{q}+\mathbf{q''})^2 Dt) \tag{32}$$

where we have replaced the time correlation functions of c_q and jq by their long time, diffusive behaviour. the collective diffusion constant D is given by $q^{-2} \int_0^\infty M(t) dt$, so than one eventually arrives at a self consistency equation for D

$$D = D_0 + \lim_{q \to 0} \sum_q{}' (1 - (\mathbf{q}.\mathbf{q'})^2/q^2 q'^2) S(\mathbf{q}-\mathbf{q'})/S(q) (D(\mathbf{q}+\mathbf{q'})^2 + \nu \mathbf{q'}^2)^{-1} \tag{33}$$

where D_0 is a "regular" contribution, corresponding to the short time decay of M This equation must in principle be solved self consistently for D. Taking into account that $\nu >> D$, one can easily show that for $q\xi >> 1$ $D = D_0 + \frac{k_B T}{6\pi\xi}$, confirming the prediction based on hydrodynamics. Of course a complete solution to the problem is obtained by writing down the coupled equations for the momentum and concentration density (in fact energy fluctuations must be included as well in a complete treatment). then all the transport coefficients are obtained by solving three coupled equations for D, η, and the thermal conductivity. η is for example found to diverge weakly, $(\eta \sim \log \xi)$ near T_c.

MODE COUPLING THEORY FOR THE SLOWING DOWN OF SUPERCOOLED LIQUIDS

"Historical" background

In the critical slowing down phenomenon described in the previous section, the essential reason for the slowing down was the divergence of the structure factor at small wave-vectors. When a liquid is supercooled, its structure factor becomes more and more structured, in the sense that its first peak (characteristic of short range molecular order) becomes more pronounced. Hence it was noticed already in 1978 by Sjolander and Turski [12] that a divergence in the structure factor at finite wave vector (similar to what is observed e.g. in diblock copolymers) could lead to a corresponding divergence in structural relaxation times. It is however quite clear both from experiment and simulation that no such divergence is observed in the structure factor upon cooling. Rather, the first peak in the structure factor increases moderately as T decreases. It was Geszti [13] in 1983 who first described, in a paper entitled "Pre-vitrification by viscosity feedback", a mechanism by which such a moderate increase could in fact yield to a very large increase of viscosity. Shortly after, Bengtzelius, Göze and Sjolander [14] and gave a more sophisticated description of this mechanism. In the next few years extensions and consequences of this more sophisticated theory were worked out by Götze and coworkers. A detail led account can be found in refs [15,16].

Before entering a detailed discussion of this mode coupling theory, I would like to summarize the essential points of the mechanism proposed by Geszti. The idea of a "viscosity feedback" consists in three steps that form the "feedback loop". the first step is to relate the viscosity to the stress-stress autocorrelation function through the exact Kubo formula, i.e. $\eta \sim \int_0^\infty <\sigma(t)\sigma(0)> dt$. Next, one uses some "projection" scheme to relate the decay of the stress stress autocorrelation function to the decay of the density-density correlation function (dynamical structure factor), i.e. $<\sigma(t)\sigma(0)> \sim <\rho_q(t)\rho_q(0)>$. Finally, assuming a diffusive behavior for $<\rho_q(t)\rho_q(0)> \sim \exp(-Dq^2 t)$, the feedback loop is closed by relating the collective diffusion constant D to the viscosity through a Stokes Einstein type relation, $D \sim 1/\eta$. the result is a closed nonlinear equation for η, with coefficients that can be related to the structure factor of the liquid. the remarkable feature of these equation is that for a finite value of this structure factor, which plays the role of a "control parameter" in the feedback loop, a divergence of the viscosity is obtained, which can be described as an "ideal glass transition".

The Bengtzelius Götze Sjolander theory

The first step in the theory [14] is the identification of the slow variables. As the phenomenon of interest is the slowing down of structural relaxation (i.e. the fact that the local structure of the liquid relaxes slowly), it is natural to choose as slow variables the Fourier components of the microscopic density, $\rho_{\mathbf{q}}$. Also included as a slow variable are the Fourier 'components of the momentum density $j_{\mathbf{q}}$ (this, however, is not compulsory; a parallel theory with equivalent results can be built for colloids, for which the momentum density is not a conserved variable). Application of the Mori Zwanzig formalism eventually leads to an integrodifferential equation for the normalized correlation function $\phi(q,t) = <\rho_{\mathbf{q}}(t)\rho_{-\mathbf{q}}(0)>/NS(q)$ ($S(q)$ is the structure factor of the fluid).

$$\ddot{\phi}(q,t) + \Omega_0(q)\phi(q,t) + \int_0^t M(q,t-s)\dot{\phi}(q,s)ds = 0 \qquad (34)$$

Here $\Omega_0(q) = q^2 k_B T / mS(q)$ is a vibration frequency. the mode coupling scheme, in which the bilinear variables of interest are density products, yields a "mode coupling" contribution of the form

$$M^{MC}(q,t) = \int d^3q' V(q,q')\phi(q',t)\phi(q-q',t) \qquad (35)$$

I will not detail here the form of the "vertices" $V(q,q')$, which can be found in the original paper. the important information is that these vertices can be expressed in terms of the structure factor of the fluid, and that they will be increasing functions of $S(q)$. Hence increasing the structure factor (either by density increase or temperature decrease) implies larger vertices.

Although the final equations are closed equations for $\phi(q,t)$, they are still quite complex due to the coupling between wave vectors. Hence it is useful to study a schematic model, which retains some of the features of the original mode coupling equations without the complexity introduced by this coupling. Such a model was formulated in [14], and corresponds to the simpler equation

$$\ddot{\phi}(t) + \Omega_0^2 \phi(t) + \lambda_2 \Omega_0^2 \int_0^t \phi(t-s)^2 \dot{\phi}(s)ds = 0 \qquad (36)$$

A possible interpretation is that the correlator ϕ in this equation corresponds to the structure at the main peak of the structure factor. the control parameter, in which all the temperature/density effects have been included, is the coefficient λ_2. I will now discuss in some detail the analysis of this simplified model.

Analysis of the schematic model

With initial conditions $\phi(0) = 1$ and $\dot{\phi}(0) = 0$, equation 36 can be written in Laplace transform form as

$$\phi(z) = 1/(z + \Omega_0^2/(z + M(z))) \qquad (37)$$

where $M(z)$ is the Laplace transform of $\lambda_2 \Omega_0^2 \phi(t)^2$. In this form, it is easy to check for the possibility of a *non-ergodic* behaviour of ϕ, i.e. a solution in which $\phi(t)$

does not relax to zero for long times. Such a solution corresponds to $\phi(z) \sim f/z$ for small z. Balancing $1/z$ singularities in 37, it is easily seen that a non-ergodic solution will be possible only if the equation

$$f/(1-f) = M(f)/\Omega_0^2 = \lambda_2 f^2 \tag{38}$$

has a nonzero real solution. Hence if $\lambda_2 < \lambda_2^c = 4$, only ergodic solutions are found. On the other hand if the "coupling" λ_2 is larger than this critical value, a non-ergodic solution is obtained. Slightly above λ_2^c, the "non-ergodicity parameter" (i.e. the long time limit of ϕ) varies as $f_c + k\epsilon^{1/2}$ where $\epsilon = \lambda_2 - \lambda_2^c$, and $f_c = 1/2$.

Close to λ_2^c, $\phi(t)$ exhibits a slowing down whose scaling properties can be studied using the powerful techniques developed by Götze and coworkers [15].

Critical correlations: $\lambda_2 - \lambda_2^c$

The first step is to rewrite the Laplace transformed equation as

$$\phi(z)/(1 - z\phi(z)) = z/\Omega_0^2 + M(z)$$

and, at the critical point, to search for a solution in the form $\phi(z) = f_c z^{-1} + G(z)$. For studying the small frequency behaviour, it is legitimate to treat $zG(z)$ as a small quantity to obtain

$$z^{-1}(f_c/(1-f_c) - \lambda_2 f_c^2 - \lambda_2 LT(G(t)^2) + zG(z)^2(1-f_c)^{-3} = \\ G(z)(2\lambda_2 f_c - (1-f_c)^{-2} + z/\Omega_0^2 + O(z^2 G(z)^3) \tag{39}$$

At this point, we have not yet taken advantage of the fact that we are at the critical point. If we now specialize to $\lambda_2 = \lambda_2^c$, cancelations occur that yield the much simpler equation

$$\lambda LT(G(t)^2)(z) - zG(z)^2 = \text{constant} \times (z/\Omega_0^2 + O(z^2 G(z)^3)) \tag{40}$$

where $\lambda = \lambda_2^c(1-f_c)^3 = 1/2$. an important remark is that the same kind of cancelation can easily be shown to take place if the "mode coupling polynomial" $M(f) = \lambda_2 f^2$ is replaced by some other, more complex, polynomial. the important feature that leads to these cancelations is that f_2^c is obtained as the first real degenerate root of $f/(1-f) = M(f)$ as some control parameter is varied. Hence the behaviour obtained within this simple model is in fact completely generic.

In the low frequency region, one can neglect the r.h.s. of 40. It is then easily shown that $G(t) = A/(\Omega_0 t)^a$, $G(z) \sim z^{a-1}$ solves 40 provided the exponent a verifies $\lambda \Gamma(1-2a) = \Gamma(1-a)^2$. This equation will have a solution $0 < a < 1$ provided $\lambda < 1$, which again can be shown to be a generic feature obtained for any "mode coupling polynomial". If $0 < a < 1$ it is then easily shown that neglecting the r.h.s. in the low frequency limit is justified, since $1 << G(z) << 1/z$.

Hence the critical correlations consist essentially (apart from a short time transient) into a slow approach to the plateau value f_c, in the form of a power law $\phi(t) = f_c + A/(\Omega_0 t)^a$

The "glassy phase": $\lambda_2 > \lambda_2^c$

Setting $\epsilon = \lambda_2 - \lambda_2^c$ and again searching for a solution in the form $\phi(z) = f_c/z + G(z)$., it is clear that $G(z)$ will now have a $1/z$ pole with a weight proportional to $\epsilon^{1/2}$, as the value for the long time plateau is now $f = f_c + k\epsilon^{1/2}$. One can essentially expect that the decay will follow the critical decay discussed above, until this decay hits the plateau value f_c. Such an argument defines a time scale τ_ϵ through $(\Omega_0 \tau_\epsilon)^{-a} \sim \epsilon^{1/2}$. At this time scale, we can expect the decay of $\phi(t)$ to depart from the critical decay and to stop at the new plateau value. This rapid analysis is born out by a more detailed construction in which one writes $G(z)$ in the form $\epsilon^{1/2} G_\epsilon(Z) \omega_\epsilon$, with $\omega_\epsilon = 1/\tau_\epsilon$ and . $Z = z/\omega_\epsilon$ In this way we expect that G_ϵ becomes a scaling function G_0 in the small ϵ limit with properties:
- $G_0(Z) \sim 1/Z$ for $Z << 1$ (corresponding to the long time plateau part.
- $G_0(Z) \sim Z^{a-1}$ for $Z << 1$ (corresponding to the critical decay.)

This is indeed the case, as can be seen by performing an analysis quite similar to that we made for the critical decay. Here the function G_0 can be shown to obey a scaling equation of the form

$$B/Z + Z G_0(Z)^2 - \lambda LT(G_0(t)^2)(Z) = 0 \qquad (41)$$

where B is a positive constant, which yields a solution with exactly the expected property. the important point at this stage is the appearance of the time scale τ_ϵ, below which the critical (power law) decay is obtained. This time scale will be described in the following as the β *relaxation* time scale. In the time domain, one can write

$$\phi(t) = f + \epsilon^{1/2} f_0(t/\tau_\epsilon) + O(\epsilon^{3/2}) \qquad (42)$$

where f_0 is a scaling function such that $f(x) \sim x^{-a}$ for $x << 1$, and $f(x) \sim \exp(-x)$ for $x >> 1$.

The liquid phase: $\lambda_2 < \lambda_2^c$

In this case, the correlation function $\phi(t)$ decays to zero for long times. It is now convenient to define $\epsilon = \lambda_2^c - \lambda_2$, and to perform the same analysis as before. Again, one writes $\phi(z) = f_c/z + G(z)$. Although the argument is perhaps less intuitive, one again introduces the time scale τ_ϵ such that $(\Omega_0 \tau_\epsilon)^{-a} \sim \epsilon^{1/2}$, and above which the behaviour of $G(t)$ is expected to depart from the critical power law decay. Making the same scaling Ansatz $G(z) = \epsilon^{1/2} G_\epsilon(Z) \omega_\epsilon$, one arrives for G_0 at the scaling equation

$$-B/Z + Z G_0(Z)^2 - \lambda LT(G_0(t)^2)(Z) = 0 \qquad (43)$$

Only the sign in the first term has changed compared to the previous case 41. This is an important change, however, which corresponds to the fact that for long rescaled times ($t \gg \tau_\epsilon$) $G(t)$ no longer decays to zero. Now, it is easily shown that for $Z \ll 1$ a solution of 43 is found in the form $G_0(Z) \sim -KZ^{-1-b}$, or in time domain $G_0(t) = -K(t/\tau_\epsilon)^b$, where the exponent $-b$ is the negative solution of $\Gamma(1-x)^2/\Gamma(1-2x) = \lambda$. Such a solution can obviously hold only as long as $G(t) = \epsilon^{1/2}G_0(t)$ remains much smaller than f_c (otherwise $\phi(t)$ would become negative), and this introduces a second time scale, τ'_ϵ, through the equality $\epsilon^{1/2}(\tau'_\epsilon\tau_\epsilon)^b = 1$. For t of order τ'_ϵ, one has to reconsider the terms that were neglected in the derivation of the scaling equation, which can be shown to become dominant in that case. Hence on this longer time scale another type of solution is looked for in the form $\phi(t) = F_\epsilon(t/\tau'_\epsilon)$. Introducing the rescaled frequency $\zeta = z\tau'_\epsilon$, one finds that F_ϵ verifies

$$F_\epsilon(\zeta)/(1 - F_\epsilon(\zeta)) = \lambda_2 LT(F_\epsilon^2)(\zeta) + \zeta/(\Omega_0\tau'_\epsilon)^2 \quad (44)$$

which in the limit of small ϵ reduces to

$$F(\zeta)/(1 - F(\zeta)) = \lambda_2 LT(F^2)(\zeta). \quad (45)$$

For short times, the solution can be shown to behave as $F(\tau) = f_c(1 - K\tau^b)$, so that the short time solution on the scale τ'_ϵ can be matched with the long time solution on the scale τ_ϵ. For long times ($t \gg \tau'_\epsilon$), $\phi(t)$ displays in that case an exponential decay. This however is a peculiar feature of the simple "mode coupling polynomial" $M(f) \sim f^2$ studied here. For more general polynomials, a decay that can be well described by a stretched exponential (but for which no analytical expression is known) solves equation 45 in the long time limit.

To summarize, describing the decay of $\phi(t)$ for $\lambda_2 < \lambda_2^c$ necessitates the introduction of three time scales. the microscopic time scale Ω_0, and two time scales $\tau_\epsilon \sim \epsilon^{-1/2a}$ and $\tau'_\epsilon \sim \epsilon^{-1/2a-1/2b}$. these time scales have been denoted in the literature as the β relaxation and α relaxation time scales, respectively. Note however that the β relaxation in this terminology is not related to the β, or secondary, relaxation observed for example on intermediate time scales in polymer glasses [17]. Very specific power law behaviors are predicted for the approach to the plateau f_c, and the decay from the plateau. It is important to notice that on a susceptibility spectrum, (the susceptibility can be defined as $\chi(\omega) = \omega Im(\phi(z = i\omega))$, the β relaxation corresponds to a *minimum* χ_{min} that scales as $\chi_{min} \sim \epsilon^{1/2}$, while the α time scale corresponds to a maximum. Specific scaling forms for the susceptibility around the β minimum can be found in the literature [15,16].

General mode coupling equations

Before we started to study the simplified model, we were dealing with large number of coupled equations for the different wave vectors. One may then suspect that the behaviour of these coupled equations is very different from that of the simple model. Fortunately, this was shown not to be the case in the very elegant

work by Götze and Sjögren [15,16]. these authors showed that the mathematical properties of the coupled equations are essentially identical to that of the equation for a single correlator, albeit with a mode coupling polynomial $M(f)$ that can be more complex than the simple f^2 considered above. Let us assume that for some critical value T_c of e.g. the temperature the structure factor becomes large enough to induce an ergodic (for $T > T_c$) to non-ergodic (for $T < T_c$) transition in the behaviour of $\phi(q,t)$. then define $\epsilon = (T - T_c)/T_c$, and $f(q)$ the long time limit of $\phi(q,t)$ for $T < T_c$. the relaxation of $\phi(q,t)$ can be described as a three step process involving
- a fast relaxation on the microscopic time scale $\Omega(q)^{-1}$
- a β relaxation on a time scale $\tau_\epsilon \sim \epsilon^{-1/2a}$. the exponent a verifies the same relation as before, with λ a parameter that can be computed from the knowledge of the vertices. In this β relaxation step, a very important and nontrivial property is the *factorization*

$$\phi(q,t) = f(q) + \epsilon^{1/2} e(q) G(t/\tau_\epsilon) \qquad (46)$$

the function e(q) describes the spatial dependence of this β relaxation, and its Fourier transform $e(r)$ is found to be short ranged in space. the function G_0 has all the power law dependencies that were found for the simplified model, i.e. $G(x) \sim x^{-a}$ for small x and $G(x) \sim -x^b$ for large x and $T > T_c$.
- an α relaxation on a time scale $\tau'_\epsilon \sim \epsilon^{-1/2a-1/2b}$. $\phi(q,t)$ in this regime is given by $\phi(q,t) = F_q(t/\tau'_\epsilon)$ where F_q is a q dependent function that can be matched with the long time β relaxation, and decays for long times like a stretched exponential. Note that this second time scale is the one that will rule the behaviour of the transport coefficients near T_c, since they are given by integrals of correlation functions. the viscosity for example is predicted to diverge like τ'_ϵ near T_c.

Extended mode coupling theory

In practice, it is well understood that the existence of an "ideal" ergodic to non-ergodic transition is an artifact of the approximations made in the derivation of the mode coupling equations, and that no such transition is present in real systems. It is therefore natural to investigate what corrections to the scaling predictions will take place if a (hopefully small) correction to the mode coupling equations is made that suppresses the transition. Let us write the exact memory function as

$$M(k,z)^{-1} = M_{MC}(k,z)^{-1} + \delta(k,z)$$

which constitutes in fact a definition of δ. the hope is that if the mode coupling approximation is not too bad, δ will be numerically small, although it has the drastic effect of suppressing the transition. Now, let us interpret this δ as another control parameter (in fact very uncontrolled!) and consider that the system is following a path in parameter space (T, δ) that approaches the critical point $(T = T_c, \delta = 0)$.

In this case, it can be shown that the scaling predictions described in the above sections will remain valid to a large extent. In particular, the properties of the β relaxation are essentially unaffected. Of course, this will be true only as long as the corresponding time scales remain smaller than $1/\delta$.

the problem obviously in this approach is the actual calculation and even more the physical interpretation of δ, which is often described as accounting for "activated" processes whose physical origin is unfortunately not understood.

SUMMARY, CONCLUSIONS

I would like to close this short introduction to mode coupling theories by some general remarks on the mode coupling approaches, and some more specific concerning their application to the glass transition. In the physics of liquids, mode coupling approaches have proved a very versatile tool to investigate various slowing down phenomena that take place in critical systems or in dense fluids (I have not discussed here the celebrated "long time tails" in velocity correlation functions that have also been explained with this type of method [4]. From the study of disordered systems, it is now becoming clearer that these approximations belong to a class of "dynamical mean-field" approximations that, in the case of disordered systems, become exact in infinite dimensions [18]. Whether this remark will help improving on the existing theories is still unclear at the moment.

Concerning the glass transition, a strong connection is also appearing between mode coupling theories for liquids and mean field theories of disordered systems. This connection opens the interesting possibility of a general "mean-field" picture that includes both the dynamical aspects covered by mode coupling theory and thermodynamic aspects very reminiscent of the earlier Adams Gibbs approach [19]. Again, going beyond mean field (or mode coupling) approximations for the description of dynamical processes seems quite difficult in this context.

Beyond this theoretical context that makes it intellectually quite satisfactory, the mode coupling theory makes a number of quantitative predictions corresponding to the fact that the system is passing nearby a "critical" point in the parameter space. I will conclude with a short list of what the main predictions are, indicating by YES or NO whether these predictions are born out by experiment or simulation. Detailed references to the relevant experimental or simulation work can be found in Ref. [20] or in the lectures by W. Kob in this volume.

- power law divergence of relaxation times at a critical T_c: NO in molecular glass formers. In colloidal systems a fast increase of viscosity as a function of density is observed, which can be fitted by a power law over a quite large range.
- two slow relaxation steps: YES
- plateau value in $\phi(q,t)$ behaves as $f_c(q) + \epsilon^{1/2} A(q)$: YES
- scaling behaviors of τ_ϵ, $\tau'_\epsilon \Lambda$ and χ_{min} give consistent values of T_c and of the exponents a and b: YES
- the β relaxation regime obeys the factorization property 46: YES

- the shape of $f(q)$, $e(q)$ is given correctly: YES, within 10% for simple systems where the calculation is possible.
- Slowing down affects all observables in the same way: YES

REFERENCES

1. see for example D. Chandler, *An introduction to modern statistical physics* (Oxford University Press, Oxford 1990)
2. N. van Kampen, *Stochastic processes in physics and chemistry*, (North-Holland, Amsterdam 1981)
3. H. Risken, *the Fokker-Planck equation* (Springer, Berlin 1996)
4. J-P. Hansen, I.R. McDonald, *theory of Simple Liquids*, (Academic Press, New-York, 1986)
5. D. Forster, *Hydrodynamic Fluctuations, Broken Symmetry and Correlation Functions*, (Benjamin Cummings, Reading, Mass., 1975)
6. K. Kawasaki, *Ann. Phys. (NY)* **61**, 1 (1970).
7. D.J. Amit, *Field theory, the renormalization group and critical phenomena*, (World Scientific, Singapore, 1991)
8. J.-P. Bouchaud, L.F. Cugliandolo, J. Kurchan and M. Mézard in *Spin Glasses and Random Fields*, Ed.: A.P. Young (World Scientific, Singapore, 1998); preprint cond-mat/9511042.
9. P.C. Hohenberg, B.I. Halperin, *Rev. Mod. Phys*, **49**, 435 (1977).
10. H.E. Stanley, *Introduction to phase transition and critical phenomena*, Oxford, Oxford University Press, 1971.
11. P.G. de Gennes, Ph. D. thesis, Paris 1957 (see reference 23, chapter 8 in P.G. de Gennes, *Scaling concepts in polymer physics* (Cornell University Press, Ithaca, 1979))
12. A. Sjölander, L.A. Turski, *J. Phys C* **11**, 1973 (1978)
13. T. Geszti, *J. Phys. C* **16**, 5805 (1983).
14. U. Bengtzelius, W. Götze, A. Sjölander, *J. Phys C* **17**, 5915 (1984)
15. W. Götze, in "liquids, freezing and the glass transition" (Hansen, Levesque and Zinn-Justin eds, North Holland, Amsterdam 1991)
16. W. Götze, L. Sjögren, *Rep. Prog. Phys.* **55**, 241 (1992)
17. J. Perez, "Physique et mécanique des polymères amorphes" (Technique et Documentation, Paris, 1992).
18. J.-P. Bouchaud, L.F. Cugliandolo, J. Kurchan, and M. Mézard, Physica A **226**, 243 (1996)
19. M. Mézard, G. Parisi, to appear in *J. Chem. Phys.* (1999). Preprint cond-mat/9812180.
20. W. Götze, review article to appear in *J. Phys. Cond. Matt* (1999)

Aging in Glassy Systems: Traps and Mode-Coupling theory

J.-Ph. Bouchaud

*Service de Physique de l'État Condensé, Centre d'études de Saclay,
Orme des Merisiers, 91191 Gif-sur-Yvette Cedex, France*

Abstract. We discuss the general link between Mode-Coupling Theory (MCT) and the motion of a particle in a random potential. This enables one to generalize the MCT equations for temperatures below the glass transition, and to describe aging effects. We compare these results with those obtained within more phenomenological 'trap' models.

I INTRODUCTION

A great variety of systems exhibit slow 'glassy' dynamics: glasses of all kinds [1], of course, but also spin-glasses [2], pinned 'defects' such as Bloch walls, vortices in superconductors, charge density waves, dislocations, etc. interacting with randomly placed impurities [3]. Another class of systems where surprisingly slow dynamics can occur are soft glassy materials, such as foams, dense emulsions or granular materials, which attracted a lot of interest recently [4–6].

Mode-Coupling Theory (MCT) [1] has had important successes in describing some features of supercooled liquids [7], in particular the two-step shape of the relaxation function or, equivalently, the behaviour of the frequency dependent susceptibility $\chi(\omega)$ around its minimum, which separates the low frequency 'α-peak' from the high frequency 'β-peak'. However, the Mode-Coupling description is based on the existence of a true dynamical transition temperature T_c (higher than the glass transition T_g) where no singularity is actually observed experimentally. Theoretically, this is assigned to the so-called 'activated processes' (or 'hopping processes'), which are not described by the MCT, and which smear out the transition. Correspondingly, the clear-cut predictions of MCT are somewhat weakened and more difficult to test unambiguously.

There are therefore several questions which one would like to answer; in particular:

• What is the physics contained in the Mode-Coupling Theory, and in what sense are activated processes left out in the description ?

- How specific are the predictions of MCT, and to what extent can these predictions be reproduced by other – possibly less sophisticated – theories?
- Is there a way to extend MCT to temperatures below T_g to describe, among other things, out-of-equilibrium *aging effects*?

II THE MODE-COUPLING THEORY: DIFFUSION IN HIGH DIMENSION

In order to shed light on the above questions, it is interesting to introduce the following simplified picture of a glass: the motion of a given particle can be thought of as taking place in a random potential created by its neighbours, creating a 'cage' within which the particle is trapped. Since the motion of the molecules is extremely slow at low temperatures, one can, as a first approximation, assume that this random potential has a static component, at least for times which are not too long. The diffusion of a particle in a (quenched) disordered potential can be studied in different ways. A first approach is to treat the problem in large spatial dimension $N \to \infty$. Assuming that the random potential is Gaussian, one can then establish *exact* equations [8] relating the two-time correlation function $C(t+t_w, t_w) = \vec{r}(t+t_w) \cdot \vec{r}(t_w)$ (where $\vec{r}(t)$ is the position of the particle at time t), and the two-time response to an external force $R(t+t_w, t_w)$ ($t_w = 0$ is defined as the moment where the system reaches the temperature T after a rapid quench from high temperatures). For temperatures higher than a certain T_c, one finds that C and R are actually *time translation invariant* (i.e. these functions only depend on time differences), and furthermore that the fluctuation-dissipation theorem $R(t) = -\frac{1}{T}\Theta(t)\frac{\partial C(t)}{\partial t}$ is obeyed. In this case, one can then eliminate $R(t)$ and find an equation for $C(t)$ which, interestingly, *is precisely the schematic Mode-Coupling equation*, with a kernel which is related to the correlation of the random potential.

Hence, the physical content of the (schematic) MCT is clear: it is a mean-field description of a single particle in a *static* random potential. The fact that MCT works well suggests that in a sense, one does not look at the motion of one particle in an $N = 3$ space, but rather at the collective motion of a set of N correlated particles. It would be very important to quantify the value of N because, as we shall discuss below, this is related to the time beyond which MCT is expected to breakdown.

From a physical point of view, it is quite important to have identified a well defined Hamiltonian which underlies the MCT equations. One should in fact note that the same equations can also be obtained starting with some mean-field models of *spin-glasses* [11]. The important point is thus that MCT implicitly assumes the presence of some *quenched disorder* which should rather, in reality, be 'self-induced' by the dynamics itself. There has been in the recent years efforts to show how 'complicated' Hamiltonians with no quenched disorder behave much as quenched disordered systems such as spin-glasses [12,2].

Coming back to the equations relating C and R, one can now investigate the 'glass' phase $T < T_c$ [9]. In this case, the correlation and response function cease to be functions of t only. More precisely, $C(t_w + t, t_w)$ can be written as the sum of an 'equilibrium' contribution $C_{eq}(t)$ which only depends on t, and an 'aging' part which depends on the ratio [1] $u = \frac{t}{t_w}$, $\mathcal{C}(u)$. The same decomposition holds for the response function; however, the aging parts of C and R are related by an 'anomalous' fluctuation dissipation theorem, where T is replaced by an effective temperature $\frac{T}{X}$, with $X \leq 1$ [14]. In more physical terms, this means that for a finite waiting time t_w after the quench below T_c, one expects that the susceptibility $\chi(\omega, t_w)$ still exhibits two peaks: a high frequency β-peak very similar to the high temperature $(T > T_c)$ one, and a low frequency α-peak which reaches a maximum at a frequency $\omega_\alpha \simeq \frac{1}{t_w}$, which thus progressively disappears as $t_w \to \infty$. An interesting prediction of this low temperature extension of MCT is that the high frequency part of the 'aging' α-peak behaves as $(\omega t_w)^{-b}$, while the low frequency 'foot' of the β-peak behaves as ω^a, with the following relation between a, b, and the 'anomaly' X [8,9]:

$$X \frac{\Gamma^2[1+b]}{\Gamma[1+2b]} = \frac{\Gamma^2[1-a]}{\Gamma[1-2a]} \quad (1)$$

This equation generalizes the famous MCT relation [1] between a and b for $T > T_c$, for which $X \equiv 1$. Aging in model Lennard-Jones systems has recently been analyzed along the lines of MCT in [15].

III THE TRAP MODEL

Another, more phenomenological approach, consists in coarse-graining the random potential, replacing it by 'traps', where the potential is locally harmonic, from which the particle can only escape through thermal activation and 'hop' to the neighbouring sites. The important quantity here is the distribution of barrier height $\rho(\Delta E)$, which in turn determines the distribution of local *waiting times* τ through the relation $\tau = \tau_0 \exp[\frac{\Delta E}{T}]$, where τ_0 is a microscopic time scale (typically the oscillation period within the trap). $\rho(\Delta E)$ describes the strength of the 'cages' created by the neighbouring particles.

Let us consider the case where $\rho(\Delta E)$ decreases for large ΔE as $\exp-(\frac{\Delta E}{T_0})^{1+\nu}$. When $\nu = 0$ (Poisson distribution of barrier heights), one can show [16–18] that there is a true dynamical phase transition at $T = T_0$, separating a high temperature phase where the correlation function $C_q(t_w + t, t_w) = \langle \exp[i\vec{q} \cdot (\vec{r}(t + t_w) - \vec{r}(t_w))] \rangle$ is time translation invariant and decays to zero for large times, from a low temperature phase where $C_q(t_w + t, t_w)$ has an 'aging' component, decaying on a time scale proportional to the waiting time t_w itself. For $T < T_0$, the asymptotic diffusion constant of the particle is zero [17]. This model has recently been generalized to account for *rheological* properties of glasses and soft glassy materials [4].

---

[1] see however refs. [13,10,2] for a more precise discussion of this point.

For $\nu > 0$ the 'glass' phase transition only takes place, strictly speaking, at $T = 0$. However, the relaxation time $\tau(T)$ beyond which $C_q(t_w + t, t_w)$ decays to zero diverges extremely strongly as the temperature is decreased:

$$\tau(T) \propto \tau_0 \exp(\frac{T_0}{T})^{1+\frac{1}{\nu}} \qquad (2)$$

For $t_w \gg \tau(T)$, the relaxation function only depends on the time difference t and is well approximated by a stretched exponential with an exponent $\beta \simeq (\frac{T}{T_0})^{\frac{1}{2}+\frac{1}{2\nu}}$. Furthermore, $C_q(t)$ obeys an approximate 'time temperature' superposition principle, i.e. $C_q(t) \simeq f_q(\frac{t}{\tau(T)})$ in the α regime [18]. Finally, when $t_w, t \ll \tau(T)$, the correlation function is aging very much like what happens in the case $\nu = 0$, $T < T_0$ [18]. In particular, the α-peak in the susceptibility spectrum shifts to low frequencies as t_w is increased, with a high frequency tail described by an effective exponent b, which is temperature and waiting time dependent.

IV DISCUSSION

Hence, these simple minded 'trap' models, describing the motion of a particle in a random potential in finite dimension, also capture many features of supercooled liquids, and lead to predictions which are very similar to those of the MCT. It is thus natural to wonder whether the above two descriptions are in fact equivalent. Surprisingly, this is not so: the motion of a particle in infinite space dimensions (which leads to the schematic MCT equations) is peculiar because the particle actually never reaches the bottom of an energy well. There is always directions along which it can escape and lower its energy [13,20,21]. Hence, aging within the low temperature MCT is not related to barrier crossing and activated processes, but rather to a slow, endless descent in phase space where the local structure of phase space changes [20]. On the other hand, in any finite dimension, the particle reaches some local minima of the potential in a finite time, and activation is crucial to leave these minima. One of the important differences between these two descriptions (besides the role of temperature) is that the late 'in cage' dynamics (described by the exponent a) is intimately related to the early 'out of cage' dynamics (described by the exponent b) within the MCT description, whereas these two regimes are a priori not related in a precise way within the trap models. This is why we believe that a precise experimental investigation of the age dependent susceptibility spectra $\chi(\omega, t_w)$ in the glass phase, in order to test Eq. (1), would be valuable to draw a consistent physical picture of supercooled liquids. From a theoretical point of view, it would be extremely interesting to understand how finite dimensional effects and the residual motion of the surrounding particles (which make the random potential time dependent) can be included in a systematic way within the Mode-Coupling framework. As a general rule of thumb, we expect mean-field, mode-coupling like theories to be valid at short times, while long time dynamics in disordered and glassy systems are better described by activated, trap-like events.

Acknowledgments. I wish to thank my collaborators on these matters, namely Leticia Cugliandolo, David Dean, Jorge Kurchan, Marc Mézard and Cécile Monthus. I also thank J.L. Barrat, M. Cates, W. Kob, P. Sollich for interesting discussions.

REFERENCES

1. For reviews, see W. Götze, in *Liquids, freezing and glass transition*, Les Houches 1989, JP Hansen, D. Levesque, J. Zinn-Justin Editors, North Holland. See also W. Götze, L. Sjögren, Rep. Prog. Phys. **55** (1992) 241.
2. For a review, see J.P. Bouchaud, L. Cugliandolo, J. Kurchan, M. Mézard, *Out of Equilibrium dynamics in spin-glasses and other glassy systems*, in 'Spin-glasses and Random Fields', A.P. Young Editor (World Scientific) 1998, and references therein.
3. For a recent review, see T. Giamarchi, P. Le Doussal, in 'Spin-glasses and Random Fields', A.P. Young Editor (World Scientific) 1998, and references therein.
4. P. Sollich, F. Lequeux, P. Hebraud, M. Cates; Phys. Rev. Lett. **70** 2020 (1997), P. Sollich, Phys. Rev. E **58**, 738 (1998).
5. E.R. Nowak, J.B. Knight, E. Ben-Naim, H.M. Jaeger, S.R. Nagel, Phys. Rev. E **57** 1971 (1998)
6. M. Nicodemi, A. Coniglio, cond-mat/9803148.
7. See e.g. H. Z. Cummins, W.M. Du, M. Fuchs, W. Götze, S. Hildebrand, A. Latz, G. Li and N.J. Tao, Phys. Rev. **E47** (1993) 4223, W. Kob, H.C. Andersen, Phys. Rev. E **53**, 4134 (1995).
8. L. Cugliandolo, P. Le Doussal, Phys. Rev.E **53** (1996) 1525, S. Franz, M. Mézard, Europhys. Lett. **26** (1994) 209, Physica **A209** (1994) 1
9. J.P. Bouchaud, L. Cugliandolo, J. Kurchan, M. Mézard, Physica **A226**, 243 (1996).
10. E. Vincent, J. Hammann, M. Ocio, J.P. Bouchaud, L. Cugliandolo, *Proceedings of the 1996 Sitges Conference of Glassy Systems*, I. Rubi Editor, Springer (1997).
11. T. R. Kirkpatrick and D. Thirumalai, Phys. Rev. **B36** (1987) 5388; S. Franz, J. Hertz, Phys. Rev. Lett. **74** (1995) 2114
12. J.P. Bouchaud and M. Mézard; J. Phys. I (France) **4** (1994) 1109. E. Marinari, G. Parisi and F. Ritort; J. Phys. **A27** (1994) 7615; J. Phys. **A27** (1994) 7647, P. Chandra, M. V. Feigel'man, L. B. Ioffe and I. Kagan, Phys. Rev. B **58** 11553 (1997), and refs. therein.
13. L. Cugliandolo, J. Kurchan, Phys. Rev. Lett. **71** (1993) 173, J. Phys. A **27** (1994) 5749.
14. L. Cugliandolo, J. Kurchan, L. Peliti, Phys. Rev E **55**, 3898 (1997).
15. W. Kob, J.L. Barrat, "Fluctuations, response and aging dynamics in simple glass forming liquids out of equilibrium", cond-mat/9905248.
16. J.P. Bouchaud, D.S. Dean, J. Physique I (France) **5**, (1995) 265.
17. T. Odagaki, Phys. Rev. Lett. **75** (1995) 3701.
18. C. Monthus, J.P. Bouchaud, J. Phys. A **29** (1996) 3847.
19. A. Barrat. R. Burioni and M. Mézard; J. Phys. A **29** (1996) 1311.
20. J. Kurchan, L. Laloux; J. Phys. A **29** (1996) 1929.
21. A. Barrat and M. Mézard; J. Physique I (France) **5** (1995) 941.

Numerical Tests of Mode Coupling Theory

Walter Kob, Tobias Gleim, and Kurt Binder

Institute of Physics, Johannes Gutenberg University, Staudinger Weg 7, D-55099 Mainz, Germany

Abstract. We discuss some results of computer simulations that have been done to study the dynamics of strongly supercooled simple liquids. By investigating various dynamical quantities, such as the intermediate scattering function, we show that mode-coupling theory is able to give a very good description of this dynamics on a qualitative as well as quantitative level.

INTRODUCTION

In the last ten years an impressive amount of work has been devoted to the investigation of the dynamics of supercooled liquids and these efforts have led to a significant improvement of our understanding of these systems. An overview on the current state of knowledge can be found in Refs. [1]. Many of these investigations have been devoted to test whether or not the so-called mode-coupling theory (MCT) [2] is able to give a correct description of this dynamics. The heart of this theory is a set of coupled nonlinear differo-integral equations for the time-and wave-vector dependence of the intermediate scattering function, equations which have been obtained in the context of normal, i.e. *not* supercooled, liquids [3]. The physical picture behind these equations is the so-called cage-effect, i.e. the fact that in a strongly supercooled liquid every particle is temporarily trapped by its surrounding neighbors, and the theory is an attempt to give a detailed description of how the particle moves inside this cage and how, at long times, this cage breaks up and thus allows the particle to escape it. Note that the particles of the cage are of course also trapped by the particles which surround them and thus MCT is a *self-consistent* theory of that process. More details on the theory can be found in Refs. [2].

Among the mentioned tests on the validity of MCT have also been several computer simulations and in this articles we will discuss some of them. Because of the limited space we will, however, restrict ourself to just a few of them and refer the reader who wants to know more about these simulations to some other review papers and the original literature [4–8]. Since in a computer simulation the position and the velocities of all the particles are known at any time it is easy to calculate all time and space correlation functions of interest, as long as the distances and times are smaller than the size of the simulation box and length of the simulation, respectively. This feature allows to make sometime more stringent tests of MCT, or other theories, than it is possible in a real experiment. Hence simulations are a very useful addition to experiments.

MODEL AND DETAILS OF THE SIMULATIONS

The most important feature of a model which one wants to use to investigate the dynamics of supercooled liquids is that it should not crystallize. One way to achieve this is to use a binary mixture of particles with interactions that are chosen carefully. An example of such a system is a binary (80:20) mixture of particles, which in the following we will call A and B particles, that interact with a Lennard-Jones potential, $V_{\alpha\beta} = 4\epsilon_{\alpha\beta}[(\sigma_{\alpha\beta}/r)^{12} - (\sigma_{\alpha\beta}/r)^6]$, with $\alpha, \beta \in \{A, B\}$ and $\epsilon_{AA} = 1.0$, $\sigma_{AA} = 1.0$, $\epsilon_{AB} = 1.5$, $\sigma_{AB} = 0.8$, $\epsilon_{BB} = 0.5$, and $\sigma_{BB} = 0.88$. To allow for an efficient simulation of this system we truncated and shifted this potential at a distance of $2.5\sigma_{\alpha\beta}$. In the following we will measure length and energy in units of σ_{AA} and ϵ_{AA}, respectively (setting $k_B = 1.0$), and time in units of $\sqrt{\sigma_{AA}^2 m/48\epsilon_{AA}}$, where m is the mass of the particles. The total number of particles is 1000 in a cubic box of volume 9.4^3 and in order to improve the statistics of the results we averaged at each temperature over eight independent runs. The temperatures investigated are 5.0, 4.0, 3.0, 2.0, 1.0, 0.8, 0.6, 0.55, 0.5, 0.475, 0.466, 0.452, and 0.446.

In the following we consider two types of dynamics: The well known Newtonian dynamics (ND) and a so-called stochastic dynamics (SD). The equations of motion for the latter are given by

$$m\ddot{\vec{r}}_j + \nabla_j \sum_l V_{\alpha_j\beta_l}(|\vec{r}_l - \vec{r}_j|) = -\zeta\dot{\vec{r}}_j + \vec{\eta}_j(t) \quad . \tag{1}$$

Here $\vec{\eta}_j(t)$ is a gaussian distributed white noise force with zero mean and an amplitude which is related to the damping constant ζ via the fluctuation dissipation theorem, i.e. $\langle \vec{\eta}_j(t) \cdot \vec{\eta}_l(t') \rangle = 6k_B T \zeta \delta(t - t') \delta_{jl}$. The value of ζ was 10, which corresponds to the strong damping limit [9]. The presence of the stochastic forces in the SD makes this microscopic dynamics quite similar to the one of a Brownian particle, i.e. the phonons are strongly damped.

RESULTS OF THE SIMULATIONS

The simplest quantity to characterize a fluid is the *static* structure factor $S(q)$, which can be measured in neutron and x-ray scattering experiments. (For a binary mixture there are *three* partial structure factors $S_{\alpha\beta}$ which describe the relative

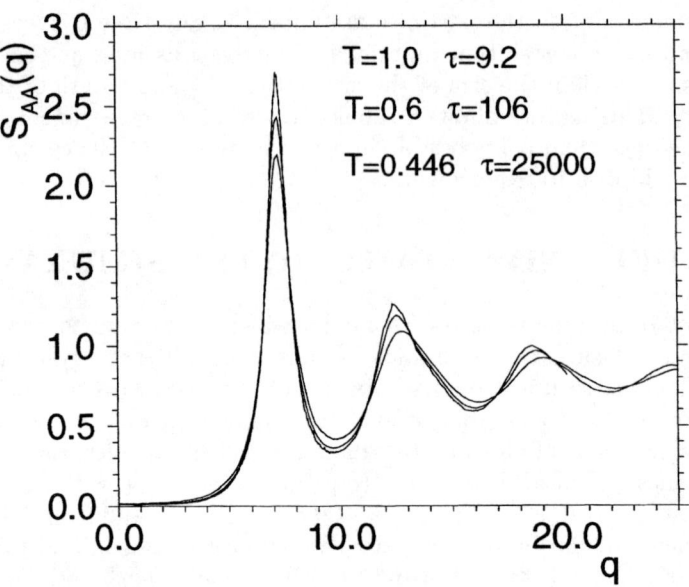

FIGURE 1. Wave-vector dependence of the static structure factor for the A-A correlation for different temperatures T. Also given are the α-relaxation times τ of the system at the same temperatures.

location of the particles of the two species [10].) MCT assumes that the form of the structure factor is a smooth function of temperature, i.e. does not show any indication of a singularity when the temperature of the system is lowered. That for the present model this assumption does indeed hold, is demonstrated in Fig. 1 where we show the static structure factor for the A-A correlation for different temperatures. In the temperature range covered the relaxation time of the system, defined below, increases by about 3.5 decades, thus showing that the temperature variation of this structural quantity is indeed very mild as compared to the one of dynamical quantities.

Since this structural quantity shows no dramatic temperature dependence, a property it shares with all other known structural quantities, we now turn our attention to dynamical quantities. The most simple one is probably the mean squared displacement (MSD) of a tagged particles, i.e.

$$\langle r^2(t) \rangle = \langle |\vec{r}_j(t) - \vec{r}_j(0)|^2 \rangle, \qquad (2)$$

where $\langle . \rangle$ denotes the thermal average. The time dependence of the MSD for the A particles is shown in Fig. 2 for all temperatures investigated. We see that at high temperatures, curves to the left, the MSD shows at short and long times a power-law with exponent 2.0 and 1.0, respectively. These two time regimes correspond to the ballistic motion at short times and the diffusive motion at long times. For the lowest temperatures, curves to the right, the same two time regimes can be

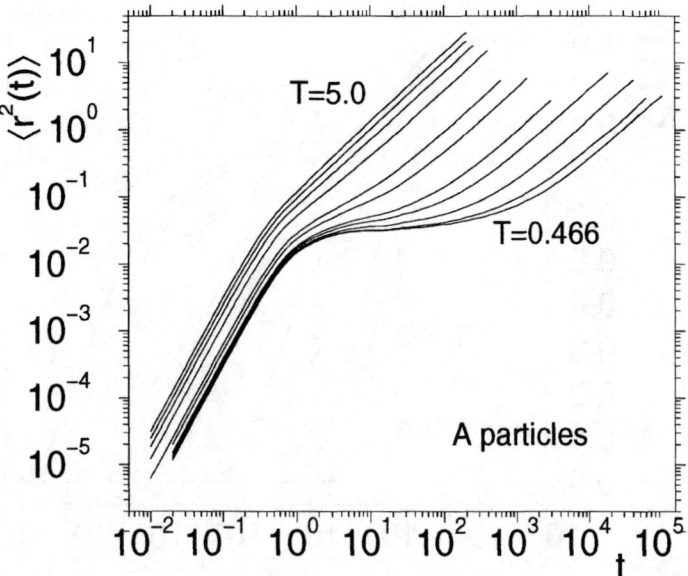

FIGURE 2. Time dependence of the mean squared displacement for the A particles for different temperatures.

observed. In addition to this we find, however, that at intermediate times there is a third regime in which the MSD hardly increases for several decades in time. The reason for the third regime is the cage-effect discussed in the introduction, i.e. the fact that on a mesoscopic time scale the particles are trapped in the cage formed by their surrounding particles. The lower the temperature is the more efficient this trapping becomes and hence the more extended the plateau in the MSD becomes.

Using the Einstein relation it is simple to calculate from the MSD the diffusion constant D and hence to discuss the temperature dependence of this quantity. For the moment we will postpone this discussion until we have investigated the temperature dependence of an other important time correlation function, the (self) intermediate scattering function $F_s(q,t)$, where q is the wave-vector. The relevance of this function lies in the fact that it can be measured in light and neutron scattering experiments and is one of the important theoretical quantities to describe the dynamics of a liquid [10]. $F_s(q,t)$ can be calculated from the particle positions via the relation

$$F_s(q,t) = \langle \exp(i\vec{q} \cdot (\vec{r}_j(t) - \vec{r}_j(0))) \rangle \quad . \tag{3}$$

In Fig. 3a we show the time dependence of $F_s(q,t)$ for the A particles for different temperatures. The value of q is 7.25, which corresponds to the location of the main peak in the static structure factor.

From that figure we see that at high temperatures this correlation function decays rapidly to zero. A more detailed analysis of the curves shows that, after the

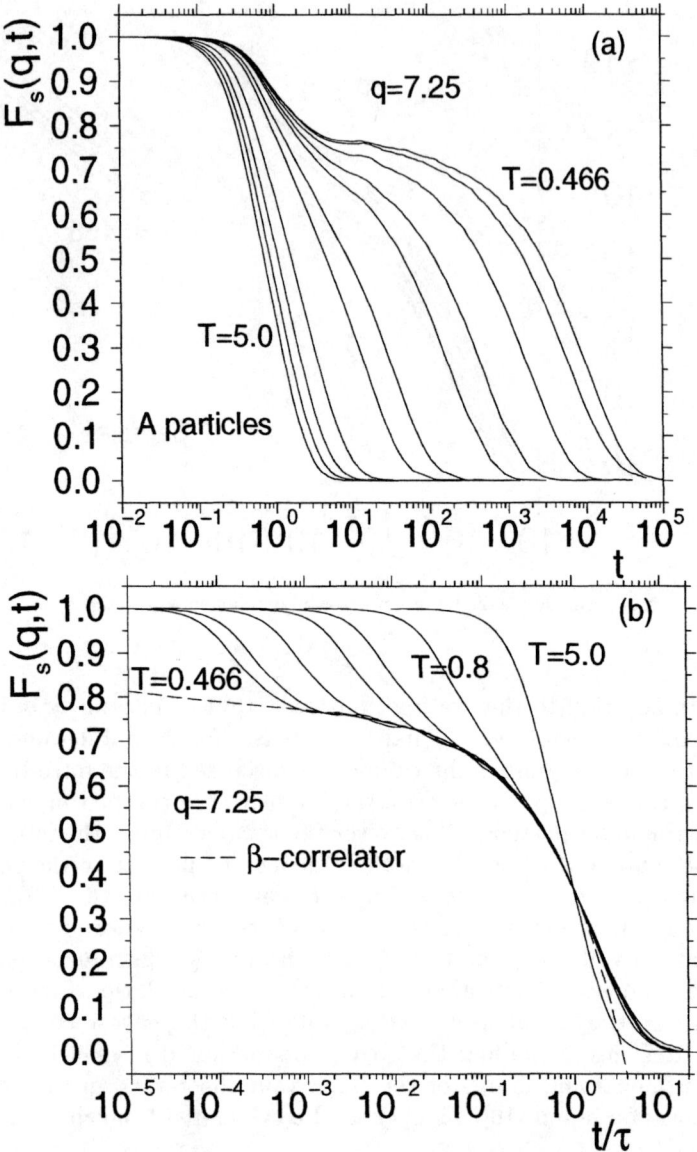

FIGURE 3. a) Time dependence of the intermediate scattering function for different temperatures. b) The same correlation functions versus the rescaled time $t/\tau(T)$, where $\tau(T)$ is the α-relaxation time of the correlation function.

ballistic regime, this decay is described very well by a simple exponential. For low temperatures, however, we see that $F_s(q,t)$ shows at intermediate times a plateau, the origin of which is again the above mentioned cage-effect. At these temperatures the correlation function decays to zero only at very long times and this decay is now given by a so-called stretched exponential, or Kohlrausch-Williams-Watts (KWW) function: $F_s(q,t) = A_q \exp(-(t/\tau_q)^{\beta_q})$. MCT does indeed predict that this final decay, also called α-relaxation, is described well by this KWW law. Furthermore the parameters A_q, τ_q, and β_q are predicted to depend on q and this is indeed what is found [6].

Having these time correlation functions it is now easy to define a typical relaxation time $\tau(T)$ of the system. This can, e.g., be done by requiring that $F_s(q,\tau) = e^{-1}$. MCT predicts that at low temperatures the so-called "time-temperature superposition principle" (TTSP) should hold which means that if a time correlation function is plotted versus the rescaled time $t/\tau(T)$ a master curve should be obtained *in the α-relaxation regime*. That this prediction of the theory is indeed fulfilled for the present system is demonstrated in Fig. 3b where she show the same time correlation functions as in Fig. 3a but this time versus t/τ. We see that in the α-regime the curves for the different correlation functions do indeed collapse nicely onto a master curve. Note that since the TTSP is supposed to hold only at low temperatures we have plotted only the data for $T \leq 0.8$. Also included is the curve for the high temperature $T = 5.0$ and we see that this curve does not fall on the mentioned master curve, thus showing that the shape of the correlation functions at low temperatures are indeed different from the ones at high temperatures.

Many of the detailed predictions of MCT concern the dynamics of the system in the β-relaxation regime, i.e. the time window in which the correlation functions are in the vicinity of the plateau. One of the main predictions of the theory is that in this time regime the so-called factorization property holds, i.e. that any time correlation function $\phi_l(t))$ can be written as

$$\phi_l(t) = f_l^c + h_l G(t) \quad , \tag{4}$$

where the temperature independent constants f_l^c and h_l are called critical non-ergodicity parameter and critical amplitude, respectively. The form of the time dependent function $G(t)$, called β-correlator, depends on one parameter, which for the present system has been calculated within the theory [11], and is thus *not a* fit parameter. That the functional form given by Eq. (4) is indeed able to describe the data very well in the late part of the β-relaxation regime is demonstrated by Fig. 3b where we have included the result of such a fit (dashed line). We recognize from that figure that at *small* rescaled times the agreement between the theory and our data is not quite satisfactory in that the approach of the β-correlator to the plateau is much slower than the one observed in the data from the simulation. However, it has been show that this discrepancy is only due to the strong interference of the short time dynamics [7]. If this phonon-like dynamics at short times is damped, as it is the case for the SD discussed above, the interference effects are much weaker and it has been shown that in that case the theory gives also a good description of the early β-relaxation.

Since we have now determined the diffusion constant D and the α-relaxation time τ we can now discuss their dependence on temperature. MCT predicts that there exists a critical temperature T_c in the vicinity of which these two quantities show a power-law dependence on T:

$$D \propto \tau^{-1} \propto (T - T_c)^\gamma \qquad (5)$$

where the exponent γ can be calculated within the theory and is for the present system 2.34 [11]. Furthermore the theory predicts that the value of T_c as well as the exponent γ are independent of the microscopic dynamics, i.e. should be the same for the ND and the SD. In Fig. 4 we show the temperature dependence of

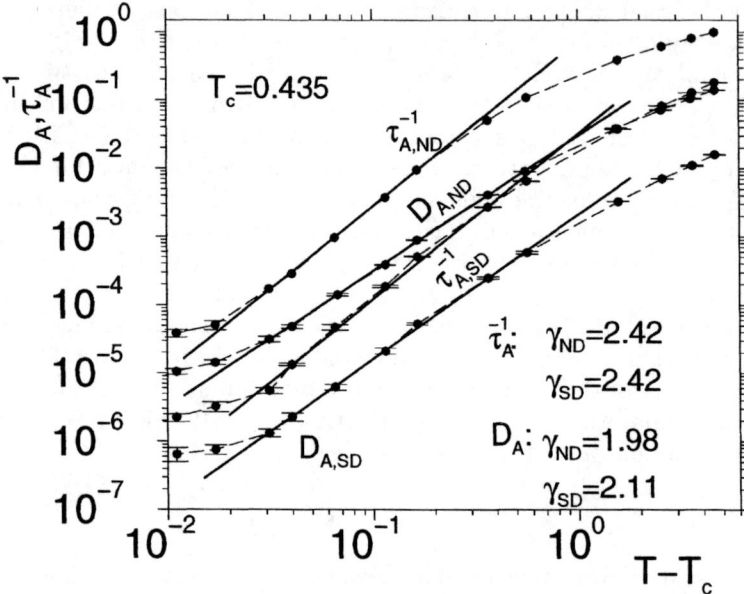

FIGURE 4. Temperature dependence of the diffusion constant D and the α-relaxation time for the A particles for the Newtonian and stochastic dynamics. Solid lines are fits with power-laws.

D and τ^{-1} for the A particles (the results for the B particles are very similar). Here T_c was determined from a fit with the power-law given by Eq. (5). We see that it is indeed possible to obtain a good fit to the data with such a functional form and that the value of the critical temperature T_c, 0.435, is independent of the quantity. Furthermore we see that the exponent for the diffusion constant and for τ are independent of the microscopic dynamics, as predicted by MCT. However, the value of γ for D and τ^{-1} are *not* the same. The reason for this is likely the fact that in this system there exist dynamical heterogeneities, see Refs. [12] for more details, that lead to an enhancement of the diffusion constant.

We now check, for the case of the SD, in more detail the factorization property, see Eq. (4). From that equation it is easy to show that for times t in the β-relaxation regime the ratio $R_l(t) = [\phi_l(t) - \phi_l(t')]/[\phi_l(t'') - \phi_l(t')]$ is independent of ϕ_l, if the times t' and t'' are also in the β-relaxation regime. To check the validity of this prediction we considered many (36) different time correlation functions (coherent and incoherent intermediate scattering function at various values of q). These correlators are show in the upper inset of Fig. 5. The functions $R_l(t)$ are shown in the main part of Fig. 5 and from it we see that in the β-regime the different correlators do indeed fall nicely onto a master curve. That the existence of this master curve is not trivial can be recognized by the fact that at early as well as at long times the different curves do *not* fall onto a master curve.

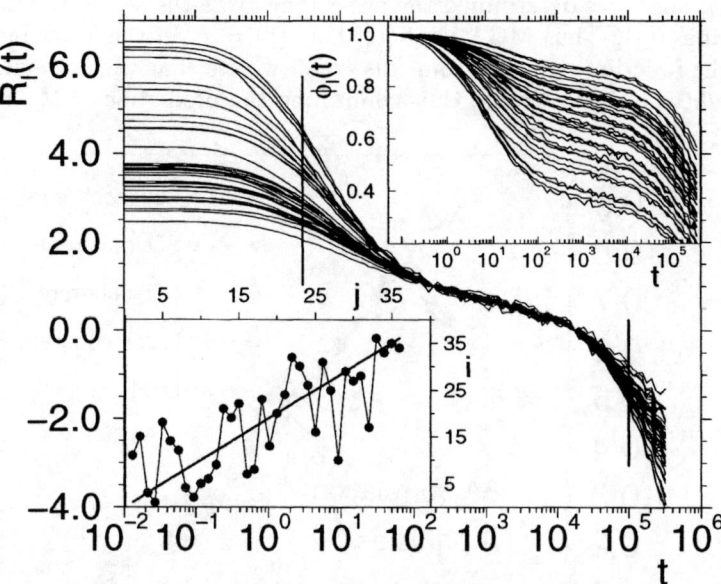

FIGURE 5. Main figure: The ratio $R_l(t)$ for various correlators (shown in the upper inset) for the SD at $T = 0.446$. Lower inset: the function $i(j)$ used to check the validity of the correction terms in Eqs.(6) and 7). See text for details.

Very recently also the corrections of the asymptotic result given by Eq. (4) have been worked out [13]. In particular it has been shown that in the *early* β-relaxation regime any correlator $\phi_l(t)$ can be written as

$$\phi_l(t) = f_l^c + h_l(t_0/t)^a\{1 + [K_l + \Delta](t_0/t)^a\} \quad , \tag{6}$$

where a and b are constants that can be calculated from γ and are for the present system 0.324 and 0.627 respectively. Δ is a l-independent constant and the constant K_l depends on l but not on temperature. In the *late* β-regime the correlation function is predicted to behave like

$$\phi_l(t) = f_l^c - h_l(t/\tau)^b\{1 - K_l(t/\tau)^b\} \quad . \tag{7}$$

The mentioned corrections are the second terms in the curly brackets in Eqs. (6) and (7). The important result about these equations is that the l-dependent part of the correction, i.e. K_l, is the same in the early and late part of the β-relaxation. From this property it is straightforward to show the following: Draw a vertical line in the early part of the β-relaxation and a second one in the late β-relaxation. The Eqs. (6) and (7) imply that the first (second, ...) correlator that intersects the left vertical line, as counted from the top, is also the first (second, ...) curve to intersect the right vertical line, also counted from the top. Thus we have labeled the correlator in the descending order they cross the first vertical line, and called this label j, and have determined the order they cross the second vertical line, and call this label $i(j)$. Thus MCT predicts that $i(j) = j$. In the lower inset of Fig. 5 we show the function $i(j)$ and from this graph we see that the correlation between the two indices is quite strong, thus confirming the prediction of MCT.

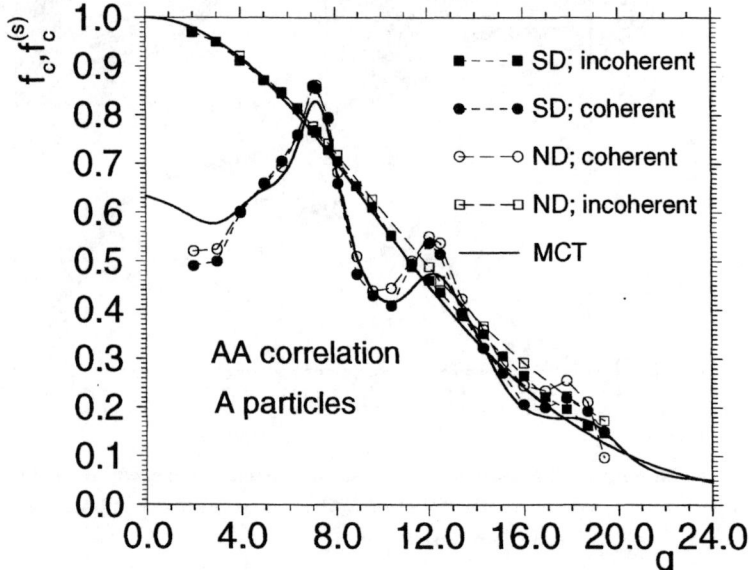

FIGURE 6. Wave-vector dependence of the non-ergodicity parameter for the SD and the ND. The solid lines are the theoretical predictions of MCT.

Before we end we demonstrate that MCT is also able to give very accurate *quantitative* predictions of the dynamics of supercooled liquids. For this we study the wave-vector dependence of f_c, the so-called non-ergodicity parameter, which is just the height of the plateau of the correlator in the β-relaxation regime. For the case of the coherent and incoherent intermediate scattering function of the A particles this quantity is shown in Fig. 6. The open and filled symbols correspond to the ND and SD, respectively. From this figure we recognize that the non-ergodicity parameter does not depend on the microscopic dynamics, in agreement with the prediction of MCT. Also included in the figure are the two corresponding curves as calculated by means of MCT [11] and we see that these curves agree extremely well with the data as obtained from the simulation. Note that for the calculation of the theoretical curves *no* free parameter was used. The only input in the calculation was the static structure factor. A similar good agreement is obtained between the non-ergodicity parameters for the AB correlation and the ones for the B particles [11,8].

SUMMARY AND DISCUSSION

We have presented the results of investigations in which computer simulations have been used to check whether or not MCT is able to give a good description of the dynamics of a simple liquids. From the results shown here and in other places it becomes clear that this theory is indeed able to give a very good description of this dynamics, and this on a qualitative level as well as on a quantitative level. This good agreement is by no means unique since similar results have also been found for the case of a binary soft sphere mixture, supercooled water, polymers, and other systems [5]. Hence we conclude that the theory is indeed able to grasp the essentials of the dynamics in these systems. Finally we mention that recent simulations of SiO_2 showed, that even many of the aspects of the dynamics of this *strong* glass-former can be understood well within the framework of MCT [14], thus giving evidence that this theory is not only applicable to fragile glass-formers.

ACKNOWLEDGMENTS

This work was supported by SFB 262/D1 of the Deutsche Forschungsgemeinschaft.

REFERENCES

1. See, e.g., Proceedings of the "Third International Discussion Meeting", Vigo 1997, J. Non-Cryst. Solids **235-237** (1998); "Second Workshop on Non-Equilibrium Phenomena in Supercooled Fluids, Glasses and Amorphous Materials", Pisa 1998, J. Phys.: Condens. Matter **10A** (1999).
2. W. Götze, p. 287 in *Liquids, Freezing and the Glass Transition* Eds.: J. P. Hansen, D. Levesque and J. Zinn-Justin, Les Houches. Session LI, 1989, (North-Holland, Amsterdam, 1991); W. Götze and L. Sjögren, Rep. Prog. Phys. **55**, 241 (1992); W. Kob, p. 28 in *Experimental and Theoretical Approaches to Supercooled Liquids: Advances and Novel Applications* Eds.: J. Fourkas, D. Kivelson, U. Mohanty, and K. Nelson (ACS Books, Washington, 1997); W. Götze, J. Phys.: Condens. Matter **10A**, 1 (1999); J.-L. Barrat, these proceedings.
3. L. Sjögren, Phys. Rev. A **22**, 2866 (1980) and references therein.
4. J.-L. Barrat and M. L. Klein, Ann. Rev. Phys. Chem. **42**, 23 (1991); W. Kob, p. 1 in Vol. III of "Annual Reviews of Computational Physics"; Ed.: D. Stauffer (World Scientific, Singapore, 1995); W. Kob, J. Phys.: Condens. Matter **11**, R85 (1999).

5. J.-N. Roux, J.-L. Barrat, and J.-P. Hansen, J. Phys.: Condens. Matter **1**, 7171 (1989); J.-L. Barrat, J.-N. Roux, and J.-P. Hansen, Chem. Phys. **149**, 197 (1990); G. F. Signorini, J.-L. Barrat, and M. L. Klein, J. Chem. Phys. **92**, 1294 (1990); J. Baschnagel, Phys. Rev. B **49**, 135 (1994); P. Gallo, F. Sciortino, P. Tartaglia, and S.-H. Chen, Phys. Rev. Lett. **76**, 2730 (1996); F. Sciortino, P. Gallo, P. Tartaglia, and S.-H. Chen, Phys. Rev. E **54**, 6331 (1996); H. Teichler, Phys. Rev. Lett. **76**, 62 (1996); F. Sciortino, L. Fabbian, S.-H. Chen, and P. Tartaglia, Phys. Rev. E **56**, 5397 (1997); C. Bennemann, W. Paul, K. Binder, and B. Dünweg, Phys. Rev. E **57**, 843 (1998); C. Bennemann, J. Baschnagel, and W. Paul, Eur. Phys. J. B (1999) (in press)
6. W. Kob and H. C. Andersen, Phys. Rev. Lett. **73**, 1376 (1994); Phys. Rev. E **51**, 4626 (1995); *ibid.* **52**, 4134 (1995).
7. T. Gleim, W. Kob, and K. Binder, Phys. Rev. Lett. **81**, 4404 (1998).
8. T. Gleim and W. Kob, preprint cond-mat/9902003 (1999).
9. T. Gleim, PhD Thesis (Mainz University, 1999).
10. J.-P. Hansen and I. R. McDonald, *Theory of Simple Liquids* (Academic, London, 1986).
11. M. Nauroth and W. Kob, Phys. Rev. E **55**, 657 (1997).
12. K. Schmidt-Rohr and H.W. Spiess, Phys. Rev. Lett. **66**, 3020 (1991); H. Sillescu, J. Non-Cryst. Solids, (1998, in press); W. Kob, C. Donati, S. J. Plimpton, S. C. Glotzer, and P. H. Poole, Phys. Rev. Lett. **79**, 2827 (1997); C. Donati, J. F. Douglas, W. Kob, S. J. Plimpton, P. H. Poole, and S. C. Glotzer, Phys. Rev. Lett. **80**, 2338 (1998).
13. T. Franosch, M. Fuchs, W. Götze, M. R. Mayr, and A. P. Singh, Phys. Rev. E **55**, 7153 (1997); M. Fuchs, W. Götze, and M. R. Mayr, Phys. Rev. E **58**, 3384 (1998).
14. J. Horbach, W. Kob, and K. Binder, Phil. Mag. B **77**, 297 (1998); J. Horbach and W. Kob, Phys. Rev. B (1999) (in press).

III GEOMETRICAL APPROACHES

Phenomenological Analysis of Supercooled Liquids

Daniel Kivelson* and Gilles Tarjus**

*Department of Chemistry and Biochemistry, University of California,
Los Angeles, CA 90095, USA
**Laboratoire de Physique Théorique des Liquides, Université Pierre et Marie Curie,
4, Place Jussieu, Paris 75005, France

Abstract. We present an overview of the phenomenology associated with supercooled liquids, in particular of the phenomena associated with the slow, and therefore separable, α-relaxations. Along with this summary, which is admittedly selective, we present an extensive analysis of the data, an analysis leading to a model of supercooled liquids based on frustration-limited domains

INTRODUCTION

Phenomenology-Based Theory

In these lectures we discuss the physical phenomena associated with metastable states often found below the normal melting point, T_m, in particular, phenomena associated with supercooled liquids. It is our goal not merely to describe a wide range of phenomena, but to interpret them with the purpose of setting a foundation upon which to develop a theory. Such a study is necessary before a physical theory can be constructed because the ultimate purpose of theory is to describe and interpret a set of physical phenomena. Of course, if one takes a truly *ab initio* approach it may, in principle, be possible to develop theory without any physical input, but when we talk about theory here we refer to a lower level theory based upon a model in which the relevant parameters are not molecular parameters but rather renormalized averages over the molecular parameters. When ultimately formulated such a non *ab initio* theory must be judged not only on its ability to rationalize the data, but also upon its predictive value and on its internal mathematical consistency.

Biased Presentation of Data.

The presentation of the data can <u>never be unbiased</u> if it is to be useful. The data-presentation always involves arbitrary ordering, correlating, and weighting. If the analysis is totally unmotivated by theory or physics, it may be motivationally impartial, but it will nevertheless be biased towards or against some viewpoint. Therefore, we unabashedly admit that although this article is a review of the relevant data, the analysis is constructed so as to motivate and test (not validate) the theory of frustration-limited domains (FLD). We will refer to other relevant models and theories, and to alternative ways in which the data might be analyzed, but we shall focus on the role of frustration-limited domains.

BASIC PHENOMENOLOGY.

Crystallization.

A liquid at temperature (T) that lies above the melting point (T_m) will be denoted as "normal," whereas below T_m as "supercooled." The supercooled liquid is metastable relative to the "normal crystal," *i.e.*, to the crystal that melts at T_m. The freezing transition is first order and can often be by-passed (the liquid supercooled) because of a high free-energy barrier of activation opposing crystal nucleation; the supercooled liquid is only metastable because with sufficient time or help, the crystal will ultimately form at T's below T_m. We shall assume throughout that normal crystallization, and hence the temperature T_m, are irrelevant to our study.

Anomalous Slowing.

The most striking feature characterizing supercooled liquids is the stupendous continuous increase in the α-relaxation time (τ_α) and in the viscosity (η) with decreasing T. (1-3) The α-relaxation is usually the slowest relaxation observed in supercooled liquids; it is often denoted "structural relaxation." For exponential relaxation there is no ambiguity in what is meant by τ_α, but in order to incorporate non-exponential relaxation we take τ_α to be an averaged time of relaxation. For a class of supercooled liquids known as "fragile," the observed increase in τ_α and η is about 15 orders of magnitude over the full temperature-range at which the substance is liquid. (1,2) The T-behavior of supercooled liquids can be expressed in terms of an effective activation free energy, E(T), where

$$\tau_\alpha = \tau_\infty \exp[E(T)/T] \tag{1}$$

and τ_∞ is a constant. If E(T) is constant, the relaxation is said to be "Arrhenius-like," whereas it is said to be superArrhenius if E(T) increases appreciably with decreasing T. Fragile liquids are extremely superArrhenius at low T. It is also of particular interest that the activation free energies of these fragile liquids at low T when measured relative to the thermal energy $k_B T$, are comparable to those of strongly-bonded network-formers such as SiO_2; this raises the intriguing question of why the effective activation energies of fragile liquids, which are often weakly bonded molecular fluids, have such high relaxation activation energies? Of course, one could replace τ_α in Eq. (1) by η, and τ_∞ by η_∞.

The Glass Temperature, T_g.

At sufficiently low temperature, below the so-called "glass transition," supercooled liquids become so viscous, and their α-relaxation times (τ_α) become so long, that the liquid-like modes prevalent in the supercooled liquid can no longer relax within the experimental times (τ_{ex}). The temperature at which the "glass-transition" occurs is denoted T_g, and it is best defined as the temperature at which $\tau_\alpha(T_g) = \tau_{ex}$. Such a change is not indicative of a true thermodynamic phase transition, but represents structural arrest in the sense that the relaxations are too slow to observe. The precise value obtained for T_g will depend upon the particular τ_{ex} in each experiment. However, in experiments carried out near the limits of current experimental technology the increase of $\tau_\alpha(T)$ with decreasing T in the vicinity of T_g is so dramatic that reasonable operational consistency in the T_g values can be obtained by many different techniques. It is, in fact, customary to specify T_g as the T at which any of the following more-or-less equivalent effects occur: $\eta(T_g) = 10^{13}$ Poise, $\tau_\alpha(T_g) = 10^3$ s, the heat capacity plunges rapidly when measured at a cooling rate of 10 K per min (the T_g obtained in this latter way being denoted the "calorimetric T_g"). (1-3) We will be primarily interested in liquids, *i.e.*, in temperatures above T_g, and not in glasses with $T < T_g$; whereas supercooled liquids are independent of path-to-formation, glasses are dependent upon such paths.

We note that T_g is not an inherent property and so cannot be central to the understanding of supercooled liquids; therefore, although scaling to T_g may often be operationally useful, it cannot lead to an underlying physical or theoretical model. It should be noted, however, that analyses based on the concept of "waiting times" may evolve about T_g. (4) Finally, it should be noted that we take supercooled liquids to be equilibrated liquids (with the crystal expunged), and that data reported just above T_g should be accepted with caution because they may refer to systems that have not been sufficiently well equilibrated to create a "supercooled liquids."[1] If the nonexponential relaxation is expressed as a distribution of exponentials, there may remain components that cannot be relaxed within a time τ_{ex}, but components with decay rates much less than the average τ_α^{-1} do not contribute appreciably.

Crossover Temperature, T^*.

Not only is the relaxation time extremely superArrhenius at low temperatures, but it appears to be quite Arrhenius-like at high temperatures. One can, therefore, specify a cross-over temperature, T^*, between these two regimes. This cross-over behavior is characteristic of many, perhaps all glass-forming liquids, and is here taken as universal. The cross-over temperature is quite clearly perceived,(1-3,5) and it is the characteristic (scaling) temperature in our analysis. Within this interpretive scheme one can write,

$$E(T) = E_\infty + BT^*f(T^*-T)\,\Theta(T<T^*) \qquad (2)$$

---

[1] There is still some controversy about whether just above T_g there is a return to Arrhenius behavior, but with a much higher effective activation energy than at high T. In our opinion, this question cannot, in principle, be resolved.

where E_∞, B, and T* are <u>species-dependent but temperature independent parameters,</u> and <u>f(T*-T) is a continuous universal function</u> with f(T*) = 1 and f(0) = 0, and $\Theta(T<T^*)$ = 1 for T < T* and = 0 for T > T*. (5) Clear perception of T* is not equivalent to precise specification, and the exact values obtained for T* will depend upon the fitting formulas used, *i.e.*, upon the form of f(T*-T). By "universal" we mean that f(T*-T) is common for all substances. We have found that all the data on all the liquids we have analyzed can be fit "adequately" to Eqs. (1) and (2) together with the expression (5)

$$f(T^*-T) = (1-T/T^*)^{8/3}, \qquad (3)$$

and so the values of the parameters obtained in this way should have reasonable physical consistency. (6) Here we take this expression as a successful empirical fitting formula, but it does, in fact, have a theoretical basis in the FLD theory to be mentioned below. (7) (For some substances the data are restricted to temperatures for which the systems are either Arrhenius or superArrhenius, although we believe that if the range could be extended, both Arrhenius and superArrhenius behavior would be observed for all substances.)

Ideal Glass Temperature, T_0.

In the vicinity of T_g, the $\ln[\tau_\alpha]$-versus-(1/T) curve increases so rapidly that it seems reasonable (but to us, not compelling) to hypothesize a divergence at a non-zero temperature T_o such that $0 < T_o < T_g$. (1-3,5) Such a divergence would signal a critical point, and models of supercooled liquid have been built about such a <u>low-T critical point</u>. (8,9) Although due to the very slow dynamics such a critical point cannot be accessed, it might nonetheless determine the dynamics at temperatures above T_g. The hypothesized inaccessible, equilibrated system below T_o is often called "an ideal glass." As with T*, the determination of T_o is dependent upon the fitting formula used. The most common way of determining T_o is by means of the Vogel-Fulcher-Tamman (VFT) relation, (also called the Williams-Landel Ferry formula in the study of polymers),

$$\tau_\alpha(T) = \tau_o \exp[DT/(T-T_o)] \qquad (T < T_{cut}) \qquad (4)$$

where τ_o, D, T_o are species-dependent, temperature-independent parameters.

Temperature Control.

The <u>α-relaxations of supercooled fragile liquids are temperature-controlled</u> and not density (volume) controlled. (6,10) This is evident in plots of $\ln[\tau_\alpha]$ versus density (ρ) rather than versus 1/T. See Fig. 4 of ref. 10. Although the constant pressure (1atm) $\ln[\tau_\alpha]$-versus-ρ and $\ln[\tau_\alpha]$-versus-(1/T) curves, along the p = 1 atm curve the temperature decreases as the density increases; when the temperature is kept constant, $\log[\tau_\alpha]$ barely increases with increasing ρ. From this we conclude that the anomalous or <u>superArrhenius behavior of fragile supercooled liquids must be temperature controlled, and this implies that the phenomenon is associated with thermal deficiency</u>

(activated dynamics in the limiting case) and not with free volume deficiency (jamming of hard spheres in the limiting case). In Fig. 4 in ref. 10, it can be seen that at temperatures well above T_g the density can be increased far above the 1 atm glass-transition density without any appreciable increase in η; it is this that leads to the conclusion that the glass transition at 1 atm is associated with thermal deficiency and not with the jamming expected from overcrowding of particles, *i.e.*, that the picture of a structural glass as a system of overcrowded hard spheres is not applicable to supercooled liquids. Note that the phrase "far above" refers to the fact that the range of densities reported extends as far above the 1 atm glass transition density as below. In contrast to the above, the anomalous slowing of the diffusion of colloidal suspensions with increasing concentration, and the consequent colloidal glass-formation above a critical colloid-concentration seems to be an example of volume (free volume) control; thus the picture of a colloidal glass as a system of overcrowded hard spheres may be applicable.

Fragility.

Fragility is associated with the superArrhenius behavior observed at low T. (1-3) Some liquids exhibit more superArrhenius behavior before reaching T_g than do others, and such liquids have been denoted as "fragile." The degree of fragility has been measured in a number of ways:[2] by the D in the VFT formula, (1) by the "steepness index" defined as $d\log[\tau_\alpha(T)]/d(T_g/T)$ evaluated at $T = T_g$, (11) and by the curvature parameter B in Eq. (2). (6,12) In many cases, but significantly not in all, these definitions of fragility are more or less equal, fragility increasing with decreasing D, increasing steepness index, and with increasing B. (10) Here we take B as the measure of fragility because it is determined by fits over a wide range of T and is independent of the choice of the non-fundamental quantity T_g. Thus superArrhenius behavior (below T*) is measured by B, large B for a "fragile" and small B for a "nonfragile" liquid, and Arrhenius behavior (above T*) is measured by E_∞/T^*, large values for "strong" and small values for "weak" liquids).(6)[3]

Why should one be concerned with the concept of "fragility?" One would like to believe that it provides a suitable categorization and set of correlations among physical

---

[2] Fragility has also been measured by the size of the sharp dip in the heat-capacity-versus-T curve as T is decreased (at 10 K per min) below T_g, but whereas the methods listed above refer to reversible dynamical phenomena in the linear response regime, the response of the heat capacity at T_g is a nonlinear, irreversible (out of equilibrium) dynamical phenomenon. Furthermore, the heat-capacity-dip could depend in a non-obvious way on which and how many modes are "frozen" out at T_g for each individual liquid studied.

[3] It has often been said, mistakenly we believe, that Lennard-Jones systems (presumably 2-component ones) are quite "fragile." The systems are "weak" in the sense that $E_\infty/T^* \approx 1$. It's fragility has been estimated (1) by placing data on an Angell-plot, *i.e.*, $\ln[\tau_\alpha/\tau_\alpha(T_g)]$ versus T_g/T, but this placement depends upon an estimate of the laboratory T_g, which has been determined from a simple Arrhenius plot to be about 0.4T*. Since the τ_α's determined by simulation are still many orders of magnitude less than $\tau_\alpha(T_g)$, the extrapolation on the 1/T axis is virtually infinite and therefore uncontrollable. In this case one must plot $\tau_\alpha(T)$ versus T, or still better E(T) versus T; it can then readily be seen that $T_g \approx 0$ K and that the Lennard-Jones system is extremely non-fragile, the most non-fragile of all systems studied. (6)

properties that lead to physical and theoretical insights. The concepts of "strong" and "weak" might intuitively be associated with strong and weak intermolecular bonding, respectively, (and indeed they were so associated initially), but as presently specified they remain dynamical and not structural concepts. The concept of "fragility" has been associated with bond-weakness since many of the most fragile liquids are found to be molecular fluids with rather weak intermolecular bonds. Systems (*e.g.*, Lennard-Jones liquids) have been found, however, in which the intermolecular bonds are thought to be weak (suggesting small E_∞/T^*), yet with fragility (B) small. In others systems (*e.g.*, H_2O) with fairly strong intermolecular bonds (suggesting large E_∞/T^*), the fragility (B) is large. (6) Below we associate fragility with a physical property, "frustration."

Heterogeneities.

The long relaxation times observed for supercooled liquids suggest the existence of long-range correlations, perhaps even correlation lengths that diverge as and if $\tau_\alpha(T)$ diverges. Curiously, although the principal features of supercooled liquids are centered on long time-scales, long length scales are not so evident. However, in recent years there has been experimental (13-28) and computational (29-33) evidence that indicates that there are indeed supermolecular correlations which might be denoted as heterogeneities, clusters, domains, or fluctuations, depending upon one's picture of the phenomenology. Although supermolecular, these correlation lengths are rather modest, perhaps five molecular diameters at temperatures near T_g, and progressively shorter as T is increased; (14,21-24) as a result by supermolecular we mean 20 to 200 molecules. (We assume that above T^* the correlations are molecular.) All the evidence appears to establish the dynamical character of these heterogeneities, *i.e.*, correlations over time and space of dynamical properties, but direct diffraction data do not seem to reflect the existence of supermolecular structural correlations. The dynamical character of these heterogeneities is supportive of the thesis that they play a dominant role in relaxation processes, but the absence of structural signatures remains an intriguing paradox because it is difficult to envisage dynamic correlations unaccompanied by structural correlations.[4] Recent studies (14) suggest that at temperatures well above T_g the lifetimes of the heterogeneities are slightly larger than the average α-relaxation time, $\tau_\alpha(T)$, and that near T_g the heterogeneity lifetime may be orders of magnitude larger; these results support, at least to some extent, the concept of a supercooled liquid characterized as a system of random, supermolecular heterogenieties.

Local Order in Amorphous Material.

Although supercooled liquids and glasses may be broken into supermolecular regions (heterogeneities or domains), they do not seem to have long-range order, *i.e.*, they are

---

[4] A correlation function, G(t), can be expanded in a Taylor series in t for which the coefficients are equal-time (structural) correlation functions, the coefficients of increasingly higher powers being dependent upon correlations between increasing numbers of molecules. In this sense G(t) is always determined by "structural correlations," but unless the dynamical correlations can be reasonably described by the first few coefficients, one may not readily find or understand the connection between dynamical and structural correlations.

amorphous. Consequently, there cannot be a regular order parameter, which means that any measure of order, which we shall denote the order variable, $O(\mathbf{R})$, must be correlated over no more than supermolecular lengths. Thus $<O> = 0$ and $<O(\mathbf{R})O(\mathbf{R'})>$ might be of the form $|\mathbf{R}-\mathbf{R'}|^{-1}\exp[-|\mathbf{R}-\mathbf{R'}|/\lambda]$, where λ is a supermolecular correlation length and $< >$ indicates an equilibrium ensemble average. The order variable or local order parameter is the structural property that distinguishes one domain or heterogeneity from the next, a necessary property if indeed the domains are structurally distinct. Often one studies correlation functions in reciprocal space, *i.e.*, one focuses on $<|O(\mathbf{q})|^2>$, where $O(\mathbf{q})$ is the transform of $O(\mathbf{R})$; if $O(\mathbf{R})$ is the number density, then $<|O(\mathbf{q})|^2>$ is the usual structure factor. Because diffraction experiments have not given evidence of supermolecular domain structure, (34-36) it appears that the order variable cannot be the number density (which is a sum of 1-particle functions), but it may be a 2-body function (such as the stress tensor) or a multibody function, and $<|O(\mathbf{q})|^2>$ could then include 4-body correlations (such as the stress modulus) or higher-body correlations. This viewpoint is appealing because these multibody correlations are more directly related to the dynamics than is the density structure factor; this could explain why it is the dynamical and not the simple structural (density) signatures that have characterized domains in supercooled liquids.

Lineshapes.

The study of the "activation free energy" gives an overview of the α-relaxations, but the study of the α-components of the "relaxation function," f(t), requires a much more detailed understanding of the relaxation processes. In lieu of f(t) one often studies the "lineshape" of the susceptibility, $X(\omega)$, in particular its imaginary part, $X''(\omega)$, where $X(\omega)$ is the transform of df(t)/dt. Typically the susceptibility, $X''(\omega)$, has a low-frequency maximum, and $X''(\omega)$ is reasonably described in the region of the maximum (called the "stretching regime") by a Cole-Davidson function, $(1-i\omega\tau')^{-\beta'}$, where $0 < \beta' \leq 1$. (In the time domain this region can be reasonably represented by the "KWW stretched exponential," $\exp[-(t/\tau)^\beta]$.) At somewhat higher frequencies (shorter times) the relaxation has been denoted "von Schweidler relaxation," and $X''(\omega)$ has been represented by a power-law, ω^{-b}, where $0 < b \leq 1$. (37-38) Although it appears that β decreases as T decreases, the exact T-dependences of the exponents β and b remain in question, but the relaxation data over both these regimes for many different liquids, over a wide range of liquids, have been placed on the single Dixon-Nagel mastercurve (a non-trivial curve specified by a relatively small number of parameters, *i.e.*, three temperature- and species-dependent parameters.) (39) Because both these regimes can be placed on a single mastercurve, they have both been taken as part of the same α-relaxation mechanism by Nagel; (40) we concur, and therefore take the α-regime as incorporating both the stretching and von Schweidler regimes. At still higher frequencies $X''(\omega)$ goes through a minimum and then increases in what is sometimes known as the "fast β relaxation regime."(38) The full spectrum is well-illustrated by Fig. 2 of ref. 41b. As explained below, here we disregard the high frequency regimes near and above the minimum in $X''(\omega)$; we also disregard a small second maximum, known as the "Johari-Goldstein β-relaxation." (42)

Stokes-Einstein Breakdown.

Whereas the rotational relaxation time τ_{rot} is nearly proportional to η/T over 15 orders of magnitude, the translational diffusion constant D_{tr}, though proportional to T/η at high T's, increases considerably above the extrapolated low-T values as T approaches T_g. (16,18) This increase in $D_{tr}\tau_{rot}$ is known as the "Stokes-Einstein breakdown" or "viscosity decoupling." This phenomenon can readily be interpreted in terms of heterogeneous relaxation; the rotational relaxation takes place within a given domain with a given relaxation rate, whereas the diffusion constant involves passage through many domains and. consequently, is determined by different averaging over domains. (43)

Boson Peak.

At still higher frequency, at tens of cm^{-1}, there is often a peak in both the Raman and inelastic neutron scattering spectra of glasses which is thought to reflect an enhancement in the density of states around this frequency. (44) This peak, denoted the "boson peak," appears to be characteristic of glasses, although it is also found in some crystals; it seems to be present in supercooled liquids although in the latter it is heavily overlaid by spectral features connected with fast relaxations that are frozen out in the glass. The boson peak may result from the presence of domains in supercooled liquids and glasses, and as such might provide a structural measure of heterogeneity, a measure, which as indicated above, is being sought. The study of the boson peak is still in its infancy and its interpretation quite controversial.

Heat Capacity Dip.

As already indicated the heat capacity, expressed as a function of T, dips sharply at T_g. (1-3) Although heat capacity is a thermodynamic quantity, this dip is a dynamic phenomenon associated with the fact that liquid-structure equilibration lags increasingly behind thermal equilibration as the system is cooled through T_g. Whereas η and τ_α are dynamic quantities in the linear response sense (and therefore specified in a given thermodynamic state), the dynamic response of the heat capacity as the system is cooled through T_g depends upon cooling rate (as does the calorimetric T_g itself). Thus the heat capacity dip is indicative of an irreversible process, not of a continuous phase transition, and this is confirmed by its cooling-rate dependence and the hysteresis observed upon heating and cooling.

Configurational Heat Capacity.

Above T_g the liquid can be equilibrated (with the crystal expunged), and the heat capacity of the supercooled liquid is then a *bona fide* thermodynamic quantity. The heat capacity is an inclusive property, perhaps too inclusive to be useful. In particular, the heat capacity depends upon the intramolecular modes, properties that are not of interest to this study. Although there are alternative ways of eliminating the contributions of the intramolecular modes, the most common procedure is to focus attention on the difference between heat capacities of the liquid and that of the crystal, the difference being infelicitously termed the "configurational heat capacity." The motivation for this procedure is that the contribution of the internal modes are similar in both phases and so are canceled in the configurational heat capacity; what is left presumably describes the relevant difference between liquid and crystal. (45,46) Although the configurational heat capacity can be measured only between T_m and T_g, extrapolated values can be obtained both below T_g and above T_m. The <u>configurational heat capacity is a classical quantity</u>, *i.e.*, T_g lies well above the Debye temperature; it should also be noted that in all systems studied, the configurational heat capacity is positive both at T_m and just above T_g, where the system is still equilibrated. (47)

Configurational Entropy.

The entropy is a less revealing quantity than the heat capacity because it is an integral over the heat capacity, but for this very reason, as explained below, it depends less upon minor contributions and so may be <u>easier to understand theoretically than higher-derivative quantities</u>. Because the configurational heat capacity is classical and positive between T_m and T_g, the extrapolated low-T values of the configurational entropy <u>must</u> turn negative below a temperature T_K known as the <u>Kauzmann temperature</u>, which, too, is presumably above the relevant Debye temperature. (48) Note that the extrapolated configurational heat capacity increases below T_g, but even if it were constant, the corresponding configurational entropy would be linear in -ln[T], and this alone would guarantee a Kauzmann temperature, one slightly below that actually obtained by extrapolation. (The data and extrapolated values of the configurational entropy fit a -ln[T] curve very well.) (47)

Differential Supercooled Entropies and Heat Capacities

All that can be gleaned from the existence and value of T_k obtained by extrapolation is that <u>the classical extrapolation must fail at or above</u> T_k. If one wishes to identify properties of supercooled liquids associated with the supermolecular heterogeneities that distinguish supercooled liquids from high-T liquids, then it is the differences between the configurational entropy and heat capacity of the high-T and supercooled liquids that are of interest. These differential properties can be estimated (but only with great uncertainty) by extrapolating high-T properties below T* and supercooled-liquid properties below T_g. (47)

Intermediate Metastable Phases.

It is known that supercooled liquids above T_g sometimes undergo a transformation to an intermediate phase which must be more stable than the liquid below the transformation temperature, T_d, but presumably less stable than the "normal crystal." Intermediate phases have been identified as plastic phases, liquid crystals, and more recently suggestions have been made that they might be defect-ordered,[35,49] modulated, and even amorphous phases distinct from the liquid (possibly a second liquid phase even for a one-component system). [50,51] The Ostwald rule suggests that since the intermediate phase lies part way between the liquid and the normal crystal in stability, it should also have a structure park way between the two. Perhaps more can be said. It is possible that the properties of such intermediate phases are closely related to those of the supercooled liquid from which they are derived; if this be so, then the study of supercooled liquids should include the study of such intermediate phases, and one might expect a theory applicable to the one to include a description of the other.

Presentation of Data.

Even before considering possible fitting formulas one must consider alternative ways of presenting the data; the presentation may strongly influence not only what fitting-formulas are used, but the very way in which the physics is perceived. Ideally the presentation should be motivated by a model or theory, and in turn the data analysis based upon such a presentation can be used to test the model or theory. Different weighting may fundamentally change the appearance of data. In the absence of theoretical motivation, one might try for "equal weighting of data;" the less weighted the data, the more likely any divergences, cross-overs (kinks), and trends perceived in the data are to be "real," and it is these features about which understanding and theory are built. For data that ranges over 15 orders of magnitude, $\ln[\tau_\alpha(T)]$ plots yield relatively unweighted presentations in which one can discern high-T Arrhenius behavior, a cross over (T*) to superArrhenius behavior, and a possible divergence at $T_o < T_g$; the values of both T* and T_o appear different if $\ln[\tau_\alpha(T)]$ is plotted versus 1/T or versus T.

Plots of derivatives heavily weight "changes of mechanism" or of "cross-overs," thereby increasing the apparent importance of mechanisms that contribute little; unless guided by theory, this is precisely what one does not wish to accomplish because it is only upon a dominant mechanism (associated with a separation of time and distance scales) that one can build a theory and achieve understanding. Plots of E(T) or $d\ln[\tau_\alpha(T)]/d(1/T)$ stress the crossover T* and the curvature associated with superArrhenius behavior, but they downplay the putative divergence at T_o; such plots are, however, motivated by theories, such as the theory of frustration-limited domains (7) and that of Adam and Gibbs, (52) which focus on the activation energy. Careful studies show that $d^2\ln[\tau_\alpha(T)]/d(1T)^2$ -plots exhibit more than one crossover, i.e., more than one change of mechanism; (53) note that such plots do not contain all the information contained in the $\ln[\tau_\alpha(T)]$ plots.

Plots of τ_α^{-n} versus $(T-T_c)$ are heavily weighted and have no chance of success over the 15 orders of magnitude studied. Such fits in the low-T region above T_g yield a T_c considerably below the T_o of the VFT theory, terrible fits at high T, and exponents n ≈ 12, which is a theory-defying value. (54) On the other hand, fits in the high-T region well above T_g yield non-universal values of n ≈ 2 to 3, values of $T_c > T_g$, and no fits at all below T_c; although as a simple fitting scheme this latter is unimpressive, it's use is strongly motivated by theory, the mode-coupling theory (MCT). (39)

Adjustable Parameters.

Fitting formulas should be formulated with a theory in mind; without such a connection, the fitting of data is merely useful as a replacement for tables and for interpolation of data. Thus, for example, if one makes use of the VFT formula one should take the putative divergence at T_o seriously despite the fact that it may be dynamically inaccessible, and that even if dynamically accessible, it might be blocked by a phase transition or simply by the onset of other mechanisms.

In the absence of an *ab initio* molecular theory, adjustable parameters are a necessity. The connection between these parameters and molecular quantities is usually not known; the parameters may be complicated renormalized averages over molecular quantities. Consequently, one has difficulty, even if the fits are quite good, being sure that a theory (or model) and the associated fitting formulas truly represent the correct physics and whether the values of the parameters actually coincide with those of the presumed underlying molecular quantities. The goodness-of-fit criterion is, unfortunately, not always useful because in the study of supercooled liquids there appears to be more than one set of adjustable parameters (corresponding to different models) that yield adequate fits. One is also interested in the connection between the adjustable parameters and independent physical properties, *i.e.*, one seeks physical identification and independent determinations of these parameters.

A requirement on fits is that they have a minimum number of adjustable parameters and a minimum number of cross-overs (changes of mechanism). One reason for this is that the uniqueness of a fit (and hence its credibility) diminishes as the number of adjustable parameters increases. Furthermore, a theory is useful only if it has a small number of adjustable parameters; a small number of parameters indicates that there are a small number of relevant mechanisms, a requirement for a theory to be more than merely a set of non-unique rationalizations. One would also like the fits to be as universal as possible, which means fits over as wide a range of variables as possible and for as many different substances as possible. All these attributes may be possible when there is good separation of time and length scales, and they are likely to be most valid in the vicinity of strong anomalies such as apparent divergences which are robust under changes of data-representation. There may well be conflict between the various suggested criteria, in which case one must have recourse to theory and judgment.

Parameter-Count.

Determination of the number of adjustable parameters is not always straightforward. There may be dimensionless parameters, especially exponents, that are "universal." So, for example, for the $\exp[E(T)/T]$ in Eq.(1) one could have $\exp[DT_o/(T-T_o)^n]$, or $\exp[D(T_o/T)^m]$, or $\exp[E_\infty/T]\exp[B(T/T^*)(1-T/T^*)^p]$; n =1 yields the VFT formula in Eq. (4), but n = 2 also gives reasonable fits; m = 2 gives a rather good superArrhenius fit at low T; (55) p = 8/3 gives a very good universal fit as indicated in Eq. (3). (5) Whether such "universal" exponents are determined theoretically or experimentally by means of a simultaneous (global) fit of all available data on all substances, they are usually not treated as adjustable parameters.

"Cut-offs" are often introduced because fits can be made only over a limited range of values of the variables. So, for example, although the VFT expression, Eq. (4), has only 3 explicit parameters, there is an important 4th parameters, a high-T cut-off, and even with these 4 parameters one can actually make fits only over a limited temperature range. <u>The choice of cut-off temperature is important because it strongly affects the values of the other parameters</u>. If one wishes to extend the temperature range over which fits can be made, one must introduce at least one more parameter (a 5th one). Not only must high-T cut-offs be considered, but in some cases so too must low-T cutoffs, not only temperature cut-offs, but both low frequency and high frequency cut-offs as well.[5]

Fits to Eq. (1)-(3) require the 4 parameters, $\{\tau_\infty, E_\infty, T^*, B\}$. The fits in terms of Eqs. (1)-(3) require only four species-dependent but temperature-independent parameters $\{\tau_\infty, E_\infty, B, T^*\}$. For substances for which data exist over a wide range of temperatures, a more demanding fit has the two parameters $\{E_\infty, \tau_\infty\}$ determined in a high-T experiment, and the two $\{B, T^*\}$ by a global fit with E_∞ and τ_∞ held constant. (The high-T analysis is carried out with a low-T cut-off sufficiently greater than T^* that the values of E_∞ and τ_∞ are unchanged by a change of cut-off.) (5)

TOWARDS A MODEL.

It is not clear just when a model earns the distinction of being a theory, but a theory must represent a mathematical formulation based upon a description of the phenomenology. We focus on descriptions to which we refer as models.

Activated Dynamics: Energy Landscape.

The fact that the α-relaxations in supercooled liquids are temperature-controlled suggests that they most likely take place predominantly via activated processes rather than via free-volume controlled processes. The other requirement for the relaxation processes to be activated is $E(T) \gg T$. The paradigm for an activated process is that of a chemical reaction where the activation energy is reasonably independent of T, the inequality given above is well satisfied, and the reaction coordinates (including specification of the initial and final states) are simple and well identified. For deeply supercooled fragile liquids the effective activation free energy satisfies the inequality quite well but is T-dependent, and the as-yet-not-understood multiple reaction

---

[5] The VFT expression has also been used to obtain fits at high-T in which case the divergence at T_o is predicted above T_g, where it is, of course, not observed. (56) Thus one needs a low-T cut-off above T_o, and the 4-parameter fit is valid only above this cut-off. The MCT formula in which $\tau_\alpha^{-\gamma} = A(T-T_c)$ is a 4-parameter fit $\{\gamma, A, T_c, T_{cutoff}\}$ which is valid only above T_{cutoff}, and possibly only below a high-T cut-off, as well. The temperature, T_g, is an adjustable parameter in those cases where its value plays a role in determining other parameters (such as those in waiting-time models) and where it plays the role of scaling parameter (such as in plots of $\ln[\tau_\alpha/\tau_\alpha(T_g)]$ versus T_g/T).

coordinates most likely involve collective motions of many molecules. Because of this collective behavior, the simple energy surfaces envisaged for chemical reactions must now be envisaged as a multidimensional energy landscapes (the potential energy in configuration space.) Although the energy landscape at constant V remains unchanged as T is lowered, the initial and final states, as well as the reaction coordinates change, and these latter changes account for the temperature-dependence (superArrhenius character) of the effective activation free energy, E(T), of fragile supercooled liquids. (57)

What is of interest in the discussion of dynamics is the activation free energy at a given T, but the E(T) that is measured is determined through the variation of the relaxation time with change of T. If the activation free energy, $E(T,\rho)$, is determined at constant ρ, then it corresponds to an energy landscape that remains unchanged as T changes, but if it is determined at constant p, the volume, and hence the energy landscape changes so that the activation free energy at constant pressure, $E(T,p)$, includes compressibility effects as well as barrier information. It can be seen that $E(T,p) \geq E(T,\rho)$, and a more refined condition than that above for activated dynamics is (6,57)

$$E(T,\rho) \gg T. \qquad (5)$$

In this picture the T-dependence of $E(T,\rho)$ arises not because of a change in the energy landscape, but because the initial and final states, as well as the paths between them, change, *i.e.*, the reaction coordinates change. Note that the discussion above is based on "energy landscapes," not on "free energy landscapes." Free energy landscapes involve local averaging (coarse-graining), and the free-energy-barrier-heights are T-dependent. The use of free energy landscapes may be useful because appropriate theories or models are likely not to be molecular, but mesoscopic or coarse-grained. Both the energy and free energy landscapes should be examined at constant V to avoid the change in relaxation rate attributable to compressibility effects.

The condition in Eq. (5), as well as those for temperature-control, appear to be well satisfied for fragile supercooled liquids for which the condition, $E(T,\rho) \approx E(T,p)$, also holds; therefore, we conclude that fragile supercooled liquids relax via activated dynamics. (57)

Flow Dynamics.

The energy landscape of hard spheres consists only of low energy passages between infinitely high walls, and the motion of the system in configuration space (on the energy landscape) can be described as flow, congested flow, or diffusional flow through the passageways. This is the extreme free-volume picture. Although this does not seem to be an appropriate model for deeply supercooled fragile liquids (with their temperature-controlled, activated dynamics) it may be a good model for colloidal suspensions (58) (for which concentration and hence free-volume is the control variable), for "weak" liquids, such as Lennard-Jones liquids[6], (for which the hard-core walls are the major obstructers of progress), (10,57) and possibly for structural glasses (in contrast to supercooled liquids) for which all relevant barriers may be so high that only diffusion around the barriers is possible. (59)

The activated picture represented by Eq. (1) envisages τ_∞^{-1} as the diffusional frequency for flow to, around, and over the peak of the barrier, (here a free energy barrier). For diffusional motion one might expect τ_∞ to have a weak T-dependence, *i.e.*, to be proportional to $T^{1/2}$. Whatever the T-dependence of the preexponential factor τ_∞ in Eq. (1), it should be negligible compared to the T-dependence that enters through activation, or else the picture of relaxations dominated by activated processes is not applicable. Thus we neglect all T-dependence arising from flow effects, and incorporate these small contributions into the effective activation free energy, thereby taking τ_∞ as a constant.

The dominant temperature-control and the strongly activated nature of the dynamics of deeply supercooled fragile liquids seem to rule out dynamic-coupling models as well as models based on deficiency of free volume. (38,61-65)

Universality.

The mastercurve for the activation energy (Fig. 1 of ref 5), together with the Dixon-Nagel mastercurve for the susceptibility (Fig. 3 of ref. 39a) indicate universal behavior; by "universal" we mean that the relevant properties of all relevant substances (molecular, H-bonded, and even networked and polymeric fluids) over the entire relevant temperature and frequency range have the same general behavior, *i.e.*, they scale similarly. Such universal behavior is suggestive of collective behavior, behavior in which the dependence upon specific molecular detail has been submerged, the molecular detail expressing itself only through the values of the fitting or scaling parameters and not in the functional dependence of the various properties. Furthermore, the properties appear to be "universal" in that they are quite similar for many different kinds of experiments: light scattering, neutron diffraction, dielectric relaxation, nmr, mechanical (viscous) measurements. All this can be interpreted as "mode-behavior," *i.e.*, in terms of some very slow and separable relaxation process which projects onto many different experimentally measured quantities and dominates their long-time relaxations. Recognition of this universality is important in setting the direction needed for formulating a model and theory, and indeed we make use of this. Of course, the degree to which one accepts this principle for supercooled liquids is a matter of judgment as to what one considers "relevant" and how significant one takes the fits to the mastercurves to be. Ultimately, the deciding factor must be the success or failure of the models and theories built upon such principles. Most of the current models {FLD, (7) MCT, (38) Adam-Gibbs, (52) low-T critical point (8,9,59)} do, in fact, incorporate such "universality."

Focus on α-relaxation.

The phenomenology on which we focus is that believed to be associated with the α-processes, *i.e.*, with the very slow, long-time, low-frequency relaxations. We therefore

---

[6] Note that it is largely on Lennard-Jones systems that computer simulation tests (33,60) of MCT have been carried out, and that the purportedly "best experimental verification" of MCT has been for colloidal systems. (58)

exclude from consideration the faster β-relaxations as well as the even faster molecular relaxations; we do so in the belief that the universality and the extreme separability of the α time-scales are indicative of a clearly distinguishable, and therefore possibly simply describable physical process. The long times and the universality suggest that the process is a collective one; this view is supported by the experimental evidence indicating that the relaxation is heterogeneous and associated with supermolecular domain-like units. The fact that the effective activation energies (when measured in units of $k_B T$) for relatively weakly interacting fragile liquids can be comparable to those of strongly networked systems such as SiO_2 is also highly suggestive of collective dynamics. All this suggests that we not seek a molecular description, but rather one that is coarse-grained or mesoscopic, *i.e.*, one based upon the existence of domains. The direct rationale for not seeking a molecular description is the fact that the long, relevant relaxation times (which are the times on which one focuses because they can be cleanly separated from all other times) must be associated with the dynamics taken over thousands of intermolecular "collisions" with momentum and energy transfer involving many molecules; a molecular-level description of such phenomena does not at present seem possible.

Despite the fact that the β-relaxations may also be a consequence of collective effects and might therefore be incorporated into a unified description of slow relaxations, it seems prudent, at least at the start, to focus on the longest time scale and the anomalies associated with it, such as superArrhenius relaxation.

Cluster and 2-Liquid Models.

We focus on models that incorporate structural domains despite the fact that the experimental evidence for domains is almost entirely dynamical in nature. Numerous models have been based on the existence of domains, clusters, fluctuations, and two liquid phases. (23,66,67) To some degree these models are rationalizations of the data, *i.e.*, they provide sensible, physically motivated fitting formulas with a considerable number of adjustable parameters, while leaving unanswered the question of how and why these systems evolve? The theory of frustration-limited domains, as well as the theories based upon a low-T critical point address these basic questions, and we expand on this point below.

Domain Size.

As indicated above, the domains constituting the supercooled liquid near T_g presumably consist of hundreds of molecules, and of fewer molecules at higher T. Are such domains sufficiently large to be treated by a coarse-grained, mesoscopic theory? One cannot be sure, but it should be noted that these domains are not isolated clusters of metal atoms but molecular aggregates in contact with a bath of similar domains, and one would, therefore, expect such systems, even though quite small, to behave mesoscopically rather than molecularly. These rather small, supermolecular domains, should not be confused with the dilute macroscopic super-clusters (thousands of molecular diameter across and consisting therefore of more than 10^9 molecules) reported by Fisher and coworkers. (68)

Domain Model.

Many domain-models envisage the α-relaxation as the sum of exponential, activated, intradomain relaxation processes, the activation energy being associated with characteristic domain size. The domains are polydisperse (they are not dilute clusters), and the characteristic size of the domains grows as T is lowered; in most, but not all models, the activation energy increases with increasing domain size. Thus one has

$$f_\alpha(t) = \int_0^\infty dL \, \rho(L,T) \exp\{-(t/\tau_\infty)\exp[E(L,T)/T]\} \quad (6)$$

where $\rho(L,T)$ is the distribution of domain diameters (L), and $E(L,T)$ is the activation energy within a domain of size L. The net relaxation, $f_\alpha(t)$, is then nonexponential and with appropriate choices of $\rho(L,T)$ and $E(L,T)$, Eq. (6) can be made to reproduce the observed α-lineshape, $X"(\omega)$. It is to theory that one must turn to obtain $\rho(L,T)$ and $E(L,T)$.[7]

HETEROGENEOUS ACTIVATED DYNAMICS MODELS

Adam-Gibbs Model.

One model that has been extensively investigated is that of Adam and Gibbs (52) in which the existence of collective relaxing regions (CRR) are assumed. A particularly interesting feature of this model is that the CRR's are dynamical entities (in accord with the phenomenology); however, the size and number of these entities are then related (by assumption) to a thermodynamic or structural quantity (the configurational entropy), and then related back again to the dynamics by relating the configurational entropy to the activation free energy (by assumption). In recent years considerable effort (45,46) has been focused on the study of the thermodynamic intermediate, the configurational entropy, in an effort to understand its behavior and its relationship to the CRR's. To date one can only show that as the configurational entropy decreases, the effective activation energy increases, and presumably the CRR's grow.[8] Studies have also related the configurational entropy to a free energy landscape. (70)

Low-T Critical Point.

One possible classical explanation for avoidance of negative extrapolated configurational entropy below the Kauzmann temperature, T_K, is the existence of a

---

[7] Chamberlin (69) was the first to show that with reasonable choices of $\rho(L,T)$ and $E(L,T)$, Eq. (6) could be used to calculate correct lineshapes for $X"(\omega)$ over the entire α-relaxation regime. His activation energy decreases with increasing domain size.

[8] In the Adam-Gibbs model, the activation energy and the configurational entropy are proportional and inversely proportional, respectively, to the size (L^3) of the domains. This is not true for the theory of frustration-limited domains. (71)

continuous transition between T_g and T_K. Support for this approach comes from the fact that the extrapolated VFT divergence at T_o (perhaps indicative of a dynamic divergence) corresponds approximately with the extrapolated T_K (perhaps indicative of a thermodynamic divergence). (1,45) The low-T critical point is not dynamically accessible, but its existence could account for the observed dynamics above T_g, possibly because of critical fluctuations. Below this low-T critical point the system is an "ideal glass," whose properties would be interesting only in the sense that they are somehow related to the observed properties above T_g. This picture may be analogous to that used in the study of spin glasses, but supercooled liquids exhibit strongly activated dynamics, whereas standard spin glasses do not; and the spectral spread, *i.e.*, $X''(\omega)$, is much broader for spin glasses which have quenched disorder. A model compatible with the required activated dynamics and with such a low-T critical point is that of Adam and Gibbs.[9,10]

Phase-Driven Domain-Growth.

How can one account for the formation and existence of the "supermolecular domains" that constitute supercooled liquids? This question is not addressed by many of the models based upon clusters or 2-liquids. It is difficult to form large clusters of molecules even if they form tetrahedrally-bonded networks, and particularly if they interact weakly. The only obvious way to form supermolecular domains of 10^2 weakly interacting molecules is to pass through a phase transition. As a consequence we envisage the growth of domains in supercooled liquids as phase-driven. What then is the phase to which the system is driven? We envisage it as a "virtual" or "reference crystal" with a local structure similar to that of the high-T liquid in its locally-preferred structure. As the temperature is lowered this locally-preferred structure is increasingly favored and so extends outward, converting the system continuously to a crystal with that structure. Of course, this scenario cannot take place because this structure is not space-filling.

Frustration-Limited Domains.

The inability to tile space is denoted as "structural frustration," and it is this frustration that prevents the phase transition from taking place, which is why we denote the phase as a "virtual" or "reference" phase. (7,75) What drives a phase transition is not a local free energy decrease, but a global one, and this global decrease, which is extensive (varies as L^3), is not sufficient to counter the local difficulties of nucleating the phase prior to formation of a critical cluster. According to classical nucleation theory, the

---

[9] Menon *et al.*, (72) have proposed that the dielectric constant, a structural (equal time) correlation function, may diverge at a T corresponding to the T_o of the VFT model. Their conjecture depends upon the invariance of the Dixon-Nagel mastercurve as T is extrapolated far below T_g, upon the divergence of $\tau_\alpha(T)$ at T_o, and upon the extrapolation of the von Schweidler exponent to $b \to 0$ in the $T \to T_o$ limit. Although the mastercurve is, we believe, significant, this extreme extrapolation based on it, is questionable; in this connection it has been argued that the master curve cannot be exact. (73,74)

[10] The $(T-T_c)^{-12}$ dependence of $\tau_\alpha(T)$, discussed above, does not fit any ordinary picture of critical fluctuations.

difficulty of building a critical cluster can be attributed to the surface free energy (which varies as L^2). This then gives an explanation of why it is difficult to build up sizable clusters or domains unless they are large enough to be associated with a nascent phasetransition. If cluster or domain growth is attributable to such a phase transition, why, once the critical cluster-size is exceeded, does the growth stop at supermolecular domain-size? This size limitation is attributed to frustration, as specified above. As the cluster or domain grows, its inability to tile (frustration) gives rise to strain which increases the free energy superextensively (more rapidly than L^3). The strain distorts the locally-preferred-liquid-structure, and the associated increase in free energy ultimately balances the decrease associated with the extension of the reference crystal structure. What is left are then frustration-limited domains with local structures that are strained versions of the locally-preferred-liquid-structure. Note that no phase transition actually takes place; one might say the transition begins, but is aborted by frustration; this virtual or reference phase is to be distinguished from the defect-ordered phase which appears at lower temperature (below T*).

Frustration.

We do not yet have a good picture of the physical basis of frustration but it seems to be associated with the lack of flexibility in extending a favored local structure outward.(76-79) The local icosahedral structure of spheres appears to be very inflexible in this sense and systems of spheres therefore exhibit large frustration, whereas more asymmetric, flexible molecules seem to find many alternate ways to extend the structure outward and therefore exhibit low frustration. In the FLD model it is the parameter B^{-1} in Eq. (2) that represents the "frustration" and in turn B represents the "fragility." This model thus gives a physical picture of fragility as inversely proportional to frustration.

Summary of Model of Frustration-Limited Domains.

We take advantage of the great separation of time scales to focus attention on α-relaxations, on the Arrhenius-superArrhenius crossover, on activated dynamics, and on the associated role of frustration-limited domains (FLD). The FLD model envisages a cross-over at T* (above which the dynamics are molecular and Arrhenius-like and below which they are collective and domain-dominated, *i.e.*, heterogeneous); it is this cross over T*, and not a low-T critical point, that is taken as the characteristic scaling temperature. The model (which is non-molecular) envisages a polydisperse, random, supermolecular domain-structure in which the average domain-size increases modestly (not divergently) with decreasing temperature (about 100 molecules near T_g), and the polydispersity decreases (rather than increasing) with decreasing T. The structure within a domain is a strained version of the high-T locally-preferred-liquid-structure. Because the domain growth is phase-driven, domains grow despite the fact that the intermolecular forces may be weak and nondirectional, and the phase growth is limited by the super-extensive strain attributed to structural frustration. The concept of structural frustration is intimately connected to that of fragility, the two being inversely proportional to each other. The negative differential entropy associated with domain-formation is primarily correlated with local order (within domains) and not with the number, size, or polydispersity of the domains, whereas the activation free energy associated with a domain depends upon its size, in particular its cross-sectional area.[11]

The normal crystal is excluded from this description, but below a transition temperature, T_d, where $T^* > T_d > 0$ K, and T_d may be either above or below T_g, a <u>defect-ordered crystal</u>, which may possibly be identified with an intermediate phase, forms; this phase may possibly also be envisaged as crystallization of the domains constituting the supercooled liquid from which it is formed. The existence of such a phase, whose properties may be strongly dependent upon those of the supercooled liquid from which it is formed, is a unique feature of the theory of frustration-limited domains (FLD). Such a phase, as discussed above, may have been detected experimentally.

The more fragile and the more supercooled the liquid, the larger the frustration-limited domains and the better the model should work.

FLD THEORY.

The theory of frustration-limited domains is designed to explain the phenomena outlined above; specifically it yields expressions for the $\rho(L,T)$ and $E(L,T)$ that enter the model. (7) It is not a molecular theory but a coarse-grained, mesoscopic theory, the kind of theory, we believe, to be appropriate here. The theory is built about a narrowly avoided critical point T^*, "avoided" because the frustration is long-range and "narrowly" avoided because the frustration is weak. If the frustration is weak, one can scale about the avoided critical point. The avoided critical point accounts for a second supermolecular critical correlation length which is associated with the domain size; the random domains grow as T is lowered below T^*, and at some temperature, T_d, a defect-ordered crystal forms. Although in principle one need only input short-time dynamics to obtain the long-time α-relaxations, at the present time one must model the activated dynamics by considering the relaxation of an order parameter in a finite environment. The theory is discussed elsewhere. (7,80) However, we note that the credibility of the theory rests in part on its close connection to the broad spectrum of phenomenology described by the FLD model, its prediction of the experimentally "indicated" non-trivial exponent (8/3) in Eq. (2), and its "apparent" mathematical self-consistency.

ACKNOWLEDGMENTS

We wish, in particular, to thank Steven A. Kivelson for his major inputs. We are grateful to the NSF, the C.N.R.S., and NATO for their support.

---

[11] This is quite different, as indicated above, than what is envisaged by the Adam-Gibbs model.

REFERENCES.

1. Angell, C. A., J. Non-Cryst. Solids **131-133**, 13 (1991),
2. Ediger, M. D., Angell, C. A., and Nagel, S. R., J. Phys. Chem. **100**, 13200 (1996).
3. Ngai, K. L., Ed., J. Non-Cryst. Solids **235-238** (1998).
4. Odagaki,T., Matsui, J., and Hiwatari, Y., Physica A, **224**, 74 (1996).
5. Kivelson, D., Tarjus, G., Zhao, X-L., and Kivelson, S. A., Phys. Rev. E **53**, 751 (1996).
6. Ferrer, M. L., Sakai, H., Kivelson, D., and Alba-Simionesco, C., J. Phys. Chem. **103**, 4191 (1999). G. Tarjus, , Alba-Simionesco, C., Ferrer, M. L., Sakai, H., and Kivelson, D., in *Slow Dynamics in Complex Systems*, AIP Books, Edit., M. Tokuyama and I. Oppenheim, (1999), pg. 406.
7. Kivelson, D., Kivelson, S. A., Zhao, X-L., Nussinov, Z., and Tarjus, G., Physica A **219**, 27 (1995).
8. Parisi, G., in *Supercooled Liquids: Advances and Novel Applications*, J Fourkas, *et al.*, Eds. (ACS Symposium Series 676, Washington, 1997), pg. 110. J. Phys. Chem. **103**, 4128 (1999).
9. Kirkpatrick, T. R., Thirumalai, D., and Wolynes, P. G., Phys. Rev. A **40**, 1045 (1989).
10. Ferrer, M. L., Lawrence, C., Demirjian, B. G., Kivelson, D., Alba-Simionesco, C., and Tarjus, G., J. Chem. Phys. **109**, 8010 (1998).
11. Böhmer, R., Ngai, K., Angell, C. A., and Plazek,D. J., J. Phys. Chem. **99**, 4201 (1993).
12. Kivelson, D., and Tarjus, G., J. Non-Cryst. Solids, **235-237**, 86 (1998).
13. Sillescu, H., J. Non-Cryst. Solids **243**, 81 (1999)
14. Wang, C-Y., and Ediger, M. D., J. Phys. Chem. B, **103**, 4177 (1999).
15. Böhmer, R., Diezemann, G., Hinze, G., and Sillescu, H., J. Chem. Phys. **108**, 890 (1998).
16. Cicerone, M. T., and Ediger, M. D., J. Chem. Phys. **103**, 5684 (1995)
17. Schiener, B., Chamberlin, R. V., Diezemann, G., and Böhmer, R., J. Chem. Phys. **107**, 7746 (1997).
18. Chang, I., Fujara, F., Heuberger, Mangel, T., and Sillescu, H., J. Non-Cryst. Solids **172-174**, 248 (1994).
19. Richert, R., J. Phys. Chem. B **101**, 6323 (1997).
20. Russell, E. V., Isrealoff, N. E., Walther, L. E., Gomariz, H. A., Phys. Rev. Lett, **81**, 1461 (1998).
21. Cicerone, M. T., and Ediger, M. D., J. Chem. Phys. **104**, 7210 (1996). Blackburn, F., Wang, C-Y., and Ediger, M. D., J. Chem. Phys. **100**, 18249 (1996).
22. U. Tracht, M. Wilhelm, A. Heuer, H. Feng, K. Schmidt-Rohr, and H.W. Spiess, Phys. Rev. Lett. **81**, 2727 (1998).
23. Moynihan, C. T., and Schroeder, J., J. Non-Cryst. Solids **160**, 52 (1993).
24. Vogel, M., and Rössler, E., J. Phys. Chem. A **102**, 2102 (1998).
25. Heuer, A., Wilhelm, M., Zimmermann, H., and Spiess, H. W., Phys. Rev. Lett.**75**, 2851 (1995).
26. Cicerone, M. T., and Ediger, M. D., J. Chem. Phys. **103**, 5684 (1995)
27. Schmidt-Rohr, K., and Spiess, H. W., Phys. Rev. Lett. **66**, 3020 (1991)
28. Miller, R. S., and MacPhail, R. A., **106**, 3393 (1997).
29. Perera, D. N., and Harrowell, P., Phys. Rev. E **52**, 1694 (1995); J. Non-Cryst. Solids (in press)
30. Muranaka, T., and Hiwatari, Y., Phys. Rev. E **51**, R2735 (1995)
31. Yamamoto, R., and Onuki, A., J. Phys. Soc. Jpn. **66**, 2545 (1997).
32 Odagaki,T., Prog. Theor. Physics, Supplement 126, (1997). pg. 9.
33. Kob, W., Donati, C., Plimpton, S. J., Glotzer, S. C., Poole, P. H., Phys. Rev. Lett. **79**, 2827 (1997). Donati, C., Douglas, J. F., Kob, W., Plimpton, S. J., Poole, P. H., Glotzer, S. C., Phys. Rev. Lett. **82**, 5064 (1999).
34. Leheny. R. L., Menon, N., Nagel, S. R., Price, D. L., Suzuya, K., Thiyarijan, P., J. Chem. Phys. **105**, 7783 (1996).
35. Kivelson, D., Pereda, J-C., Luu, K., Lee, Michelle, Sakai, H., Ha, A., Cohen, I., Tarjus, G., in *Supercooled Liquids: Advances and Novel Applications*, Ed. J. Fourkas, *et al.*, ACS BOOKS, (1997). pg. 224
36. Tölle, A., Schober, H., Wuttke, J., and Fujara, F., Phys. Rev. E. **56**, 809 (1997).

37. Knaak, W., Mezei, F., and Farago, B., Europhys. Lett. **7**, 529 (1988).
38. Götze, W., and Sjögren, L., Rep. Prog. Phys. **55**, 241 (1992). Götze,W., J. Phys. Cond. Mat. **11**, A1 (1999).
39. Dixon, P. K., Wu, L., Nagel, S. R., Williams, B. D., and Carini, J. P., Phys. Rev. Lett. **65**, 1108 (1990). Wu, L., Dixon, P. K., Nagel, S. R., Williams, B. D., and Carini, J. P., J. Non-Cryst. Solids **131-133**, 32 (1991).
40. Nagel, S. R., in *"Phase Transitions and Relaxation in Systems with Competing Energy Scales"*, T. Riste and D. Sherrington Eds. (Kluwer Academic, Netherlands, 1993), pg. 259.
41. Schneider, U., Lukenheimer, P., Brand, R., and Loidl, A., J. Non-Cryst. Solids **235-237**, 173 (1998).
42. Johari, J. P., Ann. NY. Acad, Sciences **279**, 117 (1976).
43. Tarjus, G., and Kivelson, D., J. Chem. Phys. **103**, 3071 (1995).
44. Rössler, E., Novikov, V. N., and Sokolov, A. P., Phase Transitions **63**, 201 (1997), and references therein.
45. Angell, C. A., J. Res. Natl. Inst. Stand. Technol. **102**, 171 (1997).
46. Richert, R., and Angell, C. A., J. Chem. Phys. **108**, 9016 (1998).
47. Morineau, D., personal communication.
48. Kauzmann, W., Chem. Rev. **43**, 219 (1948).
49. Cohen, I., Ha, A., Zhao, X-L., Lee, M., Fischer, T., Strouse, M. J., Kivelson, D., J. Phys. Chem. **100**, 8518 (1996).
50. Poole, P. H., Grande, T., Angell, C. A., McMillan, P., Science **275**, 372 (1997). Poole, P. H., Grande, T., Scortino, F., Stanley, H. E., Angell, C. A., Computational Materials Science, **4**, 372 (1995).
51. Dvinskikh, S., Benini, G., Seker, J., Vogel, M., Wiedersich, J., Kudlik, A., Rössler, E., J. Phys. Chem. B **103**, 1727 (1999).
52. Adam, G., and Gibbs, J. H., J. Chem. Phys. **43**, 139 (1965).
53. Stickel, F., Fisher, E. W., and Richert, R., J. Chem. Phys. **102**, 6251 (1995); *ibid* **104**, 2043 (1996). Hansen, C., Stickel, F., Berger, T., Richert, R., and Fischer, E. W., J. Chem. Phys. **107**, 1086 (1997). Rössler, R., and Sokolov, A. P., Chem. Geol. **128**, 143 (1996); Rössler, E., Hess, K-U., and Novikov,V. N., J. Non-Cryst. Solids **223**, 207 (1998).
54. Murthy, S. S. N., J. Phys. Chem. **93**, 3347 (1989). Souletie, J., Bertrand, D., J. Phys. (France) I **1**, 1627 (1991).
55. Richert, R., Bassler, H., J. Phys. Condens. Matter 2, 2273 (19??).
56. Greet, R. G., Turnbull, D., J. Chem. Phys. **46**, 1243 (1967).
57. Ferrer, M. L., Kivelson, D., J. Chem. Phys. **110**, 10963 (1999).
58. Van Meegen, W., Underwood, S. M., Phys. Rev. E **47**, 248 (1993). Phys. Rev. Letts. **70**, 2766 (1993).
59. Bouchaud, J. P., Cugliandolo, L., Kurchan,J., and Mezard, M., in *Spin Glasses and Random Fields*, A. P. Young Ed. (World Scientific, Singapore, 1998), pg. 161.
60. Kob, W., in *Supercooled Liquids: Advances and Novel Applications*, ACS BOOKS, Edit., J. Fourkas, *et al.,* (1997). pg. 28.
61. Batschchinski, A. J., Phys. Chem. Streichiom. Verwandtschaftsl **84**, 643 (1913).
62. Hilderbrand, J. H., Science **174**, 490 (1971).
63. Cohen, M. H., and Turnbull, D. J., J. Chem. Phys. **31**, 1164 (1959). Turnbull, D., and Cohen, M. H.,*ibid* **34**, 120 (1961); **52**, 3038 (1970). Doolittle, A. K., J. Appl. Phys. **22**, 1471 (1951). Grest, G. S., and Cohen, M. H., Adv. Chem. Phys. **48**, 455 (1981).
64. Ngai, K. L., Rendell, R. W., in *Supercooled Liquids: Advances and Novel Applications*, ACS BOOKS, Edit., J. Fourkas, *et al.,* (1997). pg. 45.
65. Shlesinger, M. F., and Montroll, E. W., Proc. Nat. Acad. Sci (USA), **81**, 1280 (1984).
66. Lishchuk, S. V., Malomuzh, N. P., J. Chem. Phys. **106**, 6160 (1997).
67. Fisher, E. W., Physica A **201**, 183 (1993).
68. Fischer, E, W., Geier, G., Rabinau. T., Patkowski, A., Steffen, W., Thonnes, W., J. Non-Cryst. Solids **121-133**, 134 (1991).
69. Chamberlin, R. V., Kingsbury, D. W., J. Non-Cryst. Solids **172-174**, 318 (1994).

70. Debenedetti, P. G., *Metastable Liquids*, Princeton Univ. Press, Princeton, NJ, 1996. Stillinger, F. H., J. Phys. Chem. B **102**, 2807 (1998).
71. Kivelson, D., Tarjus, G., J. Chem. Phys. **109**, 5481 (1998).
72. Menon, N., and Nagel, S. R., Phys. Rev. Lett. **74**, 1230 (1995).
73. Kudlik, A., Benkhof, S., Lenk, R., Rössler, E., Europhys. Letts. **32**, 511 (1995).
74. Schönals, A., Kremer, F., Schlosser, E., Phys. Rev. Letts. 67, 999 (1991).
75. Kivelson, S. A., Xiao, X-L., Kivelson, D., Fischer, T., Knobler, C. M., J. Chem. Phys. **101**, 2391 (1994).
76. Nelson, D. R., and Spaepen, F., Solid State Phys. **42**, 1 (1989). Nelson, D., Phys. Rev. B **28**, 5515 (1983).
77. Kleman, M., and Sadoc, J. F., J. Phys. Lett. (Paris) **40**, L569 (1979).
78. Sethna, J. P., Shore,J. D., and Huang, M., Phys. Rev. B **44**, 4943 (1991). Sethna, J. P., Phys. Rev B **31**, 6278 (1085).
79. Jullien, R., Jund, P., Caprion, D., Quitman, D., Phys. Rev. E. **54**, 6035 (1998). Jullien, R., Jund, P., Caprion, D., Sadoc, J. F., J. Non Cryst. Solids **234**, 119 (1998).
80. D. Kivelson, D., and Tarjus, G., Phil. Mag. B **77**, 245 (1998). Kivelson, D., Tarjus. G., Kivelson, S. A., Prog. Theor. Phys. Suppl. No. **126**, 289 (1997). Tarjus. G., Kivelson, D.,Kivelson, S. A., in *Supercooled Liquids: Advances and Novel Applications*, ACS BOOKS, Edit., J. Fourkas, *et al.*, (1997). pg. 67.

Geometric Frustration and the Glass Transition

Jean-François Sadoc

Laboratoire de Physique des Solides, Université de Paris-Sud, 91405 ORSAY cedex, France.

Abstract. Geometric frustration covers situations where a certain type of local order, favored by physical interactions, cannot propagate throughout the space. A classical example is that of a pentagonal, or icosahedral, order which appear in the three dimensional sphere packing problem. The concept of frustration then apply to metallic or covalent glasses. The approach developed here allows for a better definition of defects in non periodic systems. Defects will be characterised upon analysing the way to recover an euclidean space starting from the ideal structures in curved space. Among these defects, disclination lines play a major role. The complexity of the structure is then encoded into the complexity of its defect network.

TETRAHEDRAL PACKINGS

Amorphous metals can be modelled by close packing of spheres. A regular tetrahedron is the densest configuration for the packing of four equal spheres. The dense random packing of hard spheres problem can thus be mapped on the tetrahedral packing problem. It is a practical exercise to try to pack table tennis balls in order to form only tetrahedral configurations. One starts with four balls arranged as a perfect tetrahedron, and try to add new spheres, while forming new tetrahedra. The next solution, with five balls, is two tetrahedra sharing a common face. With six balls, three regular tetrahedra are built, and the cluster is incompatible with f.c.c. and h.c.p. structures. A seventh sphere gives a new cluster consisting in two "axial" balls touching each other and five others touching the latter two balls, the outer shape being an almost regular pentagonal bi-pyramid. But, the dihedral angle of a tetrahedron is not commensurable with 2π; consequently, a hole remains between two faces of neighbouring tetrahedra. As a consequence, a perfect tiling of the Euclidean space R^3 is impossible with regular tetrahedra (figure 1). So, in the case of polytetrahedral structures the frustration is due to fill Euclidean space with tetrahedra, even severely distorted, if we impose that a constant number of tetrahedra (here five) share a common edge. Then, we define an unfrustrated structure by allowing for curvature in the space, in order for the local configurations to propagate identically and without defects throughout the whole space.

Regular packing of tetrahedra: the polytope {3,3,5}

Twenty tetrahedra pack with a common vertex in such a way that the twelve outer vertices form an irregular icosahedron (figure 1-c). Indeed the icosahedron edge length l is slightly longer than the circumsphere radius r ($l \simeq 1.05r$). It is possible to make this icosahedron regular if it is embedded in curved space. One then tries to add new shells of sites while keeping the same icosahedral environment for all the sites. This procedure can be continued for several shells until one sees that no new sites are needed. Indeed a finite set of points has been generated such that each vertex has twelve neighbours in perfect icosahedral configuration. There are 120 vertices which all belong to the hypersphere S^3 with radius equal to the golden number if the edges are of unit length. The 600 cells are regular tetrahedra grouped by five around a common edge and by twenty around a common vertex.

This structure called the {3,3,5} polytope [2], provides a very dense atomic structure if atoms are located on its vertices. It allows to generate a general picture of amorphous solids, at geometrical and topological levels. The main results is probably to split the structure into ordered regions and defects.

The packing fraction of the polytope can be calculated in the spherical space [3]. One considers that hard spheres are placed at the polytope vertices. Then the filling factor is: $f = \frac{120v}{2\pi^2 r^3} = 0.774$, a very high value which exceeds the value 0.74 for f.c.c. or h.c.p. structures. Indeed the f.c.c. structure is a three dimensional regular packing of tetrahedra and octahedra while polytope {3,3,5} only contains tetrahedra.

The decurving procedure from S^3 to R^3 implies lowering the packing efficiency of the mapped structure. A simple estimation of the mapped structure density gives 0.63, a very reasonable values compared to both numerical simulation of sphere packings and experimental densities.

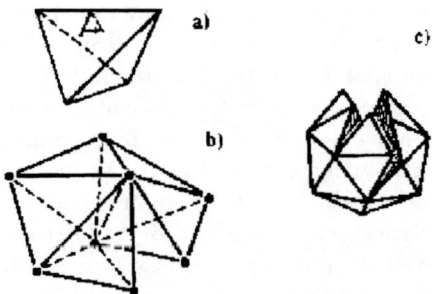

FIGURE 1. Tetrahedral packing:
-a) A tetrahedron.
-b) A hole remains between two faces of a packing of five tetrahedra with a common edge.
-c) A packing of twenty tetrahedra with a common vertex forming an irregular icosahedron

COVALENT TETRACOORDINATED STRUCTURES

In disordered covalent structures, like amorphous a-Si or a-Ge, short range order is found to be an almost perfect regular tetrahedral connectivity. But this local order can lead to numerous different structures in Euclidean or in curved spaces caraterised by the order at a scale larger than the first neighbour distance.

SiO_2 glassy structures can also, to some extent, be described as decorated tetracoordinated structures, if we consider that oxygen atoms are on edges of a tetracoordinated graph; however, oxygen atoms introduce an additional flexibility to the network.

The polytope $\{5, 3, 3\}$

This is the polytope dual to polytope $\{3, 3, 5\}$. Geometrically it is obtained by centering the 600 tetrahedral cells of polytope $\{3, 3, 5\}$ and joining two such centres by an edge whenever their tetrahedral cells share a face. This polytope has 600 tetravalent vertices, 1200 edges, 720 faces and 120 cells.

The $\{5, 3, 3\}$ is a clathrate-like "caged" structure. Indeed such dodecahedral cells are present in the clathrate crystal structure either in silicate compound or in some ice structures. The polytope $\{5, 3, 3\}$ may be useful as a template for the non crystalline versions of these materials. As far as amorphous tetracoordinated semiconductors are concerned (like a-Si, a-Ge, a-GaAs,...) the polytope $\{5, 3, 3\}$ does not seem to describe well their local order; it contains a very high number a pentagons and the expected density, is much lower than the measured one. These materials are much better described in relation to polytope "240".

The polytope "240"

This polytope shares only parts of the $\{3, 3, 5\}$ symmetries, even though it can be described as a decoration version of the latter. Let us follow a building rule similar to that which leads to the diamond structure starting from the f.c.c. structure: a new vertex is placed at the centre of some tetrahedral cells of the compact structure. In the f.c.c. case, one tetrahedron over two is centred, while in the present case, one tetrahedron over five will be centered, which has the consequence of breaking the 5-fold symmetry of the polytope $\{3, 3, 5\}$, only a ten-fold screw axis being preserved. One gets a regular structure with 240 vertices, called polytope "240", which is chiral : it cannot be superimposed to its mirror image.

In fact the polytope "240" is genrated by adding two replicas of the $\{3, 3, 5\}$, displaced along a screw axis of S^3.

A 3 dimensional graph made of vertices and edges is not necessarily a polyhedral packing : a typical example of such an impossibility is provided by the diamond structure. It becomes possible only if any pair of edges defines a face shared by two polyhedral cells. The polytope "240" is also an example of a graph (embedded here in S^3) where one encounters ambiguities while trying to define faces and cells.

This polytope contains only even-membered cycles; the smallest are hexagons in a twisted boat configuration which brings the dihedral angle into a value intermediate between the eclipsed and the staggered case without modifying the "bond" angles and the first-neighbour distances (this is possible owing to the S^3 space curvature). The polytope "240" is locally more dense than the diamond structure. This is exactly the corollary of what was said above for the $\{3,3,5\}$ compared to the f.c.c. dense structure.

DISCLINATIONS

A disclination involves a rotation operation, as opposed to dislocation, associated to a translation given by it Burgers vector. A disclination can be generated by a so-called "Volterra" process, by cutting the structure along a line and adding (or removing) a sector of material between the two lips of the cut. In two dimensions, this defect is point-like, while it is linear in three dimensions. The two lips of the sector should be equivalent under a rotation belonging to the structure symmetry group in order to get a pure topological defect confined near the apex of the cut.

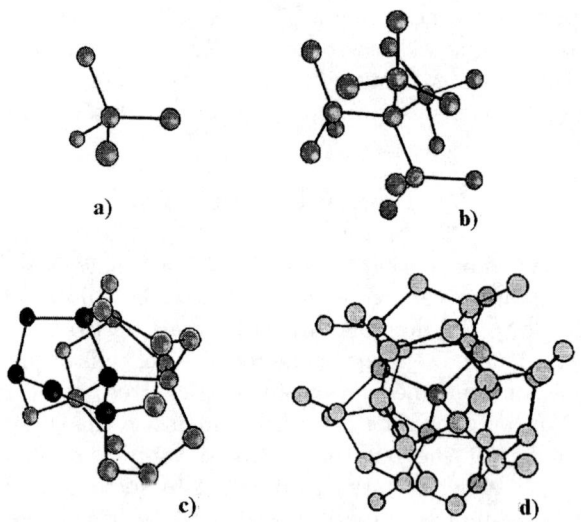

FIGURE 2. Local view of the polytope "240".
a) 5 vertices, the tetrahedral symmetry is clearly visible; b) 17 vertices;
c) 21 vertices, a sixfold ring is distinguished;
d) 39 vertices.

Wedge disclinations in two and three dimensions

It is possible to describe a 2-d disclination, and the induced deformation, as a concentration of curvature (figure 3). Let us do the Volterra construction with a sheet of paper. We first cut it along a straight segment up to its centre. Then, upon rotating around this centre, we can either add or remove a sector, and then glue again along the lips of the cut. The angle of rotation is called the weight, or the angular deficit, of the disclination. As in figure 3, one gets a non flat sheet of paper with either a conical or a saddle point singularity at its centre. Therefore, a disclination can be considered as a concentration of curvature.

A three dimensional disclination can also be generated via a Volterra process by cutting the structure along a half plane and adding (or removing) a wedge of material between the two lips of the cut. The defect is now linear. The two faces of the wedge should be equivalent under a rotation belonging to the structure symmetry group in order to allow a perfect matching between the lips and the added wedge. A pure topological defect is then confined near the axis of the cut.

Wedge disclinations can be viewed as loci of curvature concentration in a three dimensional space. Therefore, introducing negative disclinations can be used in order to decurve a positively curved space.

In three dimensions, disclinations change the network topology. For example, like in two dimensions, they change the coordination number when they go through a vertex (figure 4). So, introduction of disclinations not only allows for decurving the embedding space, but it also generates slight modifications of the local configurations.

Consider a $2\pi/5$ disclination line in the $\{3,3,5\}$ polytope (figure 4). The new coordination polyhedron is a 14-vertex triangulated structure and the central vertex

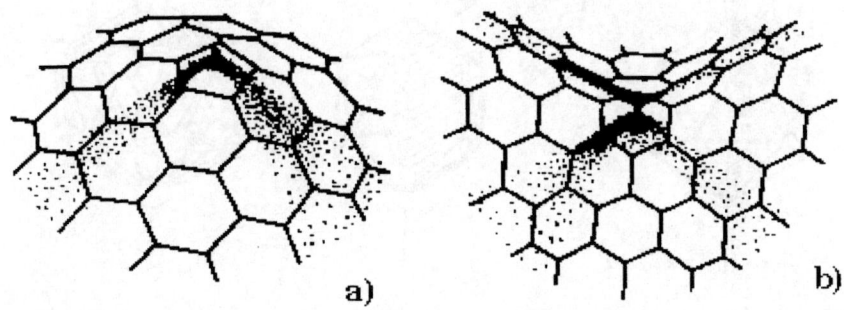

FIGURE 3. Disclinations in a hexagonal structure:
a) Positive disclination.
b) Negative disclination.

is called a Z_{14} site (Frank and Kasper notation). The central site and the two opposite vertices on the cut axis belong to the disclination line. The set of crossing points and disclination segments or lines form a so-called disclination network.

FRANK AND KASPER PHASES

In Frank and Kasper phases [4], many atoms have an icosahedral coordination shells (Z_{12} sites), but a finite proportion of atoms have a higher coordination (usually of type Z_{14}, Z_{15} and Z_{16}, the F.K. canonical cells) which form uninterrupted networks connected along the directions where the five-fold icosahedral symmetry is replaced by a six-fold local symmetry. An immediate result is that the average coordination number is higher than 12, and comparable to that of the hierarchical polytopes and of the Coxeter statistical honeycomb ($\bar{z} = 13.39$) [2,1].

A classical example of Frank and Kasper phases is the Friauf-Laves phases, as for instance Cu_2Mg or $C15$ [5] have a cubic cell containing 8 atoms of one type (Mg) arranged like carbon atoms in a diamond structure and 16 atoms of a second type (Cu) filling the free space in the voids of the diamond structure. The coordination polyhedron for the latter type of sites is a slightly distorted icosahedron (a Z_{12} site), while, for the former type, it has 16 vertices which can be obtained from an icosahedron by inserting 4 half disclinations with tetrahedral symmetry, this site being therefore a Z_{16} site.

The structure of this cubic Laves phase can be described as a stacking of sheets made of Friauf-Laves polyhedra and small tetrahedra. All Cu atoms are on the

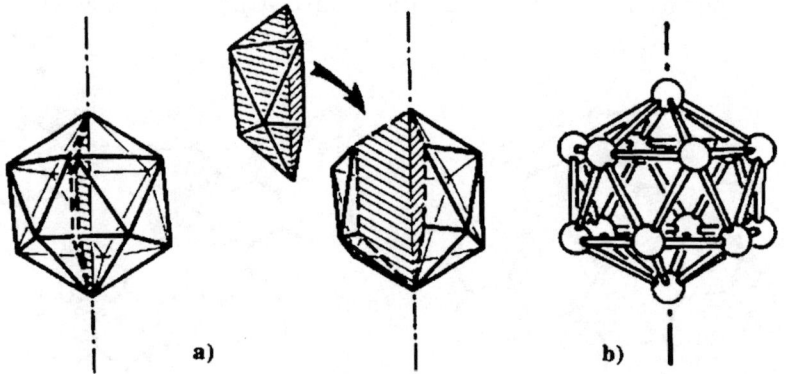

FIGURE 4. a) Procedure to insert a disclination.
b) Effect of a disclination on a icosahedral configuration: a Z_{14} coordination polyhedron has been generated.

vertices of these polyhedra, and Mg atoms are in the centre of the Friauf-Laves polyhedra. The Laves phase structure can be understood as originating from a $\{3,3,5\}$ polytope flattened by a network of disclination lines organized like the bonds of a diamond structure. The mean coordination number is $\bar{z} = 13.33$.

HIERARCHICAL DISCLINATION NETWORKS

Introducing disclinations step by step is impossible. It is however possible to achieve a complete flattening of the polytope and get an infinite structure in R^3. The key idea consists in introducing at each step a complete disclination network, whose symmetry group is contained in the polytope symmetry group [6].

The disclination configuration is obtained by a decoration procedure of the tetrahedral cells of successive polytopes. After each decoration step, a rescaling is done in order to keep a constant average first neighbour distance. Consequently, the radius of the hypersphere containing the polytopes increases upon iteration, so the space curvature is decreased.

Friauf-Laves decoration

Consider a tetrahedral cell in the $\{3,3,5\}$ polytope and add two new vertices on each of its edges, dividing them into three equal segments. The solid tetrahedron has been decomposed into four smaller tetrahedra and one truncated tetrahedron, the Friauf-Laves polyhedron(figure 5). All $\{3,3,5\}$ cells are thus decorated, leading to a decomposition of the polytope into tetrahedral and Friauf-Laves cells. New vertices are then added at the centre of Friauf-Laves polyhedra.

The proposed hierarchical flattening procedure is based on the iterative transformation of a tetrahedrized structure into another tetrahedrized structure containing more tetrahedra and vertices. From the set of new vertices, we define new tetrahedral cells, which are smaller than the initial cells. The coordination polyhedron of all sites remain icosahedra, except for those at the centre of Friauf-Laves polyhedra. Disclination lines go through the hexagonal faces of the Friauf-Laves polyhedra, so that the disclination network is formed by the edges of the dual $\{5,3,3\}$ polytope. The above decoration yields a new structure containing 1560 atoms 12-fold coordinated (Z_{12}) and 600 atoms on Z_{16} sites. This structure is called the P_1 polytope in table 1. The P_1 polytope is a packing of tetrahedral cells, five or six sharing a given edge.

The decoration has been described for one regular tetrahedral cell of the $\{3,3,5\}$ polytope, but nothing prevents to apply it again to a P_1 cell even if these tetrahedra are not regular. This is the starting point of the iterative method generating hierarchical defects. The resulting structure (P_2 polytope) is very similar to P_1, except that a new kind of coordination polyhedron now appears. These sites correspond to vertices added on an edge common to six tetrahedral cells of the P_1 polytope, which are now decorated. Thus, Z_{14} sites appear in-between two previous Z_{16}.

TABLE 1. Informations about the iterative polytopes P_i ($\lambda = 3$): N_p is the number of p-fold coordinated vertices, N the total number of vertices, T the number of tetrahedra and $\bar{z}$ the mean coordination number.

	N_{12}	N_{14}	N_{16}	N	T	$\bar{z}$
P_0	120	0	0	120	600	12
P_1	1560	0	600	2160	12000	13.111
P_2	27480	2400	12600	42480	240000	13.299
P_3	537240	57600	252600	847440	4800000	13.328
P_4	10706520	1183200	5052600	16942320	96000000	13.332
P_5	214014360	23760000	101052600	338826960	1920000000	13.333

Polytope P_2 contains two dislocation networks, interlaced in a way schematized on figure 6. It is possible to iterate *ad infinitum* this decoration procedure and obtain larger and larger structures. At each iteration, a scaling factor $\lambda = 3$ is generated between each P_i polytope and, consequently, between each interlaced disclination network. When the process is repeated again and indefinitely, only Z_{12}, Z_{14} and Z_{16} are generated.

Other scaling factors

The above decoration has been described by stating that, upon dividing the edges into three equal parts ($\lambda = 3$), each $\{3, 3, 5\}$ tetrahedral cell has been cut into one Friauf-Laves polyhedron and four smaller tetrahedra. But the above procedure can be extended to any scaling with λ an odd number, leading to the splitting of a tetrahedron into smaller tetrahedra and Friauf-Laves polyhedra.

Consider tetrahedra presented on the figure 7. There are obtained using layers of tetrahedra and Friauf-Laves polyhedra. Their edges are divided by vertices using

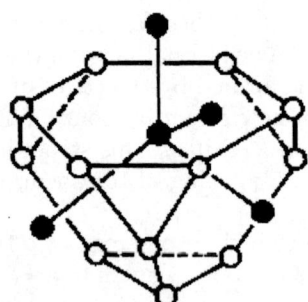

FIGURE 5. A Friauf-Laves polyhedron. It is a truncated tetrahedron: its faces are equilateral triangles or hexagons. Its vertices, denoted M_1 in the text represented by open circles; black circles represent Z_{16} sites placed at the centre of the Friauf-Laves polyhedra.

different λ factors.

By putting atoms on the vertices and at the centres of every Friauf-Laves polyhedra, a new close-packed structure is obtained in curved space. The crystalline cubic Laves phase corresponds to a stack of infinite layers formed by Friauf-Laves polyhedra and tetrahedra. The decorated tetrahedron can therefore be considered as part of a cubic Laves crystal. Thus, we have obtained an infintely large family of possible decorations parametrized by the odd integer λ.

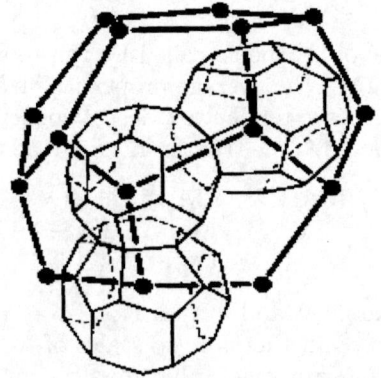

FIGURE 6. Local view of two interlaced disclination networks corresponding to two successive decurving operations. A dodecahedron, which is the elementary cell of one network ($\{5,3,3\}$ edges), is represented by heavy lines. A second network is shown interlaced with the first one.

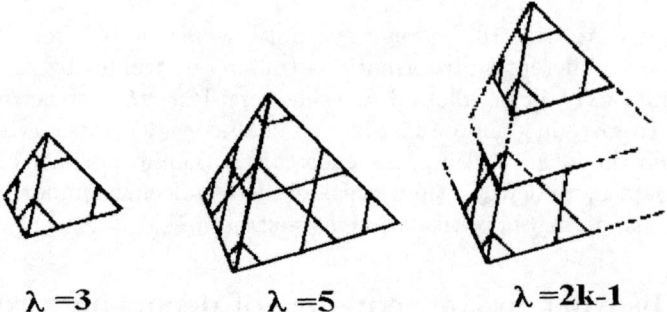

FIGURE 7. filling of tetrahedra by Friauf-Laves polyhedra and tetrahedra. The scaling factors are $\lambda =3$, 5 and $2k-1$.

Matrix formulation of the hierarchical structures

We have seen how the hierarchical structure can be described geometrically by the disclination network; in this section we use an algebraic approach which leads to the numbers of all Z_{12}, Z_{14} and Z_{16} sites at each step of the decoration (Sadoc et Mosseri 1985). Let us gather the three numbers n_{12}, n_{14} and n_{16} of Z_{12}, Z_{14} and Z_{16} sites after the ith iteration into a three-dimensional vector $\mathcal{N}^{(i)}$. We can then write a matrix Ω, which acts as a transfer matrix, and relates the representative vectors of the structure before and after one decurving operation: $\mathcal{N}^{(i)} = \Omega \mathcal{N}^{(i-1)}$

The Ω matrix reads

$$\Omega = \begin{pmatrix} 13 & 12 & 12 \\ 0 & 3 & 4 \\ 5 & 6 & 8 \end{pmatrix}$$

To the largest eigenvalue of Ω (the so-called Perron root) corresponds an eigenvector whose components gives the average properties of P_∞.

When the above procedure is extended to a decoration associated to a generic odd value for the scaling parameter λ, the transfer matrix reads, with $A = \lambda(\lambda^2 - 1)/24$.

$$\Omega_\lambda = \begin{pmatrix} 10A + \lambda & 12A & 14A - (\lambda - 1) \\ 0 & \lambda & 2(\lambda - 1) \\ 5A & 6A & 7A + 1 \end{pmatrix}$$

The Ω_λ three eigenvalues are $1, \lambda$ and $v(\lambda) = 17A + \lambda$, the latter being the Perron root; the eigenvalues correspond therefore to scales of $\lambda^0 = 1$, λ^1 (length) and λ^3 (volume) respectively. The Perron eigenvalue gives the multiplicative factor for the number of tetrahedra upon iteration (and therefore decurving) by Ω. Note that $v(1) = 1$, $v(3) = 20$, $v(5) = 90$, ..., so that $\lambda/v \ll 1$ and the Perron root dominates the decurving at every stage. From the third eigenvector it is possible to obtain a mean coordination number $z = 40/3$, independent of λ.

DISORDER AND NON-COMMUTATIVE DEFECTS

The procedure will still be repeated until decurving is complete, but a random choice among different transformations (different λ values by mixing Friauf-Laves like decorations) will be allowed at each step. But, such structures are still quite ordered. To introduce more disorder, we shall consider a decurving operation that depends on the local position and leads to a non-uniform spatial disordering. This will generate more defects, than those needed to decurve uniformly, extra positive disclinations, giving more realistic glassy structures.

Disorder in the sequence of decurving processes.

In the limit of an infinite sequence of decurving operations, $\prod \Omega$, (Euclidean limit) the main structural characteristics are completely given by the Perron eigenvector related to the Perron eigenvalue $\mu_3 = \prod v_i$. All eigenvalues are independent of the order of the decurving operations. The structure of the completely decurved polytope depends only weakly on the order of the sequence of decurving parameters $\{\lambda_i\}$ and mainly of the last operation. In practice a small number of decurving operations already yield enormous structures with very low curvature (for example, after two decurving operations Ω_3 and Ω_5 the structure contains 190800 atoms).

Some quantities of physical interest are independent of the order of the decurving operations, the total number of atoms, the elastic energy which, to first order, is proportional to the total length of disclination segments, the average coordination number $\bar{z} = 13.333$. Spatial disorder, absent here where the transformations act globally, is essential to make $\bar{z}$ fluctuate (by introducing anisotropy or fluctuations in the size of coordination polyhedra [7]).

Non-uniform decoration and spatial disorder

Up to now, we have supposed that, at a given stage of the decurving, all tetrahedral cells of the structure are identically decorated. But different decoration procedures, at different places in the structure would constitute a very efficient way to generate more disorder. It is clearly impossible to divide two neighbouring tetrahedra, by two different decorations associated to different scaling factor λ_1 and λ_2. But it is possible to commute the order of two successive decurving operations at different points in space. The scaling is conserved if one tetrahedron is divided using $\Omega(\lambda_1)$ at the first step and $\Omega(\lambda_2)$ at the second step, while another nearby tetrahedron is divided by first using $\Omega(\lambda_2)$ and then $\Omega(\lambda_1)$. The two structures are locally different, but the edge length is divided by $\lambda_1 \lambda_2$ everywhere. As the number of points generated by a sequence of operations $\prod_i \Omega(\lambda_i)$ does not depend on the order of the operations $\Omega(\lambda_i)$, decorating a tetrahedron using $\Omega(\lambda_1).\Omega(\lambda_2)$ gives the same number of sites as decorating with $\Omega(\lambda_2).\Omega(\lambda_1)$.

As far as the defect network is concerned, mixing randomly in space such pairs of transformations amounts to moving some disclination lines and creating a few closed disclination loops.

A complete description of all new coordination polyhedra and new disclination segments appearing in these random structure is rather tedious (and not unique since there are several choices on the interface). Then, the disclination network generated by the iteration $\Omega(3)$ and the disclination network generated by the iteration $\Omega(5)$ are now connected by new disclination segments (positive or negative). These new segments form a "pelote" (wool ball) close to the common interface separating two differently decorated tetrahedra.

Thus, using a non-uniform decoration, a new type of disorder is introduced, with new disclination segments. The density of defects increases, but these new defects can be called "extrinsic", since their occurence was not required in order to decurve the space, in contrast with the hierarchical networks generated by the uniform iterative decoration, which can therefore be called "intrinsic". The energy cost of these new defects is not large, they may, in some cases, be entropically stabilized.

HIERARCHY AND TWO-LEVEL SYSTEMS

An important property of the non-uniform decorations is that they generate two-level systems (TLS, or maybe better called "tunneling level systems"). As already mentioned, the distortion of the structure necessary to accommodate two different decorations is not unique. For instance, there is an ambiguity in the choice of the connection of points at interfaces separating to decoration of the kind $\Omega(3).\Omega(5)$ or $\Omega(5).\Omega(3)$.

A classical transformation between the two configurations require cuts and reconstructions of bonds, but a tunneling mode between two configurations can also be envisaged: this defines a TLS. Furthermore, it can be shown in the $\Omega(3).\Omega(5)$ example that, even if the interface configuration is fixed, the flow of points that are displaced in one tetrahedron have a rotational part relatively to the axis orthogonal to the interface: so there are still two possibilities, and a small number (1 to 4, approximately 2) of TLS can be associated to an interface separating two configurations.

We therefore have in hand a microscopic model for TLS: let us try to learn something with it, by supposing on the one hand that the ideal order is well described by the polytope, and on the other hand that the source of disorder is this non homogeneous decoration. In the above example, decorating one over five of the 600 tetrahedra of the $\{3,3,5\}$ with $\Omega(5).\Omega(3)$ in place of $\Omega(3).\Omega(5)$, leads to $480 \times 2 = 960$ TLS in a structure containing 190800 atoms. Note that this concentration of 0.005 TLS per atom is rather high compared with experiment: ($10^{-4} \sim 10^{-5}$), which may indicate that real structures are far from being saturated with these defects. We can also estimate in this example the approximate number of atoms that are moving in every TLS, by looking at those points that have their their coordination number changed: about 50 atoms per TLS. It is also possible to define TLS involving a larger number of atoms. They will have a larger associated energy. Using the structures generated from the $\{3,3,5\}$ polytope with $\Omega(3).\Omega(5)$ in some places and $\Omega(5).\Omega(3)$ in other places, it is possible to derive new structures by doing uniform decorations at later iterations. This is always possible because the structure is still decomposable into tetrahedra. In the matrix formulation, one should then take into account new sites, of type Z'_{10}, Z'_{13} and Z'_{15} (These non-canonical Frank and Kasper sites are threaded by positive and negative disclinations). In fact, the number of Z'_{13} and $'Z_{15}$ sites do not change upon subsequent iterations, and so only those sites introduced by the non-uniform operations remain (with an increase of their respective distance upon subsequent iterations). On the other hand, new Z'_{10} sites are appearing at each iteration, as intermediate points lying on the positive disclination segments introduced by the non-uniform operation. These sites contribute to the energy proportionally to the length of these lines.

Defects associated to the interfaces become more and more diluted in the structure as new uniform iteration are performed. They are still related to TLS, but which involve a very large number of atoms: $\prod_i v_i$, where v_i are Perron roots associated with the next iterations. The TLS associated with these "extended" defects are related to collective displacement of atoms. The energy barrier is higher, the tunneling rate longer, and the energy splitting between the tunneling mode smaller than with less extended defects. By considering all the non homogeneous decorations which consist in local commutations in the order of decurving processes described by $\prod_i \Omega(\lambda_i)$, on then gets a hierarchy of TLS, at all scales in the structure, which would lead to a broad density of TLS per energy range.

We can now propose the following qualitative picture. The glass transition is defined according to the change of the specific volume, which is approximately constant below T_g (neglecting the asymmetry of the pair potential) and increase linearly above T_g. Below T_g, we suppose the structure to be almost perfect, that is containing only the intrinsic defect density required by the decurving to flat space, and having no commutation defects (or a few frozen such defects which slightly increase the specific volume). Above T_g, local commutation defects are first thermally activated, while above T_m, extended defects are now generated. If we consider that energy is proportional to the specific volume, we can relate T_m and T_g, and estimated the relation : $1.3T_g < T_m < 1.8T_g$. With this simple model, we therefore find a rather good agreement with experimental results, showing that in many glasses melting temperature and glass transition temperatures are empirically related by: $T_m \simeq 1.5T_g$

Energy landscape in configuration space

We then discuss of the energy landscape in configuration space, for a structure, containing a commutation defect at an interface, generated at the end of the series of iterations. ΔE_g is the energy needed to decurve the ideal polytope, which is related to an intrinsic density of disclinations (for instance a perfect hierarchical structure), while ΔE_f characterises the interface.

There are two kinds of energy barriers. One is between the homogeneous state without disorder (ΔE_g) and a state containing an interface defect, $\Delta E_g + F\Delta E_f$ (F is a small integer). The other corresponding to a flip between the different choices for the interface structure. It is this kind of barrier which is responsible for the TLS behaviour. These energy barriers can be estimated

The separation between tunneling levels is related to ΔE_{bTLS}: the higher the barrier the smaller the separation. The energy of more "extended" defects, corresponding to earlier stages in the decurving process, reads

$$\frac{\Delta E_f^*}{\Delta E_f} \simeq \prod_i \lambda_i$$

It scales with length, whereas the barrier ΔE_b^*, involving collective displacements of atoms, scales with the volume:

$$\frac{\Delta E_b^*}{\Delta E_b} \simeq \prod_i v_i$$

Consequently, barrier heights increase much more rapidly than energy levels. Valleys, created at early stages of the decurving process become very deep, and correspond to "frozen" configurations.

Hence, the above hierarchical model leads to a rich structure in the configuration space. Homogeneous decurving yields well separated regions, with very close energies. Inhomogeneous decurving greatly complexify the energy landscape, multiplying valleys with a hierarchy of barriers heights. Thus, the kind of disorder (associated with this inhomogeneous decurving) should play an active role in a breaking ergodicity phenomenon. Note that this purely geometrical model displays the main features expected for two level systems, the latter being introduced phenomenologically in order to describe both the glass transition and the low temperature properties of glasses.

REFERENCES

1. Sadoc, J.F., Mosseri, R., *Geometrical Frustration*, Cambridge University Press, 1999; or in french *Frustration Géométrique*, Eyrolles, 1997.
2. Coxeter, H.S.M., *Introduction to Geometry*.
3. Sadoc, J.F., *C.R.Acad. Sc. Paris*, **292** *(1981) série II 435*.
4. Frank, F.C.,Kasper, J.S., *Acta Crystallogr.* **11** *(1958) 184* ; **12** *(1959) 483*.
5. Sadoc, J.F., Rivier, N., *Phil. Mag.*, **B 55**, *(1987) 537*.
6. Mosseri, R., Sadoc, J.F., *J. de Physique lett.*, **45** *(1984) L827.* , *J. de Physique*, **46** *(1985) 1809*.
7. Rivier, N., *J. de Physique Colloque*, **C9-43** *(1982) 91*.

Two-level systems/tunnelling modes in glass

Nicolas Rivier

Laboratoire de Dynamique des Fkuides Complexes
Université Louis Pasteur, F 67084 Strasbourg cedex, France

Abstract. Glass has a complicated structure (a random, regular graph of degree 4), but very simple elementary excitations (decoupled tunnelling modes between two valleys degenerate in energy). There is one tunnelling mode per odd loop (a necklace through odd rings in the network). Tunnelling is imposed by gauge invariance of the structure, the symmetry of disorder.

INTRODUCTION

Silica glass is a disordered structure of silicon atoms connected by chemical bonds (decorated by oxygen atoms which play no topological part). Silicon is tetravalent chemically, so that the structure of glass is represented by a *random, regular graph of degree 4* (dangling bonds are very rare and can be ignored). The position of the atoms (ground state), and the various modes of oscillation (elementary excitations) in glass are given by an energy, which depends on the relative orientations of and distance between connected tetrapods representing silicon atoms, on the given graph. One would expect this energy to be a very complicated function of a large number of degrees of freedom.

Experiments (carried out in the 1960-70's) show that the elementary excitations in glass are in fact very simple, independent tunnelling modes, general and specific to disordered (topologically frozen) condensed matter (1) (Fig.1).[1] It has also been confirmed that glass is the most disordered and homogeneous of all solids ("...One of the most interesting discoveries made in the comparatively early history of X-ray analysis was the fact that silk or even paper are more crystalline than glass" (Kathleen Lonsdale)). Maximal disorder and tunnelling modes provide a blueprint for the geometrical structure of glass, and for the decomposition of its low energy excitations into independent modes.

---

[1] The experimental demonstration of tunnelling modes has a long history, beginning in 1959 (2). The four crucial sets of experiments ((i) specific heat, (ii) thermal conductivity, (iii) saturation or (non-linear) ultrasonic attenuation and (iv) echoes) were first performed by (i) Hornung et al. (3), and Zeller and Pohl, (ii) A.C. Anderson, Reese and Wheatley, (iii) Hunklinger, Arnold, Stein, Nava and Dransfeld, and Golding, Graebner, Halperin and Schutz, and (iv) Golding and Graebner, and Arnold and Hunklinger. The tunnelling model was proposed independently by Anderson, Halperin and Varma, and by W.A. Phillips, before the saturation (iii) and echo (iv) experiments which confirmed it. See (1).

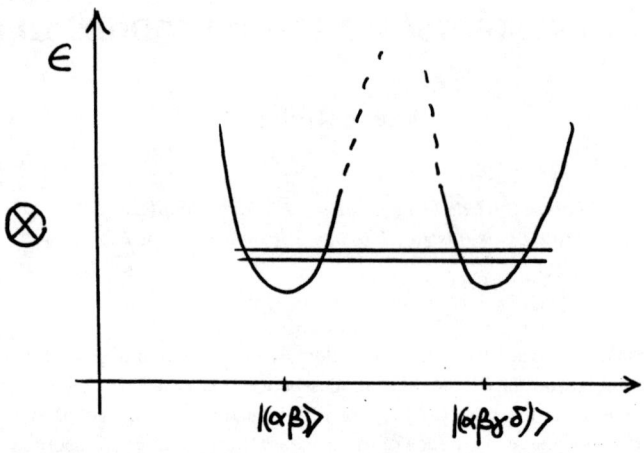

FIGURE 1. Elementary excitations in glass are (almost) decoupled tunnelling modes between two valleys (nearly) degenerate in energy. Disorder decomposes naturally the configuration space into a direct product of 2LS. This is a good example of "complicated dynamics as generic member of a disordered ensemble" (J.-P. Bouchaud, this volume).

We will show that a generic random structure (a regular graph of degree 4, with linear elasticity of its edges) has the following ground state and excitations:
  (i) It has several pairs of degenerate ground states (minima of the potential (elastic) energy), as represented in Figure 1.
  (ii) The two alternative ground states are not physically distinguishable; they are related by a (large) gauge transformation. Gauge invariance imposes tunnelling and lifts the degeneracy.
The degenerate ground states are a result of combinatorics on a random, regular graph representing the structure of the glass. Tunnelling is a consequence, through gauge invariance (4), of the randomness of the structure.

The elementary excitations are called indifferently tunnelling modes or two-level systems (2LS). Neither name is entirely satisfactory. While the two states involved are classical, there is only one excitation path from one to the other, and the potential barrier separating them cannot be overcome classically. Each 2LS is thus highly non-linear. If the ground and lowest excited states are close in energy, higher excitations are entirely out of reckoning.

TUNNELLING MODES: EXPERIMENTAL EVIDENCE

Glass has extraordinary physical properties at low temperatures (1), ascribed to tunnelling modes (or 2LS). Phonons, normal modes of sufficiently long wavelength to ride over any inhomogeneity in the structure (i.e., > 4 nm) are elementary excitations in any solid, crystalline or otherwise. We hear them as the glass rings. But consider what happens in glass:

(i) The specific heat is approximatively linear in temperature below 1K, where it exceeds the T^3 contribution of phonons in a three-dimensional system. There are therefore additional elementary excitations in glass besides phonons. A somewhat flat distribution of 2LS yields a linear specific heat (as for electrons in metals: Boltzmann's distribution on a two-level system looks like Fermi-Dirac's, with the distribution of energy spacing playing the part of the density of electronic states).

(ii) Heat is carried by phonons. Glass has a lower thermal conductivity than its corresponding crystalline material (quartz), so the phonons are absorbed by the additional two-level excitations.

(iii) When the power of the transducer is turned up, most 2LS will have absorbed a phonon and be in their excited state. They can no longer absorb phonons, and thermal conductivity of the crystal is recovered.

(iv) An echo can be set up on the 2LS, as in any quantum two-level systems (spin 1/2, but this is a real 'acoustic' echo).

Saturation and echo show that the two-level excitations are due to quantum tunnelling (classical, thermal excitations are at much higher energy, out of reach at these low temperatures). The first two experiments could also be explained by classical jumps over potential barriers but not the last two.

(v) Observation of echo (analogous to spin echo, but generated by sound pulses) indicates that different 2LS are *decoupled* enough to preserve phase coherence over the long echo time (10 µs at 20 mK). In fact, the interaction energy between 2LS is very small (<1µeV). It is only recently (5) that the coherent (albeit complex) state of the interacting 2LS has been identified and studied, at temperatures below 6mK. It responds to magnetic fields of tens of µT or less.

These low temperature properties are common to all glasses, covalent oxides, amorphous elemental semiconductors and metallic glasses (6). Only their intensity and characteristic energy and temperature scale differ (1,7). (Exceptions, like amorphous arsenic (z=3), or III-V compounds which have no odd rings, or hardly any (they imply the presence of wrong (Ga-Ga or As-As) bonds, which are energetically costly), confirm the model).

The anomalous physical properties of glass (i)-(iv) below 1K (1) are modelled by *tunnelling* between independent pairs of potential valleys in configuration space. Thus, phase space is represented (Fig.1) as a direct product of pair of classical configurations, nearly degenerate in energy. The ground state and first excited state of each pair are tunnelling modes, superpositions of the classical configurations. Excitations in each potential well are at much higher energy, inaccessible at low temperatures below 10K (as would be classical transitions over a saddle point between potential wells). The local, tunnelling modes and the long wavelength phonons constitute therefore the elementary excitations in glass.

The energy splitting between the two levels is very small.[2] This splitting is therefore entirely the contribution of tunnelling Δ_0 between degenerate potential wells. Any deviation from degeneracy Δ yields an additional energy difference (>Δ and Δ_0). This also implies that there are many atoms involved in the tunnelling, moving very little. (Inversion tunnelling of *one* atom in a tetrahedron (a symmetrical ammonia molecule) would occur at an energy higher by orders of magnitude).

---

[2]Linear specific heat is observed between <10^{-2} K and 1 K in silica, which implies a continuous range of splittings between <10^{-6} and 10^{-4} eV.

The presence of nearly *degenerate* classical ground states (potential minima) in a system with no obvious symmetry to impose the degeneracy is astonishing. Indeed, bulk condensed matter usually has one single ground state, and its potential energy, one single minimum in a many-dimensional configuration space.

We present here a simple model which explains clearly the nature of tunnelling modes, is predictive and falsifiable, generic but non-universal: it applies to covalent glasses, for example, to window glass SiO_2. Tunnelling modes are set on odd rings (circuits) of the network. These odd rings are threaded through by closed loops (8), topological "defects" of glass. There is one tunnelling mode per odd loop. The density of odd loops, thus of tunnelling modes, is determined by maximum entropy. The maximum entropy configuration of the loops is semi-dilute, with only one single length scale (the distance between different loops is equal to the average loop size), thereby taking maximal advantage of mixing (small loops) and configuration (longest loops) degrees of freedom. A tetravalent network is also the generic scaffolding spanning a disordered elastic continuum representing polymeric glasses.

TUNNELLING MODES IN COVALENT GLASSES

What then are these tunnelling modes, why are they decoupled, and why are they degenerate, in a strongly correlated system of many atoms with trivial space group ?

The answers are provided by an analysis of the structure and elasticity of covalent glasses. The argument is in several steps listed below. Details and proofs have been published elsewhere (4,9).

Generic geometry of disorder

1. Glass structure is modelled as a *continuous random network*, that is, as a regular graph. Each vertex of the network represents a Si atom. It has z=4 incident edges (z=3 in the case of arsenic and boron), and each edge of the network represents one (in elemental glasses like a-Si) or two covalent bonds separated by an oxygen atom (in silicate glasses). The oxygen atom only decorates the edge and plays no topological part. We count edges and call a ring odd if it has an odd number of edges. The vertices and edges of the network are physical objects. The constant vertex coordination z is imposed by chemistry, and the only connection between two vertices, an edge of the network, is a covalent bond (whether decorated by oxygen or not), which has some rigidity: It costs some energy to stretch, bend ot twist it, in decreasing order.

2. The long-range homogeneity of glass is a gauge invariance. Glasses have perfect short-range order, imposed by chemistry (regular z=4 graph). Each silicon atom can therefore be represented as a perfect tetrapod (Vierbein). Beyond 4 nm, i.e., beyond a few rings, glass is completely homogeneous, but this homogeneity is not the generative symmetry, exact superposability of crystals. The translated glass is locally different, for example in the orientation of the nearby Si tetrapods, but this difference is physically irrelevant. This is the symmetry of disorder (4). It describes a forest in which one is lost: one tree in the forest differs slightly different from another, but this difference is physically, globally or objectively irrelevant. It does not help us to find a way out. Thus, each Si atom in the network has, besides its position, a *gauge* parameter which connects it to its immediate environment (at this stage, it is easiest to

think of the orientation of the tetrapod, but this will be refined as the group of permutations of its bonds). The gauge parameter changes slightly from atom to atom but the overall structure and physical properties of the glass are independent of these changes. These local variations are together obligatory and macroscopically irrelevant in a disordered material. The variations, albeit local, are transmitted from atom to atom by a connection, the elasticity of the covalent bond linking them. These two remarks are generic properties of gauge symmetry, of which glass is the most concrete if not the simplest example.[3]

The continuous random network is a discrete, random scaffolding. This has two consequences. The first is that its physical properties will be gauge invariant, by symmetry. The second is the existence of topological defects.

3. Glass has "defects" as its structural signature. These defects are odd lines, necklaces threading continuously through the odd rings of the network, while avoiding even rings (8). Odd lines either close in as loops, or exit at the boundary of the material. The proof of this theorem is elementary: On any closed surface in the network there is an even number of odd rings, thereby providing an exit for any odd line entering the surface.[4] Thus, odd rings are never isolated, but form loops. The theorem holds also for crystalline structures, but there, odd rings are rare or absent (only three-fold rings are crystallographic), odd loops are rare or very small, and there is neither gauge invariance, nor maximum entropy to mould their dynamics. What are the physical effects of these odd lines? They are the seat of tunnelling modes.[5].

Gauge invariance and odd loops are the only essential ingredients of the structure of network glasses. I stress that they are natural ingredients, both in the physical (physical atoms and bonds in a continuous random network) and in the mathematical (simplest and generic) senses of the word.

Elasticity of $z=4$ random networks

We now describe the classical elasticity of continuous random networks. The elastic energy is carried by the edges of the network, that is by the chemical bonds connecting two tetrapods. This potential energy consists of two terms, a strong, bond-stretching V_1, and a weaker, bond-bending contribution V_2 (10). V_1 can be written in terms of the relative displacement of two neighbouring atoms only. V_2 requires also the direction towards a third atom. Its precise nature is not important for tunnelling modes.

---

[3]The geometric representation of gauge invariance is a *fibre bundle*. Its *base space* is the regular graph. The total space is the position and relative orientation of the tetrapods. The set of all points in the total space that are mapped onto the same point in the base is called the *fibre*.

[4]Associate (-1) to any edge on the (arbitrary, closed) surface S. Define a ring index as the product of its edges. An even/odd ring has index +1/-1. The product of all ring index on S equals +1, simply because each edge is counted twice as it borders two rings.

[5]Normally, the gauge parameter returns to its original value when one returns to the origin after some closed path in the network. Not so if the closed contour is odd, in which case the gauge parameter must be changed: it will be 'entangled' about odd rings, in one of two equivalent fashions (as the vector potential is wound about a solenoid in electromagnetism. It cannot be eliminated by a gauge transformation, even though the physical magnetic induction is zero outside the solenoid). Tunneling between the two, physically equivalent fashions must occur to restore physical symmetry. Once again, we see the double role - obligatory and physically irrelevant - of gauge symmetry, the double agent of disorder. The network itself is not entangled.

4. Neglect V_2 in a first stage. The normal modes are a band of N phonons flanked by two sets of N degenerate modes (11). One set has zero frequency ($\omega = 0$) and energy. The network is underconstrained (wobbly) in the absence of bond-bending forces. *The ground state configuration(s) and the lowest energy excitations of the material* (besides phonons) *are the N floppy modes at* $\omega = 0$.

The motion of an unstretched edge is given by one single number, the projection on the edge of the displacement of the two vertices bounding it. This number plays the part of a current in an electrical network (12), and Kirchhoff's current (incidence) and voltage (closure) laws are obviously satisfied (13). The N floppy modes are *independent* currents. They lie on edges not on a spanning tree of the network, so that each such edge closes an independent circuit or ring.[6] The independent floppy modes form a basis for displacements leaving all bonds unstretched. They are the N ($\omega = 0$) normal modes.

5. Let us now stiffen the network by bond-bending forces $V_2 \neq 0$. The floppy modes are reorganised, separated into ground states with elementary excitations, and modes of higher energy. We will see that even rings are trivial (they have a single ground state). Odd rings have two stable configurations.

Let us measure the energy of a ring configuration. By definition, floppy modes have only bond-bending energy. This energy is measured by comparing the orientations of the two tetrapods (at i and iα) connected by bond α, or, equivalently, through a *congruent transformation* of the tetrapod, from its orientation at i to that at iα.[7] Accordingly, the chemical bond (of label α) connecting the two neighbouring tetrapods imposes a mirror reflection fixing its midpoint. So, the congruent transformation is a rotatory-reflection. (The other 3 non-shared bonds may rotate)

In a given configuration, the tetrapod must be returned to its original orientation (or an equivalent one) after being carried around the ring. Configuration of a n-sided ring is the product of n rotatory-reflections. The product of rotatory reflections is therefore a covering transformation of the tetrapod, namely a *permutation* of the labels (colours) of its legs. If the ring is even, the permutation, a product of n reflections, is even. If the ring is odd, the permutation is odd.[8] Permutation is the gauge parameter indicating how the Si atom is connected to its neighbours through a ring of covalent bonds. To this permutation can be associated a small quantitative energy, involving V_2 only, since bonds remain unstretched.

6. In fact, it is not the permutation which labels the configuration, but only its conjugacy class: Let us go around two different rings in succession, starting from a common vertex, recording permutation **R** through one ring, permutation **Q** through the other. The total permutation is **P = Q.R**, or **P' = R.Q**, depending on the order of the circumnavigations, and, in general, **P** and **P'** are different. They are related by

$$\mathbf{P' = R.P.R^{-1}} \tag{1}$$

---

[6] A tree reaching all vertices of the network is called spanning. It is, of course, not uniquely defined, and selecting one spanning tree partly fixes the gauge.

[7] Consider the bond energy $S_i J_{ij} S_j$ in Ising magnetism, not as an interaction between two identical objects (S_i and S_j), but as a *connection from* S_i *to* S_j. The first point of view regards J as the standard for comparing directions of S_i and S_j; the second, adopted here, through the congruent transformation (flip or identity) J imposes on the spin.

[8] This is because the number of legs, z=4, is even. Elementary rotation about the shared bond, a cyclic permutation of the (z-1) others, has parity $(-1)^z$.

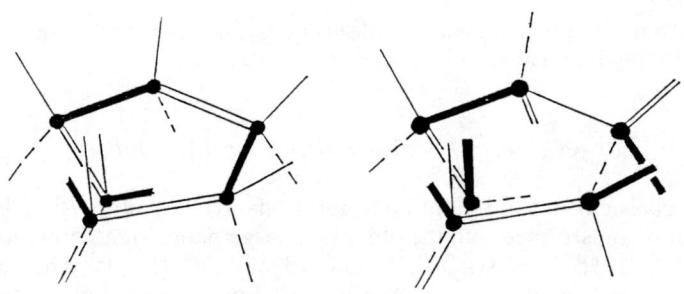

FIGURE 2. The two classical configurations (edge labellings) about an *odd* ring, resulting *inevitably* in odd permutation $\{(\alpha\beta)\}$ or $\{(\alpha\beta\gamma\delta)\}$. Change of labelling is a gauge transformation, because the edges are physically indistinguishable.

and belong to the same class of the permutation group. The physical configuration, which must be independent of the order of circumnavigations made to measure it, is labelled by the set which includes **P**, **P'**, etc., namely by the *class* of the permutation group to which they belong. Permutations belonging to the same class are identified physically.

The ground state of even rings clearly belongs to the identity class of the permutation group of degree z=4, S_4. It is nondegenerate, and even rings are dynamically trivial. They can be coloured consistently (labels are not permuted).

The configurations of odd rings are labelled by the two classes of odd permutations of S_4 (each containing six elements), $\{(\alpha\beta)\}$ and $\{(\alpha\beta\gamma\delta)\}$, so that odd rings have two distinct lowest-energy configurations (characterized by one permutation in each class, selected by choosing and colouring a spanning tree). It is easy to label consistently (identity permutation) all even rings, and impossible to do so for odd rings. Change of labelling of the latter does not destroy consistency in the former (the physical meaning of equation (1)). Moreover, one odd ring is the sole, independent representative of a whole odd line, because all the other odd rings of the loop are linked to the first one by even rings and their bonds are permuted in concert.

This shows that each odd line has two distinct classical ground states, separated by a large gauge transformation. The two classical ground states are represented in Fig.2. Permutation class $\{(\alpha\beta)\}$ on the left has been realized by a connection algorithm which labels eclipsed bonds by the same colour. This gives the sequence of transformations $\alpha\beta \rightarrow \beta\alpha \rightarrow \alpha\beta \rightarrow \beta\alpha \rightarrow \alpha\beta \rightarrow \beta\alpha$, hence permutation $(\alpha\beta)$ for the ring. On the right, bonds of the same colour are rotated by by $2\pi/3$ about the shared bond, yielding the sequence $\alpha\beta\gamma\delta \rightarrow \beta\gamma\delta\alpha \rightarrow \gamma\delta\alpha\beta \rightarrow \delta\alpha\beta\gamma \rightarrow \alpha\beta\gamma\delta \rightarrow \beta\gamma\delta\alpha$, i.e. permutation $(\alpha\beta\gamma\delta)$ for the ring. Any colouring algorithm must lead to one or the other permutation class, but it is not a physical attribute since bonds are identical and odd rings cannot be coloured consistently.

The physical attributes of the random network architecture are edges shared between two neighbouring atoms, and the impossibility (for z=4) of edge-colouring an odd cycle. Change of algorithm, and of class, is therefore a gauge transformation, leaving invariant the physical properties of the system, notably its energy, (maximum) entropy and defects (odd loops). Colouring is just a metaphor for bonds shared between two tetrapods. It is not the actual colouring which matters physically (bonds are identical), but whether consistent colouring is possible, and, when not, in how many essentially distinct ways.

Large gauge transformations and tunnelling

7. Neither classical ground state configurations $\{(\alpha\beta)\}$ or $\{(\alpha\beta\gamma\delta)\}$ are gauge-invariant. Each is transformed into the other by a large gauge transformation or change of labelling $\mathbf{G}$, $\mathbf{G}\,|\{(\alpha\beta)\}> = |\{(\alpha\beta\gamma\delta)\}>$, $\mathbf{G}|\{(\alpha\beta\gamma\delta)\}> = |\{(\alpha\beta)\}>$. They are therefore *degenerate* in energy. However, the physical configurations are given by the *gauge-invariant* linear combinations

$$|\pm> = (1/\sqrt{2})[|\{(\alpha\beta)\}> \pm |\{(\alpha\beta\gamma\delta)\}>] \,, \qquad (2)$$

with one sign for the ground state and the other for the excited state. Tunnelling, however slow, *must* take place to restore gauge invariance. Tunnelling modes must exist in z=4 glasses, and also in disordered elastic continua or in polymeric or mesoscopic glasses for which the tetrapod is the generic local frame of reference spanning the three spatial dimensions.

Tunnelling modes do not occur in trivalent network glasses (a-As or borate glass), for two reasons:

(i) For z=3, an elementary rotation about the shared bond is an odd permutation, so that odd rings can accommodate the identity permutation and be edge-coloured.[9]

(ii) Odd permutations of S_3 belong to one class only.

Interaction between 2LS ?

Topologically, two 2LS on different odd loops are decoupled since the bonds closing their representative odd rings are independent. The even rings inbetween are in the ground state (identity permutation (α)). Gauge transformation (a permutation) on one loop does not affect the class of permutation representing the structure of another (eq.(1)). Non-trivial permutations on even rings can couple two odd loops, but these are excitations at higher energies.

However, two odd loops interact in the ground state through the deformation (virtual phonons) which they impart to the network. This interaction involves the real energy cost of bond-bending, the rotation part of the rotatory-reflection connection. The two permutation states of a loop cause slightly different deformations, so that the 2LS of two loops interact, very weakly (6 mK $\approx 0.5.\mu eV$) and randomly (5).

This interaction reveals itself as a transition from incoherent tunnelling of individual 2LS at higher temperatures, to coherent behaviour of an assembly of weakly coupled 2LS, minimizing their energy at low temperatures (5). This collective dielectric response of coupled 2LS sets in at very low temperatures ($T<T_c = 5.84$ mK). It is a nonlinear response to extremely low magnetic fields ($B\approx\delta B\approx 10\mu T$), whereas a glass with incoherent, individual 2LS is a linear dielectric (its dielectric constant is independent of B). Deformations of a network of elecronic bonds are affected by a

---

[9] If this was not the case, odd rings in planar z=3 graphs would constitute the simplest counter-example of the 4-colour theorem. (Vertex and edge-colouring of graphs are closely related (14)).

magnetic field, and thus the collective state of the 2LS is that of a very fragile (classical) spin glass.

At the higher temperatures ($T>T_c$), thermal fluctuations overwhelm the very weak interaction energy between 2LS, and coherence is destroyed. Incoherent, individual behaviour by 2LS is also restored at 'high' magnetic fields $B>10mT(\approx kT_c/\mu_B)$. The "Zeeman" energy of individual 2LS is larger than their interaction energy (one Bohr magneton has been used arbitrarily as the conversion factor).

TOPOLOGICAL ENTROPY OF GLASS AND DENSITY OF TUNNELLING MODES

The topological entropy, remaining in the glass at T=0, is that (configuration and mixing) of odd loops. We make the maximum entropy assumption that every face (smallest ring) of the continuous random network can be odd or even, independently of other rings, without restriction apart from the very continuity of the odd lines. Odd loops can move freely and be generated in the glass-forming liquid.[10] The maximum entropy distribution of odd loops is semi-dilute.

The number of equivalent configurations Ω of an arbitrary number of odd loops in any position, shape or length, is simply two configurations (odd or even) per face of the network, with one face per cell as parity control to insure continuity, providing an exit for an odd line entering that cell. There are thus $\Omega=2^{F-C}$ configurations in the network (16). The (maximal) topological entropy is therefore

$$S/k = \ln \Omega = C\,(<f>/2 - 1)\ln 2 = N\,(z/2 - 1)\ln 2 , \qquad (3)$$

since a facet always separates two cells, and $<f>$ is the average number of facets per cell, $2F = <f>C$. k is Boltzmann's constant, and N is the number of Si atoms.

For a froth (z=4, eclipsed edges), $<f> = 13.4$. At the other extreme (random diamond, z=4, staggered edges), $<f> = 4$. There are on average $<f>/2$ odd faces per cell, thus $<f>/4$ odd lines per cell, which varies between 3.35 for all eclipsed edges and 1 for all staggered edges.

In the last expression (3), the topological entropy is given per atom, using Euler's formula for a three-dimensional network and the valence relation $zN = 2E$ (an edge is bounded by two vertices on which z edges are incident) (16). If cooling has been too rapid for full exploration of phase space, the zero point entropy is smaller than (3) and the odd loops are no longer in a semi-dilute configuration.

Thus, odd loops are what count as configurations in the partition function of glasses and liquids. Indeed, they account (16) for the topological entropy of melting of simple, atomic substances (Ar, Na, Cs) (17).

CONCLUSIONS

Tunnelling modes are located on odd loops in z=4 network glasses. See Fig.1 and equation (2) (gauge invariance imposes tunnelling between nearly degenerate classical

---

[10]This has been suggested as a thermodynamic mechanism for the glass transition (15).

sectors). Each classical configuration involves at least all the Si atoms on the rings constituting one odd loop. That is, a considerable number of atoms, but moving very little.

Topological disorder is essential: The space group is trivial, and even the involution associated with the bonds (reflection) never adds up to a consistent action on odd rings. It is indeed the action of this involution which differentiates odd and even rings. Gauge transformations, the permutations between bonds are universal, like their sources, the odd loops in the network.

The chemical bond is the main actor in a gauge-invariant, disordered solid: As the energy-carrying connection, and as the geometrical constituent of the odd rings and loops responsible for tunnelling modes.

Thus glasses, with their somewhat complicated ground state and very simple elementary excitations (direct product of tunnelling modes or ring modes), are exact opposites from crystalline solids.

REFERENCES

1. Hunklinger, S. and Raychaudhari, A.K., *Progr.Low T.Phys.* **IX**, North Holland, 1986, p.265.
  Phillips, W.A., *Amorphous Solids. Low-Temperature Properties*, Springer, Berlin, 1981.
2. Flubacher, P., Leadbetter, A.J., Morrison, J.A., and Stoicheff, B., *J. Phys. Chem. Solids* **12**, 53 (1959).
3. Hornung, E.W., Fisher, R.A., Brodale, G.E., and Giauque, W.F., *J. Chem. Phys.* **50**, 4878 (1969).
4. Rivier, N., *Geometry in Condensed Matter Physics*, ed. J.F. Sadoc, World Scientific, 1987, pp.1-88.
5. Strehlow, P., Enss, C., and Hunklinger, S., *Phys. Rev. Letters* **80**, 5361-4 (1998).
  Enss, C., and Hunklinger, S., *Phys. Rev. Letters* **79**, 2831-4 (1997). The physical picture of a coherent state of interacting 2LS at very low temperatures, is given in the '98 Letter.
6. Weiss, G., and Golding, B., *Phys. Rev. Letters* **60**, 2547 (1988).
  Black, J.L., *Glassy Metals I*, ed. H.-J. Güntherodt and H. Beck, Springer, Berlin 1981, pp.167-90.
7. Rivier, N., *J. Non-Cryst Solids* **182**, 162-71 (1995).
8. Rivier, N., *Phil.Mag.*A **40**, 859-68 (1979)
9. Rivier, N., *J. Math. Chem.* **13**, 1-14 (1993).
10. Keating, P.N., *Phys.Rev.* **145**, 637 (1966).
11. Alben, R., Weaire, D., Smith, J.E. Jr., and Brodsky, M.H., *Phys. Rev.* B **11**, 2271 (1975).
  Sen, P.N., and Thorpe, M.F., *Phys. Rev.* B**15**, 4030 (1977).
12. Rivier, N., *Adv.Phys.* **36**, 95-134 (1987).
13. Kirchhoff, G., *Pogg. Ann. Physik u. Chemie* **72**, 32 (1847).
  Biggs, N., *Algebraic Graph Theory*, Cambridge Univ.Press, 1974.
14. Fiorini, S. and Wilson, R.J., *Edge Colouring of Graphs*, Pitman, London, 1977.
15. Rivier, N., and Duffy, D.M., in *Numerical Methods in the Study of Critical Phenomena*, eds. J. Della Dora, J. Demongeot, B. Lacolle, Springer, 1981, pp. 132-42.
  Anderson, P.W., in *Ill-Condensed Matter*, eds. R. Balian, R. Maynard, G. Toulouse, North Holland, 1979, Ch. 3, § 2.1.
16. Rivier, N., and Duffy, D.M., *J. Phys. C: Solid St. Phys.* **15**, 2867-74 (1982).
17. Tallon, J.,L., *Phys. Lett.* **76**A, 139-42 (1980), and ref. therein.

Frustration in model glass systems : numerical investigations

Rémi Jullien[a], Philippe Jund[a], Didier Caprion[a] and Jean-François Sadoc[b]

[a] Laboratoire des Verres, Université Montpellier II, Place Eugène Bataillon, 34095 Montpellier, France

[b] Laboratoire de Physique des Solides, Bât. 510, Université Paris-Sud, Centre d'Orsay, 91405 Orsay, France

Abstract. Numerical Voronoï tessellation is used to investigate the mechanisms of frustration in some model glass systems. First, random packings of 8192 hard spheres of increasing volume fraction c are built using an efficient computer algorithm. Their Voronoï statistics evolves with c as if the system would like to reach a pure icosahedral order when extrapolating the volume fraction above the Bernal limit $c_b \simeq 0.645$. Second, super-cooled liquid and glass samples of 1000 atoms are generated at different temperatures T after a quench from the liquid state, using classical micro-canonical molecular dynamics with a simple soft-sphere potential. When decreasing T, the ideal icosahedral order appears again as an extrapolated situation which cannot be realized due to geometrical frustration. Third, a model silica glass of 648 atoms is studied using the potential of van Beest, Kramer and van Santen and a quite similar quenching procedure is performed. As in the soft-sphere case the structural freezing following upon the glass transition is noticeable in all the geometrical characteristics of the Voronoï cells and again a possible interpretation in terms of geometrical frustration is proposed.

I INTRODUCTION

In this paper, we would like to report on the mechanisms of geometrical frustration in some model glassy systems analyzed numerically via the Voronoï tessellation. We first consider random packings of hard spheres since they were extensively used throughout the last decades to represent the structure of liquids, amorphous solids and glasses [1-6]. One of the most fascinating features of random packings, which was first evidenced by Bernal [1], is that there exists an upper limit of the volume fraction $c_b \simeq 0.645$ which cannot be exceeded and which is significantly smaller than the one $c_m = 0.7405$ of the ordered close packings (hexagonal-closed-packed and face-centered-cubic). It is generally believed that the existence of such an upper limit is related to the degeneracy between hexagonal-closed-packed and face-centered-cubic structures leading to the so-called geometrical "frustration" [7] associated with the impossibility to tile the space with perfect tetrahedra only. The similarities between the local tetrahedral order found in random packings and the one observed in many metallic glasses has been stressed since a long time [8]. Here, using numerical Voronoï tesselation, we show that, when increasing the volume fraction up to the Bernal threshold, the system evolves as if a perfect tetrahedral order would like to take place. All the characteristics of the Voronoï cells vary with c in such a manner that, when extrapolated above c_b, they converge to those of a perfect dodecahedron circumscribed to a sphere. However, one conceptual problem when trying to use random packings to simulate the structure of glasses is that one does not consider neither any realistic potential nor the true thermodynamics, and therefore all the features associated with the existence of a glass transition cannot be accounted for. This is why, for comparison, we have performed in this work the same kind of geometrical analysis on atomic configurations, but of more modest size, generated by using classical molecular dynamics (MD) of a model glass. By determining the histograms for the distribution of volumes and total surface areas of the Voronoï cells, we are able to analyze the general trends for the evolution of the local order as a function of the temperature T. The same tendency to develop a perfect tetrahedral order, with dodecahedral Voronoï cells and five-fold symmetry, is observed when decreasing T from the supercooled liquid state. Our results strongly support the idea that the perfect tetrahedral order would be an ideal situation which can not be reached due to geometrical frustration, as postulated in a recent theory for the glass transition [9]. While it appears that the frustration ideas apply quite well to soft sphere systems, which are the archetypes of "fragile glasses", such as metallic glasses, not so many attempts have been tried to apply them to "strong glasses", such as silica-based vitreous systems, in which the interactions are more likely of covalent nature. Therefore we have combined molecular dynamics calculations with the Voronoï tessellation scheme to study a model silica glass using the two-body potential proposed by van Beest et al. [10]. Our results as a function of temperature show the structural freezing consecutive to the glass transition in all the geometric characteristics of the Voronoï cells. Moreover when extrapolating the high temperature data to lower temperatures, it seems that the system behaves as if it would like to reach locally an ideal underlying structure satisfying the basic energetic requirements imposed by the potential but which cannot extend to infinite distances in the usual three dimensional space.

In section II, we present the calculations for random packings, in section III we present the calculations for the soft-sphere system, in section IV we present the calculations for the model silica glass and in section V, we conclude. Some accounts of the work presented here have been published elsewhere [11-14].

II RANDOM PACKINGS OF HARD SPHERES

A The Jodrey-Tory algorithm

To generate sphere packings in a square box of edge length L, with periodic boundary conditions (PBC), we have followed an efficient numerical recipe which was introduced by Jodrey and Tory (JT) [15] more than ten years ago. The JT algorithm proceeds by an iterative sequential resorption of overlaps of -fictitious-spheres which consists in successive displacements of pairs of nearest neighboring points (sphere centers) starting at iteration $i = 0$ from a set of N points randomly located in the simulation box. At each iteration i the set of points (the sphere centers) is characterized by the list of coordinates and also by a list of distances (between pairs of points) in increasing order together with some other tables necessary to identify the points in the list. To the minimum distance d_m^i corresponds a minimum packing fraction $c_m^i = N(\pi/6)(d_m^i/L)^3$. Along the iterative procedure, one also carries a maximum distance d_M^i, related to a maximum packing fraction $c_M^i = N(\pi/6)(d_M^i/L)^3$, which is set to $d_M^0 = L(6/\pi N)^{1/3}$ (i.e. $c_M^0 = 1$) at $i = 0$. After the identification of the pair of points M_1^i and M_2^i realizing the minimum distance $d_m^i = M_1^i M_2^i$, these points are spread apart symmetrically along the $M_1^i M_2^i$ line to new positions M_1^{i+1} and M_2^{i+1} such that $M_1^{i+1} M_2^{i+1} = d_M^i$. Then, before going to the next iteration, the list of distances and the related tables are updated, the new minimum distance d_m^{i+1} is determined, and the maximum distance is set to a lower value given by:

$$d_M^{i+1} = d_M^i - \frac{K}{N}(c_M^i - c_m^i)^\alpha \tag{1}$$

where the "rate" K and the exponent α are two input parameters of the algorithm in addition to L, N and d_M^0. Note that formula (1) is slightly different (simpler) than the original one used by Jodrey and Tory [15] and consequently our definition of the rate K is different. The process stops at iteration n when one finds $d_M^n < d_m^n$. Then the final minimum distance d_m^n is taken as the particle diameter d for the resulting packing. The final packing fraction is then given by:

$$c = N\frac{\pi}{6}(\frac{d}{L})^3 \tag{2}$$

Note that the value taken for the box edge length L does not play any role as it only fixes the unit of length. In practice L has been set to an integer value $L = 20$ because we have used an underlying cubic lattice of 8000 cells to label the spheres in order to accelerate the search for their neighbors. The number N of spheres has been set to $N = 8192$, almost an order of magnitude larger than the JT value $N = 1000$ [15]. We have checked on a few other N values that the results are size insensitive. While the parameter d_M^0 is not very important (it should be taken sufficiently large however to guarantee that the spheres of diameter d_M^0 do overlap in the initial configuration), the remaining parameters K and α are essential not only to fix the final packing fraction, but also to determine the rate at which it is reached. Here we have taken $\alpha = 0.33$ and we have varied K to obtain a set of samples of various volume fractions ranging from $c = 0.4235$ to $c = 0.6430$, a value close to the Bernal's threshold c_b. When extrapolating these volume fractions to $K = 0$, we obtain $c_b \simeq 0.645$ in good agreement with previous estimates. To build the most dense packing with $K = 10^{-4}$, of volume fraction $c = 0.643$, we used a few days of IBM RISC 6000/580 computer time.

As soon as they have been built, all the packings are rescaled to get a particle diameter of 1 and, consequently, in all what follows, the distances will be expressed in diameter units, A detailed analysis of the long range correlations in these packings obtained from a calculation of the pair correlation function $g(r)$, up to distances r of about twenty sphere diameters, have been reported elsewhere [11].

B Determination of the Voronoï cells

We recall that a Voronoï cell is the extension of a Wigner-Seitz cell for a disordered structure: it is defined as the ensemble of points closer to a given sphere center than to any other and it is characteristic of the local environment around this sphere. The positions of the sphere centers being known at a given time, the Voronoï tessellation has been numerically performed by first determining the Delaunay tetrahedral simplicial cells which are, among all the tetrahedra formed with four sphere centers, the ones such that no other sphere center lies inside their circumscribed spheres. After determining the simplicial tetrahedra, all the elements, faces and edges, of the Voronoï cell of a given sphere can be determined knowing that its vertices are the centers of the circumscribed spheres of all the simplicial tetrahedra sharing this sphere center. We have checked that the cell vertices have always three edges in common, as expected in random structures. Therefore for each cell, the total number of edges E and the total number of vertices V are related by $2E = 3V$. Knowing that these two quantities are related to the total number of faces F through the well-known Euler formula, $V - E + F = 2$ [16,17], one has also $F = 2 + V/2 = 2 + E/3$. These relations have been verified by our numerical results. Another formula, $< F > = 12/(6- < e >)$, has been verified. It relates the mean number of faces $< F >$, also called coordination number z, to the mean number of edges per face $< e >$ [17]. Not only V, E, F and their mean over the N atoms have been calculated but also, the volume v of each cell and its total area s, which is the sum of the face areas.

C Results

In Fig. 1a and 1b we report on the distribution functions h_v and h_s of the Voronoï cell volumes and surfaces for some of our sphere packings with various volume fractions. They are defined such that $h_v dv$ is the proportion of cell volumes lying between v and $v + dv$ (and similarly for h_s). Both histograms become more

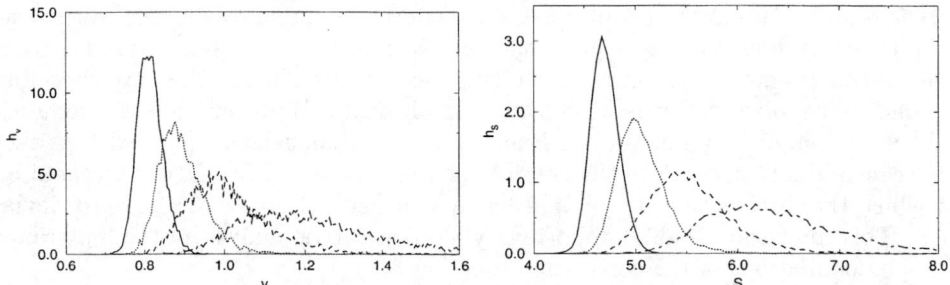

FIGURE 1. Cell volume $h_v(v)$ (a) and cell surface $h_S(S)$ (b) distribution functions for sphere packings. Dot-dashed, dashed, dotted, and solid lines correspond to $c = 0.424, 0.518, 0.585$ and 0.643, respectively.

and more peaked as the volume fraction increases. As a measure of the widths of these histograms we have calculated the standard deviations σ_v and σ_s of v and s which have been reported as a function of c in Fig.2.

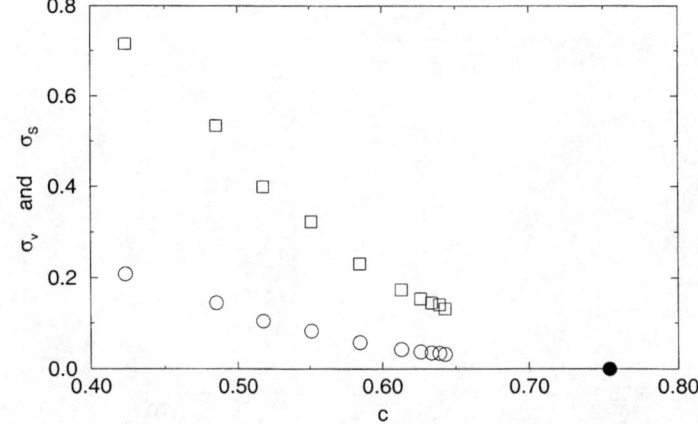

FIGURE 2. Standard deviations σ_v (open circles) and σ_S (open squares) as a function of c for sphere packings. The filled circle indicates their expected extrapolations for $c = c_0$.

Both quantities extrapolate to zero at about the same value of the packing fraction. The black dot indicated on the figure corresponds to $c_0 = 0.754$ a value obtained by dividing the volume of a sphere by the volume of the regular dodecahedron with all faces tangent to it. It is worth noticing that c_0 is larger than the packing fraction $c_m = 0.7405$ of the FCC structure.

The results of Fig.2 can be interpreted as follows: when increasing the concentration, the mean local arrangement of the neighboring sphere centers around a given one evolves progressively towards a regular icosahedron while the corresponding Voronoï cell evolves towards a perfect dodecahedron. However, due to geometrical frustration, one cannot tile the Euclidean three dimensional space with perfect dodecahedra only. Therefore the volume fraction c cannot exceed the threshold c_b, at which the distribution of volumes has a well defined non-zero standard deviation. This conclusion is also supported by a quantitative analysis of the long range damped oscillations of the correlation function $g(r)$ [11].

It is worth noticing that the mean cell volume v_m is trivially related to the packing fraction through the formula (in which the sphere diameter is set to unity):

$$v_m = \frac{L^3}{N} = \frac{\pi}{6c} \qquad (3)$$

As a consequence the maximum of the distribution function h_v is shifted towards low values as c increases. However the evolution of the mean cell surface s_m is non trivial and can give some information on the evolution of the mean cell shape. In Fig.3 we have reported the quantity $S_m/v_m^{\frac{2}{3}}$ as a function of c. It extrapolates

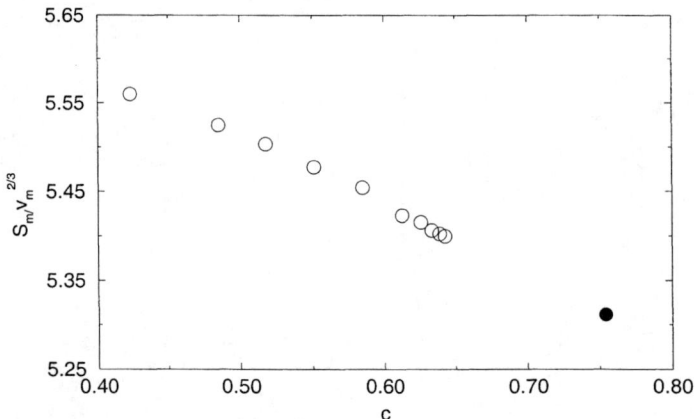

FIGURE 3. Dimensionless ratio $S_m/v_m^{2/3}$ as a function of c for sphere packings. The filled circle indicates the expected extrapolation for $c = c_0$.

nicely at $c = c_0$ to the corresponding quantity calculated for a perfect dodecahedron. This provides another support to the above analysis. In Fig.4 we report the mean coordination number $z = <F>$ as a function of c. Unless one accepts a very slow convergence, there is no clear evidence for an extrapolation to $z = 12$ (dodecahedron) for $c \to c_0$. In fact this is not inconsistent with the above reasoning because extra faces might persist, the areas of which vanish only at $c = c_0$. In

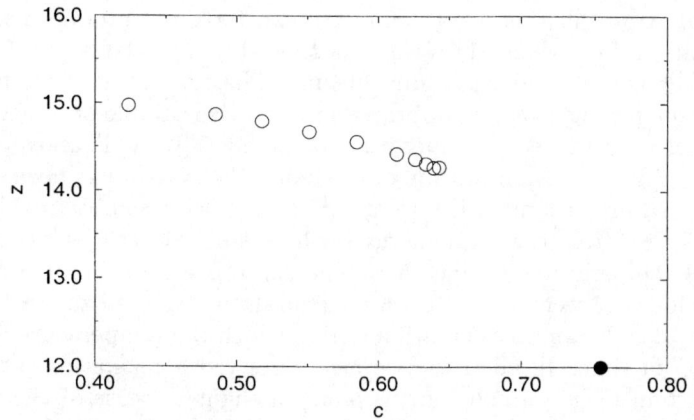

FIGURE 4. Mean coordination number z as a function of c for sphere packings. The filled circles indicates the expected extrapolation for $c = c_0$.

that case z might be discontinuous at $c = c_0$. The same kind of behavior occurs when disturbing any degenerate regular structure, for example when including an infinitesimal temperature in an FCC crystal.

III SOFT SPHERE SYSTEM

A Principles of the calculations

In our molecular dynamics calculations of a soft-sphere system we have used the repulsive potential of Laird and Schober [18] which has been proven to be appropriate to describe the vibration spectrum of glasses [19,20]:

$$U(r) = \epsilon\left(\frac{\sigma}{r}\right)^6 + Ar^4 + B \quad \text{for} \quad r < 3\sigma \tag{4a}$$

$$U(r) = 0 \quad \text{for} \quad r \geq 3\sigma \tag{4b}$$

In this expression σ is a typical length fixing the range of the potential and the constants A and B have been chosen so that both the potential and the force become zero for $r = 3\sigma$. In our simulations we have used $N = 1000$ atoms with a density such that $(N/L^3)\sigma^3 = 1$. This would correspond to a packing fraction of 52.4% for hard spheres of diameter σ. A few simulations have been performed on larger systems with the same density to verify that our results are size independent. In order to give some physical meaning to the simulations and to be able to give the temperatures in Kelvin, we have chosen $\sigma = 3.405 \text{Å}$ and $\epsilon = 0.0103\text{eV}$, which are the Lennard Jones (LJ) values of Argon [21] (of course we do not pretend to simulate real Argon in this way). Similarly we took the atomic mass of Argon ($m = 40$ amu) and consequently we used a time step $\Delta t = 0.004\tau = 10^{-2}$ps where $\tau = (m\sigma^2/\epsilon)^{1/2}$ is the standard LJ unit of time. The glass configurations have been obtained by quenching a well equilibrated initial liquid sample obtained by melting a simple cubic crystal at a temperature of about 50K, well above the melting temperature. After full equilibration of the liquid the system has been cooled down to zero temperature at a quench rate of 10^{12} K/s which was obtained by removing $\Delta E = 8.6 \ 10^{-7}$eV from the total energy of the system at each iteration.

At several temperatures during the quenching process the configurations (positions and velocities) were saved. Each configuration was used to start a constant-energy molecular-dynamics calculation during which the temperature was recorded as a function of time. In all cases, we have observed a relaxation process typical of such a system (Fig.5 in [22]) during which a slight increase of the temperature was observed before a saturation regime was taking place. We found that typically 10000 iterations (i.e. 100 ps) were enough to insure a well defined constant temperature for each sample. After these 10000 relaxation steps we have started to calculate all the physical quantities reported below which have been monitored during 20000 additional steps. It is worth mentioning that during the total run time of 30000 steps (= 300 ps) we have never observed a strong tendency to crystallization. We now know that this comes from the choice of the quench rate which insures a maximum stability of the resulting glass samples [23].

Standard physical quantities such that the two-point correlation function $g(r)$ and its Fourier transform $S(q)$, as well as the diffusion coefficient have been reported elsewhere [24]. They all exhibit the expected general behavior for such amorphous systems [25,26]. The glass transition was determined as usual by following the temperature evolution of the diffusion coefficient (calculated over the 30000 steps mentioned above). These results show a clear change of behavior between a very small constant ($< 5.10^{-3}$Å^2.ps^{-1}) and a linearly temperature dependent diffusion coefficient at a temperature $T_g = 10.5$K (as estimated by extrapolating the high-T linear regime) in perfect agreement with a previous study using the same potential [18].

B Results

In Fig. 5a and 5b, we report on the cell-volume and cell-surface distribution functions (in reduced units $\sigma = 1$) h_v and h_s, respectively, resulting from an average over 20000 timesteps after the relaxation period of 10000 timesteps at each temperature, as explained in section II-B.

With decreasing temperature, the distribution h_v is more and more peaked around the mean cell volume v_m, which is here precisely equal to 1 (i.e. σ^3), due to our particular choice of the density. The same qualitative behavior is found for h_s, except that, now, the mean surface S_m (close to the location of the maximum) is slowly varying with temperature.

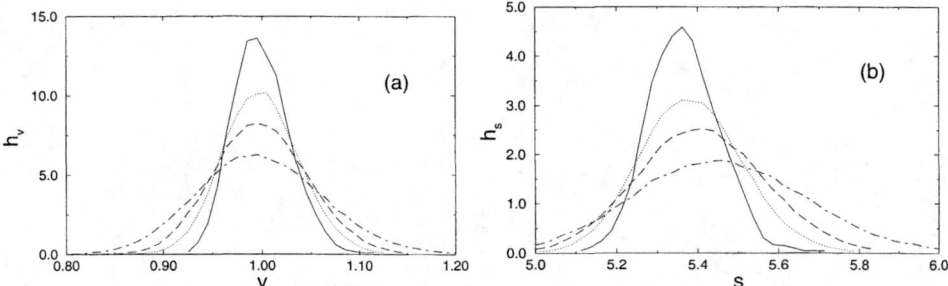

FIGURE 5. Cell volume $h_v(v)$ (a) and cell surface $h_S(S)$ (b) distribution functions for the soft-sphere glass. Dot-dashed, dashed, dotted, and solid lines correspond to $T = 29.7, 14.5, 9.0$ and 0.3 K, respectively.

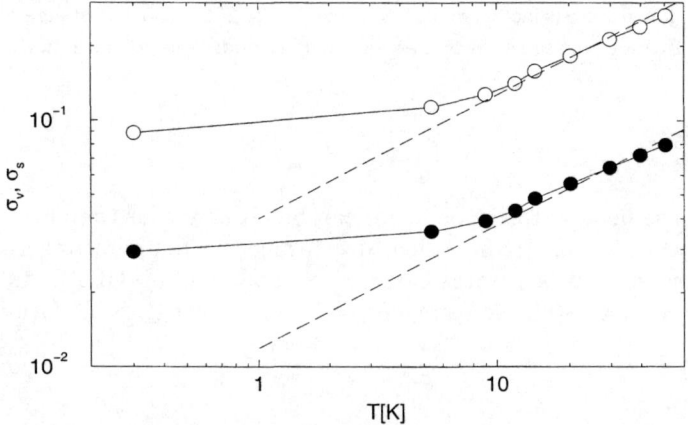

FIGURE 6. Standard deviations σ_v (filled circles) and σ_S (open circles) as a function of T for the soft-sphere glass. The dashed lines indicate $T^{1/2}$ behaviors.

The corresponding standard deviations σ_v and σ_s have been plotted as a function of temperature in the log-log plot of Fig.6. It is worth noticing that both σ_v and σ_s are varying like $\sqrt{T}$ in the liquid phase, as expected for classical thermally activated density fluctuations. But both quantities are much more slowly varying in the glass phase, and, as T vanishes, they tend to non-zero values, characteristic of spatial disorder. Such a low temperature saturation of the density fluctuations which occurs around T_g is the signature of the glass transition [25,26]. The temperature dependence of the mean cell surface s_m is shown in Fig.7. For comparison, the surface s_d of the regular dodecahedron of volume 1 is indicated by the filled circle. The data can be interpreted as if, when decreasing the temperature from the liquid phase, the mean surface would like to reach the one of the dodecahedron, but, since it is impossible to fill the space with dodecahedra only, a glass transition takes place and below T_g the mean surface saturates towards a value higher than s_d as T goes

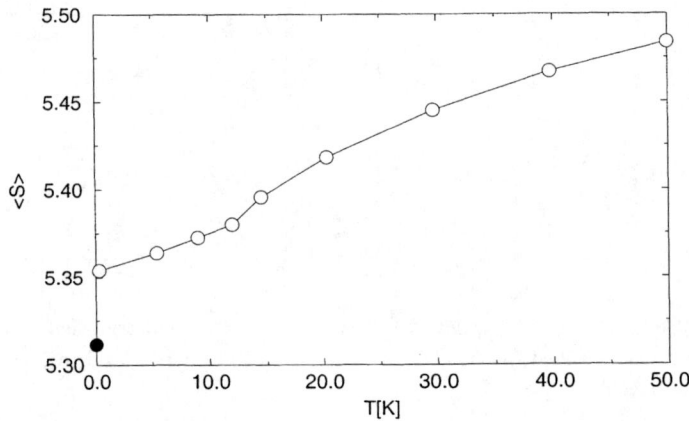

FIGURE 7. Mean cell surface area S_m as a function of T for the soft sphere glass. The filled circle at $T = 0K$ indicates the surface area for a perfect dodecahedron of unit volume.

to zero. The same qualitative behavior can be observed for the mean coordination number $z = <F>$ in Fig.8. When decreasing the temperature from the liquid phase z decreases, but saturates below T_g to a value of about 14. As in Fig.4, one observes however that the convergence to $z = 12$ of the $T > T_g$ data is quite slow.

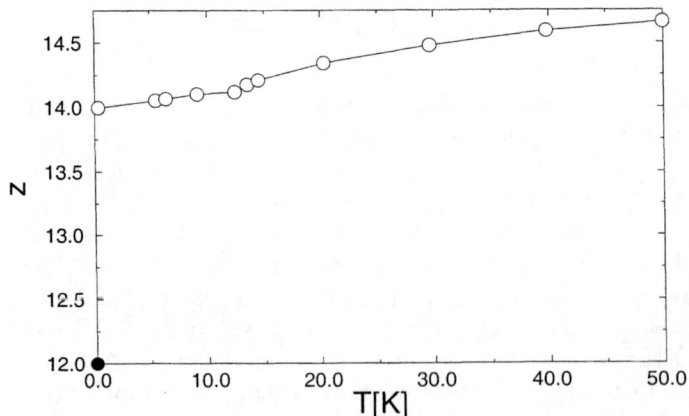

FIGURE 8. Mean coordination number z as a function of T for the soft sphere glass. The filled circle at $T = 0$ indicates $z = 12$.

IV   MODEL SILICA GLASS

A   Principles of the calculations

To perform realistic simulations on silica glasses we have used the so-called BKS potential developed by van Beest et al. [10]. Though designed originally from the crystalline phases of silica, it has been shown that it also describes very well the structural [27], vibrational [28] and transport [29] properties of amorphous silica. We performed molecular dynamics simulations for microcanonical systems containing 216 silicon and 432 oxygen atoms confined in a cubic box of edge length $L = 21.48$ Å, which corresponds to a mass density of $\approx 2.18 \text{g/cm}^3$ very close to the experimental value of 2.2g/cm^3. Periodic boundary conditions were used to limit surface effects. In order to insure energy conservation even at high temperature a timestep of 0.7 fs was necessary. The 4-th order Runge-Kutta algorithm was used to integrate the equations of motion. The glass configurations were obtained by quenching well equilibrated initial liquid samples obtained by melting β-cristobalite crystals at a temperature around 7000 K. After full equilibration of the liquid (≈ 40000 timesteps), the system was cooled to zero temperature at a quench rate of 2.3×10^{14} K/s which was obtained by removing the corresponding amount of energy from the total energy of the system at each iteration. At several temperatures during the quenching process the configurations (positions and velocities) were saved. Each configuration was used to start a constant-energy molecular dynamics calculation during which the temperature was recorded as a function of time. To avoid transient configurations we allowed for each temperature and for each sample, 10000 relaxation steps followed by 50000 supplemental time steps (for a total simulation time of 42 ps) during which all the calculations were done. Apart from the calculation of standard quantities, we included also in our molecular dynamics code a Voronoï tessellation scheme similar to the one that we have developed for monocomponent soft-sphere glasses (see section 2). This scheme has been modified to take into account several types of atoms and thus it permits to follow the local structure around the silicon atoms and the oxygen atoms as a function of temperature during the quenching procedure. Here the Voronoï cell is always defined as being the region of space closer to a given atom center than to any other and no dissymetry between the two components has been introduced. The Voronoï cell characteristics as well as all the other quantities have been averaged over samples obtained from 5 independent starting configurations in order to improve the statistics of the results.

B   Results

As said earlier this potential has already been used in other studies of amorphous silica and since we did not modify the BKS parameters our results concerning the radial pair distribution [27,28] or the diffusion constant [30] are exactly identical to the referenced results and therefore we do not come back to these standard results here. Our aim is to localize the glass transition temperature T_g through the study of the structural characteristics (via the Voronoï tessellation) of our model silica system. In fact T_g can already be evidenced by a slight change of slope in the behavior of the potential energy as a function of temperature, which, in our system, occurs between 3000 and 4000 K. Due to the fast cooling rate the value of T_g is much higher than the experimental value (1446 K [31]) but it is coherent with the value of $\approx$ 3500 K obtained from the fit of T_g versus the quenching rate proposed by Vollmayr et al. [27]. Here we observe a non negligible increase of the potential energy in the glass phase contrarily to the results we have obtained in a recent investigation of the "inherent structures" [32] of amorphous silica [33].

Once we know approximately the value of T_g (to our purpose this level of accuracy is sufficient) we can tackle the study of the geometrical characteristics of the Voronoï cells in order to follow the local structure as a function of temperature. The first characteristic is the variation of the volume of the Voronoï cells. This variation is represented in Fig.9 for the silicon (a) and the oxygen (b) atoms together with the corresponding standard deviations Fig.10.

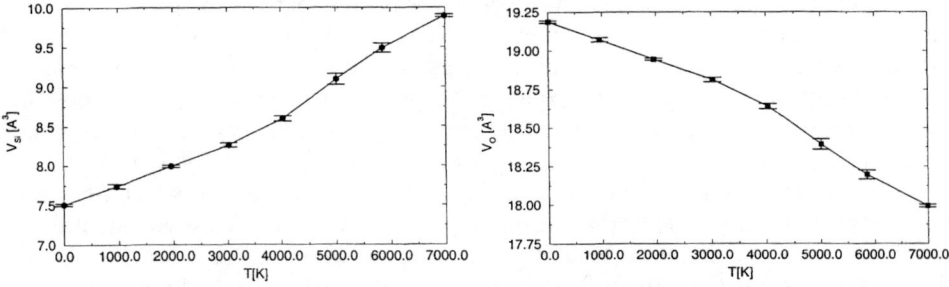

FIGURE 9. Variation of the cell volume versus temperature: (a) silicon; (b) oxygen.

With decreasing temperature the volume of the silicon cell decreases while the volume of the oxygen cell increases (these opposite variations are a consequence of our constant-volume calculations). Again a change of behavior is visible and corresponds to a slowing down of the evolution below the glass transition temperature. An even more striking behavior is observed in Fig.10 where the standard deviation σ_V is plotted as a function of temperature. This is a quantity of physical interest since it measures the local density fluctuations around the particles. For both types of atoms, σ_V decreases with decreasing temperature as if it would like to tend to zero at $T = 0K$ but then below 4000 K this trend is stopped and finally σ_V saturates around non-zero values characteristic of spatial disorder. This is a direct observation of the low-temperature saturation of the density fluctuations which is a signature of the glass transition. To investigate further the structural evolution during the quench, we looked at the angle distributions. First we studied the tetrahedral O-Si-O angle which should be ideally equal to 109°.47 in a perfect tetrahedron. As can be seen in Fig.11a, this angle varies between 110°.5 at low

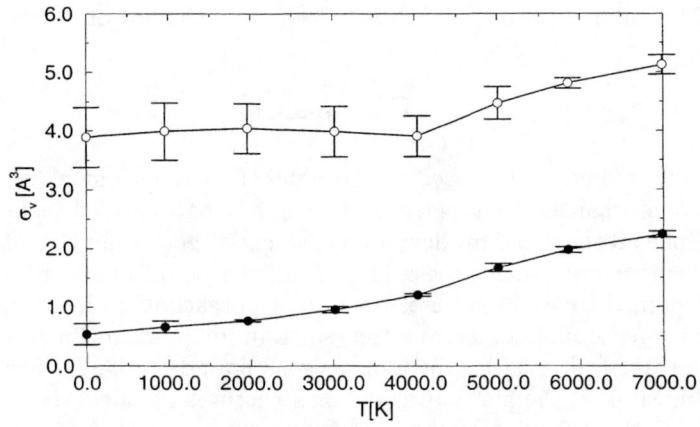

FIGURE 10. Standard deviation σ_V versus temperature: •: silicon; ○: oxygen.

temperature and 117° at 7000 K, with a slight change of behavior around T_g.

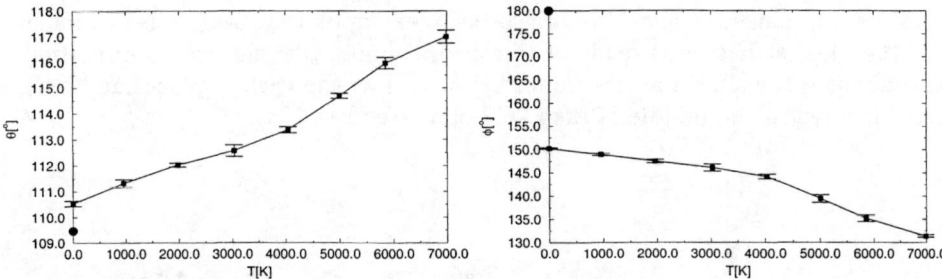

FIGURE 11. Variation as a function of temperature of: (a) θ, the O-Si-O angle, (b) ϕ, the Si-O-Si angle.

This shows firstly that with increasing temperature the SiO_4 tetrahedra survive even in the liquid phase but become more and more distorted and secondly that the glass transition does not strongly affect the local environment around the silicon atoms. On the contrary the glass transition is more clearly visible in the angle Si-O-Si, which measures the relative position and orientation of two neighboring SiO_4 tetrahedra, as can be seen in Fig.11b. With decreasing temperature within the liquid phase, the Si-O-Si angle increases and seems to converge towards 180° (a least square quadratic fit of the four points in the liquid phase gives 175°), but again below 4000 K this decrease slows down and finally the angle converges towards a value close to 150°, a value coherent with previous simulations [27,28], but slightly higher than the value 144° found in X-ray diffraction experiments [34]. Since this angle measures the relative orientation between two neighboring tetrahedra, this decrease corresponds to a decrease of the effective volume of the oxygen atoms with increasing temperature and since we work at constant volume it implies an expansion of the silicon volume, which is indeed the behavior observed in Fig.9.

C Discussion

All these results can be discussed in the light of the conclusions drawn for the analogous geometrical analysis performed in a monoatomic soft-sphere glass in section III. Since all the standard deviations presented here seem to go to zero with decreasing temperature, when extrapolated from the liquid phase, it is reasonable to assume that in the case of silica also a $T = 0$ unreachable ideal local structure exists. It is also reasonable to assume that such an ideal arrangement corresponds to a perfect tetrahedral order for the four oxygens bounded to a given silicon atom, as it is for almost all of the known crystalline structures of silica. This assumption is supported by the behavior of the coordination number of silica (not reported here), which extrapolates to 4 when T is lowered from the liquid phase, and is not incompatible with our results for the variation of the angle O-Si-O with temperature (see Fig.11a). Even if this angle does not show a major change of behavior at T_g, an extrapolated value of 109.°47 at $T = 0$ is not inconsistent with the data above T_g. Moreover, since the angle Si-O-Si seems to extrapolate to 180°, one can imagine that the ideal structure is made of tetrahedral units, like the sp_3 coordination of carbon where the silicon atoms would be located at the carbon places and oxygen atoms located in the middle of the C-C bonds (see Fig.12a).

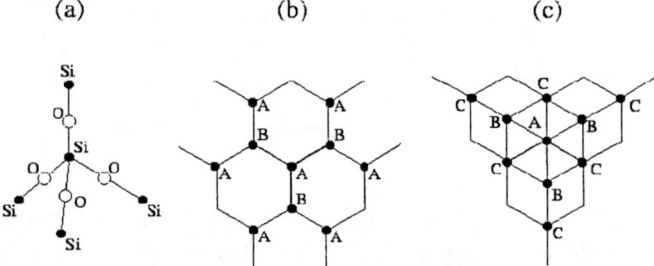

FIGURE 12. The ideal tetrahedral unit around a silicon atom (a) is shown together with the trydimite (b) and β-cristobalite (c) structures viewed from the top. In (b) and (c) two successive layers connected to a single top unit are represented (the positions of the oxygen atoms have been omitted).

Indeed the tendency to build such a local structure should result from the form of the potential. In particular the tetrahedral arrangement of the oxygens around a silicon atom results from a combination of the Si-O attraction and the repulsion between the oxygens. The tendency to align the Si-O-Si bridges between neighboring tetrahedra is more subtle however, since the long range nature of the ionic part of the potential certainly plays a role.

It is interesting to notice that, among all the known crystalline structures of silica, two particular structures (at least) fully satisfy these criteria, namely the β-cristobalite and the tridymite structures. In these two structures the above defined tetrahedral units are stacked with sequences ABCABC...(Fig.12b) and ABABAB...(Fig.12c), respectively, like in FCC and HCP structures. In fact these structures can be simply built from FCC and HCP structures, by adding to the original structure another one shifted by a fourth of the diagonal of the cubic cell (in the FCC case) and by 3/8 of the c-axis of the hexagonal cell (in the HCP case). They correspond respectively to the diamond and wurtzite structures of carbon. When these two structures are considered with the same density, they have exactly the same Si-O distance d_{SiO}, and therefore, in the two cases the Voronoï cells for the silicon atoms are regular tetrahedra with the same volume $V_{Si} = \sqrt{3}d_{SiO}^3$. The volume of the oxygen cells, V_O, is also the same and therefore the two structures are characterized by the same Si/O volume ratio $R = V_{Si}/(2V_O) = 9/55 = 0.164$, independent on the density. In Fig.13 we have plotted $R = V_{Si}/(2V_O)$ as a function of T, as calculated from our simulations, and reported the value $R = 0.164$ at $T = 0$ (open circle).

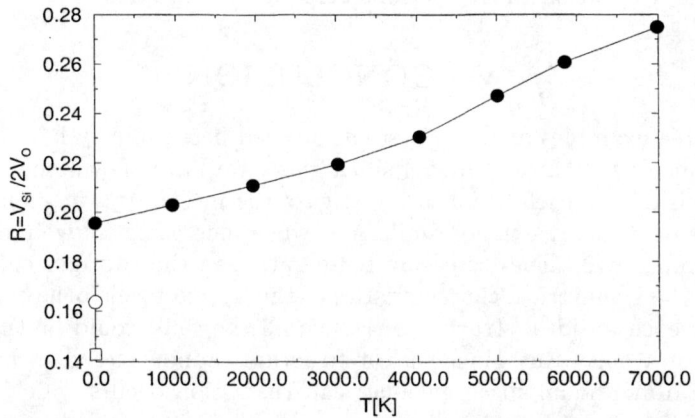

FIGURE 13. Variation of $R = V_{Si}/2V_O$ as a function of temperature. o: value of R obtained in tridymite or β-cristobalite; □: value of R obtained in the ideal structure built on S_3.

When decreasing the temperature from the liquid phase R decreases as if it would like to reach a value quite close to (or even lower than) 0.164. Therefore, one can interpret the glass transition as a result of local frustration effects. Indeed the system tries to establish a local order similar to the one of tridymite or cristobalite, but these two structures are different, and therefore, due to spatial incompatibilities, no long range order can be generated.

But let us pursue this analysis a little bit further, like it has been done for soft sphere glasses or hard sphere packings in section III. Similarly to these model systems, the frustration can be resolved by considering a curved space with positive curvature, namely the sphere S_3 [7]. Consider a regular tetrahedron with a silicon atom in the middle and oxygen atoms at the centers of the faces. This tetrahedral unit contains one silica molecule. One can exactly tile an S_3 space with 600 tetrahedra like this, and the resulting SiO_2 structure (which contains 600 silicons and 1200 oxygens) satisfies all the local requirements defined above. The 120 vertices of the unit tetrahedra are located on the so-called {3,3,5} polytope [7]. A good approximation of the ratio R for this structure can be obtained by considering the tetrahedron unit in the regular three dimensional space: the silicon Voronoï cell, limited by the bisector planes of the Si-O bonds, is a tetrahedron of volume 1/8 of the volume of the unit. Therefore the remaining volume for the two oxygen atoms is 7/8 and consequently $R = 1/7 = 0.143$. This value is represented by the open square in Fig.13. It appears that this value corresponds to a better extrapolation of the four points above 4000K than the one obtained from tridymite or cristobalite (the second order fit of the four liquid points leads to a value of 0.136). Therefore one is tempted to conclude as in the soft sphere case: when lowering the temperature from the liquid phase the system evolves as if it would try to build locally such an ideal structure, but since this structure cannot be realized in the regular three dimensional space, the systems gets frozen in a glassy state below T_g. Obviously, the above interpretation should be considered as a suggestion.

V CONCLUSION

In the three examples of glassy systems studied here, namely hard sphere packings, soft sphere system and model silica glass, we have shown that the Voronoï tessellation is a very useful tool to investigate the mechanism of geometrical frustration. We have found striking similarities when increasing the volume fraction in sphere packings and when decreasing temperature in the two molecular dynamics systems, al the geometrical characteristics of the Voronoï cells behave as if the system would reach an ideal structure in which all the cells would be the same for a given type of atoms. But, since such a structure cannot extend to infinity in the usual three dimensional space, geometrical frustration occurs which explains the existence of the Bernal's threshold in the hard sphere case and the glass transition in the molecular dynamics systems. In these systems our results show that the Voronoï cell characteristics not only give useful informations on the local structure but also can be used to determine the glass transition unambiguously. This means that even if nothing *dramatic* happens at the glass transition concerning the local structure, *something* happens, which is basically a dynamic freezing of the natural evolution of the structure towards an unreachable ideal structure. These conclusions are consistent with the ideas developed by D. Kivelson in these proceedings.

Part of the numerical calculations were done on the IBM/SP2 computer at CNUSC (Centre National Universitaire Sud de Calcul), Montpellier, France.

REFERENCES

1. Bernal J., *Proc. Roy. Soc.* **A 280**, 299 (1964).
2. Finney J. L., *Proc. Roy. Soc.* **A 319**, 479 (1970).
3. Bennett G. H., *J. Applied Phys.* **43**, 2727 (1972).
4. Adams D. J. and Matheson A. J., *J. Chem. Phys.* **56** 1989 (1972).
5. Cargill G. S., *J. Applied Phys.* **41**, 12 (1972).
6. Bernal J. D. and Mason J., *Nature* **188**, 910 (1990).
7. Sadoc J.-F. and Mosseri R., *Frustration Géométrique"*, Eyrolles ed. and Alea Collection, Saclay, France, 1997.
8. Sadoc J.-F., Dixmier J. and Guinier A., *J. Non Cryst. Solids* **12**, 48 (1973).
9. Kivelson D., Kivelson S.A., Zhao X., Nussimov Z. and Tarjus G., *Physica* **A219**, 27 (1995).
10. van Beest B.W.H., Kramer G.J. and van Santen R.A., *Phys. Rev. Lett.* **64**, 1955 (1990).
11. Jullien R., Jund P., Caprion D.and Quitmann D., *Phys. Rev. E* **54**, 6035 (1996).
12. Jund P., Caprion D. and Jullien R., *Europhys. Lett.* **37**, 547 (1997).
13. Jullien R., Jund P., Caprion D. and Sadoc J.-F., in *Foams and Emulsions*, ed. by J. F. Sadoc and N. Rivier, NATO ASI series E, Vol. 354 (Kluwer Academic, 1999), p. 571.
14. Jund P. and Jullien R, *Phil. Mag. A* **79**, 223 (1999).
15. Jodrey W. S. and Tory E. M., *Phys. Rev. A* **32**, 2347 (1985).
16. Sadoc J.-F. and Mosseri R., in *Aperiodidicity and order 3, Extended icosahedral structures*, (Academic Press, New York, 1989) p. 163.
17. Rivier N., in *Disorder and granular media*, ed. by D. Bideau and A. Hansen, (Elsevier Science Publishers, North Holland, 1993), p. 55.
18. Laird B. and Schober H., *Phys. Rev. Lett.* **66**, 636 (1991).
19. Schober H. and Oligschleger C., *Phys. Rev. B* **53**, 1 (1996).
20. Caprion D., Jund P. and Jullien R., *Phys. Rev. Lett.* **77**, 675 (1996).
21. Gazzillo D. and Della Valle R., *J. Chem. Phys.* **99**, 6915 (1993).
22. Steinhardt P. J., Nelson D. R. and Ronchetti M., *Phys. Rev. B* **28**, 784 (1983).
23. Jund P., Caprion D. and Jullien R., *Phys. Rev. Lett.* **79**, 91 (1997).
24. Jund P., Caprion D. and Jullien R., *Mol. Sim.* **20**, 3 (1997).
25. Gaskell P.H. in *Materials science and technology*, ed. by R. W. Cahn, P. Haasen and E. J. Kramer, **9**, 175 (1991).
26. Cusak N.E., *The Physics of structurally disordered matter: an introduction* (Adam Hilger, Bristol and Philadelphia, 1987)
27. Vollmayr K., Kob W. and Binder K., *Phys. Rev. B* **54**, 15808 (1996).
28. Taraskin S.N. and Elliott S.R., *Europhys. Lett.* **39**, 37 (1997).
29. Jund P. and Jullien R., *Phys. Rev. B* **59**, 13707 (1999).
30. Guillot B. and Guissani Y., *Phys. Rev. Lett.* **78**, 2401 (1997).
31. Angell C.A., *J. Chem. Phys. Solids* **49**, 863 (1988).
32. Stillinger F. H. and Weber T. A., *Phys. Rev. A* **25**, 978 (1982).
33. Jund P. and Jullien R., *Phys. Rev. Lett.* in press.
34. Neuefeind J. and Liss K.D., *Ber. Buns. Ges. Phys. Chem.* **100**, 1341 (1996).

IV NUMERICAL SIMULATIONS

Multimillion Atom Molecular Dynamics Simulations of Glasses and Ceramic Materials

Priya Vashishta, Rajiv K. Kalia and Aiichiro Nakano

Concurrent Computing Laboratory for Materials Simulations
Department of Physics & Astronomy and Department of Computer Science
Louisiana State University, Baton Rouge, Louisiana 70803-4001, U.S.A.

Abstract. Molecular dynamics simulations are a powerful tool for studying physical and chemical phenomena in materials. In these lectures we shall review the molecular dynamics method and its implementation on parallel computer architectures. Using the molecular dynamics method we will study a number of materials in different ranges of density, temperature, and uniaxial strain. These include structural correlations in silica glass under pressure, crack propagation in silicon nitride films, sintering of silicon nitride nanoclusters, consolidation of nanophase materials, and dynamic fracture. Multimillion atom simulations of oxidation of aluminum nanoclusters and nanoindentation in silicon nitride will also be discussed.

INTRODUCTION

Recent advances in computing technology - hardware, especially parallel computer architectures, software and development of robust $O(N)$ algorithms - have revolutionized the field of computer simulation. Some of the reasons for doing computer simulation studies are:

- Computer simulations bridge the gap between theory and experiment.
- Nonlinearities, lack of symmetry, and a large number of degrees of freedom can be handled much more easily by computer experiments than by analytical methods.
- The hypothetical universe available to computer experimentation is not limited to the processes occurring in nature.

Some scientific disciplines where computer simulations have been used widely are

- Astrophysics

- Biology
- Chemistry
- Condensed matter physics - especially disordered materials which lack symmetry
- Lattice gauge theory
- Statistical physics [Classical and quantum liquids, plasmas, superionic conductors, glasses, high temperature ceramics, *etc.*].

In these lectures we will deal with the simulation of classical systems, *i.e.*, systems in which the dynamics of particles is governed by classical mechanics. For simulations of time-independent and time-dependent quantum systems we suggest the reader to look into other review articles and research papers.

In many materials, such as semiconductor binary oxides (SiO_2, GeO_2), chalcogenides ($GeSe_2$, $SiSe_2$), III-V semiconductors (GaAs, InI, InSb), fast-ion conductors (AgI, CuBr, Ag_2S, Ag_2Se), and high temperature ceramic materials (Si_3N_4, SiC, Al_2O_3, AlN, GaN) one of the dominant interaction is the Coulomb potential arising as a result of charge-transfer effects. As the Coulomb potential is a long range interaction, each particle interacts with all the other particles in the system. Therefore, the computing time involved in the evaluation of interaction increases as N^2, where N is the number of particles in the system. This makes the large-scale simulation of Coulombic systems difficult.

Special techniques have been developed to reduce the computational complexity from $O(N^2)$ (1-4). Greengard and Rokhlin have proposed the Fast Multipole Method (FMM) (5). The far-field contribution to the Coulomb interaction is calculated with the multipole expansion for the Coulomb potential. Computational complexity of the FMM is $O(N)$.

The goals of this article are: (i) to discuss the molecular dynamics (MD) method and to design and implement a parallel MD algorithm including calculation of long range Coulomb interaction for large scale systems; (ii) to investigate the structural and dynamical correlations in a variety of systems under extreme conditions of density, temperature, and external strain using MD simulations.

MOLECULAR DYNAMICS METHOD

In order to set up a computer simulation, we first need to formulate a mathematical model to describe the physical phenomenon of interest. The mathematical model underlying a molecular-dynamics (MD) simulation is the Newton's equations of motion in classical mechanics, which state that the acceleration of an atom is proportional to the force exerted on the atom by the other atoms (6). In MD simulations, a physical system consisting of N atoms is represented by a set of coordinates, $\{\vec{r}_k = (x_k, y_k, z_k) \mid k = 1, ..., N\}$, and we trace the atomic trajectories, $\vec{r}_k(t)$, by integrating the Newton's equations numerically with respect to time, t.

The above mathematical model is formulated as differential equations where time has continuous values. To perform computer simulations, however, this continuous law must be cast into a discrete algebraic form which is amenable to numerical solution on a digital computer. The physical system is sampled at times $t_n = n \, \Delta t$ ($n = 1, 2, 3, ...$). For example, the most common discretized form is

$$m_k \frac{\vec{r}_k(t_{n+1}) - 2\vec{r}_k(t_n) + \vec{r}_k(t_{n-1})}{(\Delta t)^2} = \vec{F}_k(\vec{r}^N(t_n)),$$

where $\vec{F}_k = -\partial V(\vec{r}^N)/\partial \vec{r}_k$ is the force acting on the k-th atom, $V(\vec{r}^N)$ is the interatomic potential energy, and $\vec{r}^N = (\vec{r}_1, \vec{r}_2, ..., \vec{r}_N)$ is a $3N$-dimensional vector representing the positions of all the atoms (6).

In order to write a simulation program, solutions to the discretized algebraic model must be translated to a sequence of computer instructions based on some numerical algorithms. For example, the commonly used velocity-Verlet algorithm translates the above equation into iterated operations of a time-stepping procedure (6):

Repeat

1. Compute the forces, $\vec{F}_k(t_n)$, as a function of $\vec{r}^N(t_n)$ for $k = 1, ..., N$

2. Update the velocities, $\vec{v}_k \leftarrow \vec{v}_k(t_n) + \vec{F}_k(t_n)\Delta t/2m_k$, of the atoms ($m_k$ is the mass of the k-th atom)

3. Obtain the new atomic positions, $\vec{r}_k(t_{n+1}) \leftarrow \vec{r}_k(t_n) + \vec{v}_k \Delta t$

4. Compute the new forces, $\vec{F}_k(t_{n+1})$, as a function of $\vec{r}^N(t_{n+1})$

5. Obtain the new velocities, $\vec{v}_k(t_{n+1}) \leftarrow \vec{v}_k + \vec{F}_k(t_{n+1})\Delta t/2m_k$

6. Update the time, $n \leftarrow n + 1$

until n reaches the maximum number of time steps, Max_steps

The choice of numerical algorithms is crucial for efficient simulation. For example, the velocity Verlet algorithm is time reversible, i.e., a simulation can be played back to recover the starting state. In addition, the solution satisfies a certain symmetric property called symplecticness which is related to the conservation of phase-space volume along the trajectory (7). These properties are essential for the long-time stability of a simulation.

Accurate atomic force laws are essential for realistic materials simulations of sintering, consolidation, dynamic fracture, nanoindentation, and oxidation on metallic nanoclusters. Mathematically, a force law is encoded in the interatomic potential energy, $V(\vec{r}^N)$. In the past years, we have developed reliable interatomic potentials for a number of materials, including ceramics such as silica (SiO_2) (8), silicon nitride (Si_3N_4) (9), and silicon carbide (SiC) (10), and semiconductors such as gallium arsenide (GaAs) and indium arsenide (InAs) (11). Interatomic potential energy for these materials consists of two- and three-body terms. The two-body potential energy is a sum over contributions from $N(N + 1)/2$ atomic pairs, (i, j). The contribution from each pair depends only on their relative distance, $|\vec{r}_{ij}|$. Physically, the two-body terms are steric repulsion and electrostatic interaction due to charge transfer between atoms, and charge-dipole and dipole-dipole interactions that take into account the large electronic polarizability of negative ions. The three-body potential energy consists of contributions from atomic triples (i, j, k), and takes into account covalent effects through bending and stretching of atomic bonds, $\vec{r}_{ij}$ and $\vec{r}_{ik}$.

Multiresolution algorithms

Efficient algorithms are key to extending the scope of simulations to larger spatial and temporal scales that are otherwise impossible to be simulated. These algorithms often utilize multiresolutions in both space and time.

The most computationally intensive problem in an MD simulation is the computation of the electrostatic energy for N charged atoms. Direct evaluation of all the atomic-pair contributions requires $O(N^2)$ operations. In 1987, Greengard and Rokhlin discovered an $O(N)$ algorithm called the Fast Multipole Method (FMM) (5). The FMM groups distant atoms together and treats them collectively. Hierarchical grouping is facilitated by recursively dividing the physical system into smaller cells, generating a tree structure (see Fig. 1). The root of the tree is at level 0, and it corresponds to the entire simulation box. A parent cell at level l is decomposed into $2 \times 2 \times 2$ children cells of equal volume at level $l + 1$. The FMM uses the truncated multipole expansion and the local Taylor expansion of the electrostatic potential field. By computing both expansions recursively for the hierarchy of cells, the electrostatic energy is computed with $O(N)$ operations.

The discrete time step, Δt, in MD simulations must be chosen sufficiently small such that the fastest characteristic oscillations of the simulated system are accurately represented. However, many important physical processes are slow and are characterized by time scales that are many orders-of-magnitude larger than Δt. Molecular-dynamics simulations of such "stiff" systems require many iteration steps, and this severely restricts the applicability of the simulation. We have used an approach called the multiple time-scale (MTS) method (12) which uses different Δt for different force components to reduce the number of force evaluations. To further speed up simulations, we have also used a hierarchy of dynamics including rigid-body motion of atomic clusters (13).

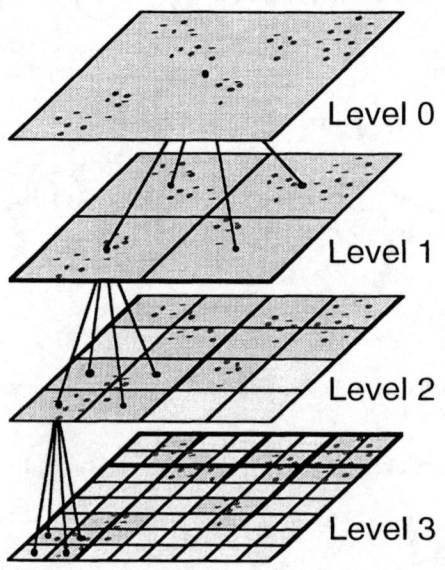

FIGURE 1. Schematic of the far-field computation in a two dimensional system in the fast multipole method. The multipoles of a parent cell at level l is obtained by shifting the multipoles of its 4 children cells at level $l + 1$ and summing them.

Our multiresolution molecular-dynamics (MRMD) algorithm (14) combining the FMM and MTS has been implemented on a number of parallel computers using a spatial decomposition. The MRMD program is highly scalable: For a 100-million-atom SiO_2 system, one MD step takes only 24 seconds on 256 Cray T3E nodes. The parallel efficiency of this system, defined as a speedup divided by the number of processors, is 0.94.

INTERATOMIC POTENTIALS

Our MD simulations are based on an effective interatomic potential which includes both 2-body and 3-body interactions. The 2-body potential function consists of steric repulsion due to atomic sizes, screened Coulomb interaction caused by the charge transfer effect, and charge-dipole interaction resulting from the large electronic polarizability of O^{2-}, N^{3-}, and Se^{2-} (15,16). The 3-body covalent contribution takes into account the bond bending and bond stretching effects, shown schematically in Fig. 2.

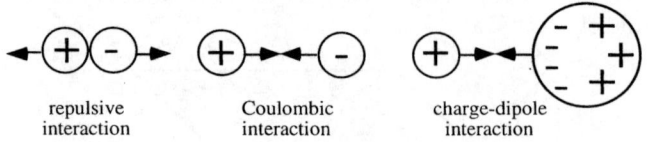

(a) two-body interaction

(b) three-body interaction

FIGURE 2. Schematic diagram of interatomic interactions.

For a system with N particles, the 2-body interaction takes the form of

$$V_2 = \sum_{i<j}^{N} V_{ij}^{(2)} \tag{1}$$

where

$$V_{ij}^{(2)} = \frac{H_{ij}}{r_{ij}^{\eta_{ij}}} + \frac{Z_i Z_j}{r_{ij}} e^{-r_{ij}/r_{1s}} - \frac{P_{ij}}{r_{ij}^4} e^{-r_{ij}/r_{4s}}, \tag{2}$$

$$H_{ij} = A_{ij}(\sigma_i + \sigma_j)^{\eta_{ij}}, \tag{3}$$

$$P_{ij} = (\alpha_i Z_j^2 + \alpha_j Z_i^2)/2, \qquad i, j = 1, 2, ..., N. \tag{4}$$

This 2-body term is a short-range pair-additive potential with a cutoff radius r_c. The 3-body interaction takes the form of,

$$V_3 = \sum_{i<j<k}^{N} V_{jik}^{(3)}, \tag{5}$$

where

$$V_{jik}^{(3)} = B_{jik} f(r_{ij}, r_{ik}) (\cos \theta_{jik} - \cos \bar{\theta}_{jik})^2, \tag{6}$$

and

$$f(r_{ij}, r_{ik}) = \exp\left(\frac{l}{r_{ij} - r_{c3}} + \frac{l}{r_{ik} - r_{c3}}\right), \qquad r_{ij}, r_{ik} < r_{c3}. \quad (7)$$

$$\cos\theta_{jik} = \frac{\vec{r}_{ik} \cdot \vec{r}_{ij}}{r_{ik} r_{ij}}. \qquad r_{ij}, r_{ik} < r_{c3}. \quad (8)$$

This potential has been used in the molecular-dynamics simulations to study structural correlations and dynamical properties of many covalent materials under various conditions of densities and temperatures. The MD simulation results for structural correlations, and static structure factors for SiO_2 (8,17-19), $GeSe_2$ (20), Si_3N_4 (9), and $SiSe_2$ (21,22) are in good agreements with corresponding experimental results.

Interatomic Potential for SiO_2 Glass

The parameters defined in the potential, Eqns. (1)-(8), such as ionic radii σ_i, repulsive exponents η_{ij}, repulsive strength A_{ij}, effective charges Z_i, polarizabilities α_i, strength of 3-body interaction B_{jik}, average bond angles $\bar{\theta}_{jik}$, and the cut-off distances for 2-body and 3-body interactions are tabulated in Table 1.

TABLE 1 Parameters in the interaction potential for SiO_2 glass.

$A_{ij}(erg)$	$r_{s1}(Å)$	$r_{s4}(Å)$	$r_c(Å)$	$l(Å)$	$r_{c3}(Å)$
1.242×10^{-12}	4.43	2.5	5.5	1.0	2.6

		$\sigma_i(Å)$	$Z_i(e)$	$\alpha_i(Å^3)$
	Si	0.47	1.20	0.00
	O	1.20	-0.60	2.40

	η_{ij}		$B_{jik}(erg)$	$\bar{\theta}_{jik}$
Si-Si	11	Si-O-Si	3.2×10^{-11}	141.00
Si-O	9	O-Si-O	0.8×10^{-11}	109.47
O-O	7			

Interatomic Potential for Si_3N_4

The parameters defined in the potential, Eqns. (1)-(8), are tabulated in Table 2.

TABLE 2. Parameters in the interaction potential for Si_3N_4.

$A_{ij}(erg)$	$r_{s1}(Å)$	$r_{s4}(Å)$	$r_c(Å)$	$l(Å)$	$r_{c3}(Å)$
2.00×10^{-12}	2.5	2.5	5.5	1.0	2.6

	$\sigma_i(Å)$	$Z_i(e)$	$\alpha_i(Å^3)$
Si	0.47	1.472	0.00
N	1.30	-1.104	3.00

	η_{ij}		$B_{jik}(erg)$	$\bar{\theta}_{jik}$
Si-Si	11	Si-N-Si	2.0×10^{-11}	120.00
Si-N	9	N-Si-N	1.0×10^{-11}	109.47
N-N	7			

RESULTS OF MOLECULAR DYNAMICS SIMULATIONS

In this section we will discuss the results of molecular dynamics simulations on a number of systems to demonstrate its applicability on a variety of materials and phenomena. Specifically we will discuss the following seven topics.

- Structural correlations in silica glass;
- Universal morphology of fracture surfaces;
- Branching instabilities of crack front;
- Fracture toughness of gallium arsenide;
- Nanophase ceramics with high fracture toughness;
- Interfacial fracture, dislocation formation and its motion;

- Nanoindentation of silicon nitride;
- Environmental effects on fracture.

Structural Correlations in Silica Glass

Structure of binary chalcogenide glasses (e.g. SiO_2, $GeSe_2$, $SiSe_2$) has been widely studied by neutron and X-ray scattering measurements (23). These measurements reveal many similarities in the static structure factor, $S(q)$, when expressed in terms of the dimensionless quantity $Q_0 = qr_0$, where r_0 is the bond length. The most celebrated structural feature in these glasses is the first sharp diffraction peak (FSDP) observed between 1.0 and 1.5 Å^{-1}, which is believed to be a signature of intermediate range correlations. Although the chemical constituents and topologies of these glasses are widely different, the dimensionless quantity Q_0 for the FSDP is nearly the same. So the key issues are: (i) What is the nature of structural correlations that give rise to the FSDP? (ii) How are these correlations related to the connectivity of glasses when described in terms of n-fold rings.

Structural Correlations and First Sharp Diffraction Peak in SiO$_2$ Glass

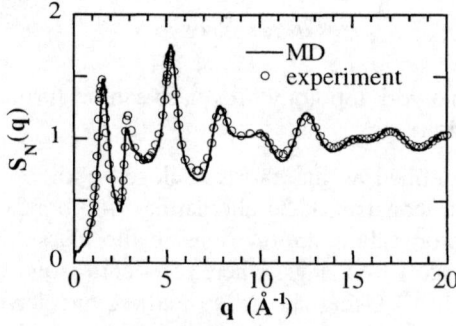

FIGURE 3. Neutron-scattering static structure factor, $S_N(q)$, of SiO_2 Glass. Solid curve, the MD result for a 41,472 particle system at 300 K; open circles, neutron diffraction experiment at 10 K [P. A. V. Johnson, A. C. Wright, and R. N. Sinclair, J. Non-Cryst. Solids 58 (1983) 109].

The structure factor from MD simulations at the normal density (2.2 g/cm^3) is in excellent agreement with neutron diffraction experiments, including the height of the FSDP (Fig. 3). The dominant contribution to the FSDP at $q \sim 1.5$ Å^{-1} is related in r-space to density-density correlations in the region of 4-8 Å (20, 24). The FSDP provides information about the intermediate range order in glasses, although the nature and how the structural entities give rise to the FSDP is still controversial. The FWHM is small and remains nearly constant with the density. From FWHM we can infer the correlation length of the intermediate range order, $\sim 2\pi/(FWHM/2)$, which is in the range 15-25 Å (20,24). The agreement between the MD and experimental results can be quantified in terms of the R_c factor introduced by Wright (25). The R_c factor for the comparison of Fig. 4 between $1 < r < 10$ Å is 4.4 %.

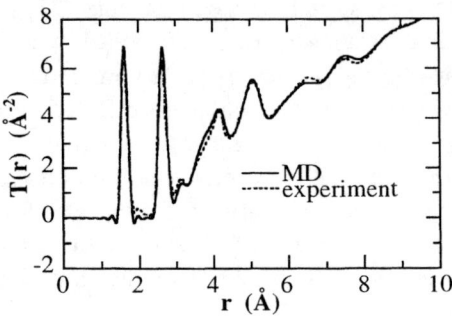

FIGURE 4. Neutron-scattering pair correlation function, $T(r)$, of SiO_2 Glass. Solid curve, the MD result for a 41,472 particle system at 300 K; dotted curve, neutron diffraction experiment at 298 K (26).

Distribution of n-Fold Rings and Height of the FSDP in SiO_2 Glass Under Pressure

The structure and network topology of a glass can be further characterized through the distribution of *n*-fold rings.

An *n*-fold ring is defined as the shortest closed path of alternating Si-O bonds. Therefore, an *n*-fold ring consists of $2n$ alternating Si-O bonds. For a system with N 4-fold coordinated Si atoms there are $6N$ rings in the glass. For example, α- and β-cristobalite have only six 6-fold rings, whereas α- and β-quartz have four 6-fold and two 8-fold rings (27,28). Using the MD method, we have studied the effect of pressure on the distribution of rings and their relationship to intermediate range correlations manifested as the FSDP for SiO_2 glass. A systematic analysis of the modifications observed in the FSDP for densities ranging from 2.0 to 3.2 g/cm^3 and temperatures from 0 to 1500 K has been carried out (28).

We have calculated the distribution of rings from $n = 2$ to 12. The distribution of distances and angles for each *n*-fold ring are also calculated (28). We have explored the relationship between the height of the FSDP and the number of 6-fold rings per Si atom in SiO_2 glass. In Figs. 5 (a) and (b) we plot the density dependence of the number of 6-fold rings and the height of the FSDP. The ratio of the height of the

FSDP and the number of 6-fold rings is shown in Fig. 5 (c). It is quite remarkable that even though the number of six-fold rings and the height of the FSDP vary substantially with density, their ratio is almost constant for the entire range of density, from 2.0 g/cm^3 to 3.2 g/cm^3.

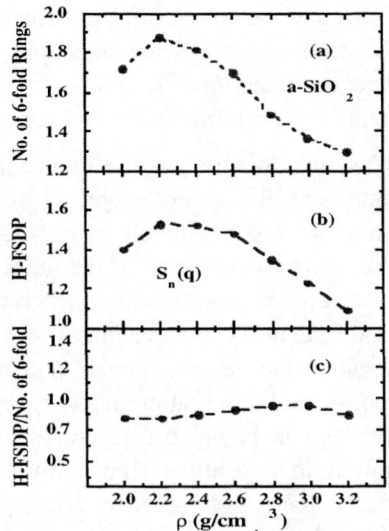

FIGURE 5. (a) Number of 6-fold rings, (b) height of the FSDP, and (c) ratio of height of the FSDP and the number of 6-fold rings as a function of the density.

Correlations in SiO$_2$ glass span two domains: (i) the short range order which extends to 4 Å, and (ii) intermediate range order between 4 Å and 25 Å, where dominant contribution to the FSDP comes from 4-8 Å region which is related to the diameter of six-fold rings. It has been shown by MD simulations that beyond this length ($L/2 > 25$ Å, L being the length of the MD box) there is no contribution to the FSDP (20,24,28).

Universal morphology of fracture surfaces

In recent years roughness of fracture surfaces has drawn a great experimental attention. Measurements on a variety of brittle and ductile solids have revealed that the widths of fracture profiles scale with the lengths as $w \sim L^\zeta$. For out-of-plane fracture profiles, the roughness exponent, ζ, above a certain length scale has a "universal" value of 0.8 for a three-dimensional system and 0.7 for a two-dimensional system. Many experiments also indicate that these values of ζ may be independent of the material (metals, semiconductors, and ceramics) or the mode of fracture. Furthermore, there are two regimes for dynamic fracture—at small length scales or when the crack front propagates quasi statically, the roughness exponent has a value of ~ 0.5 which crosses over to the larger value of 0.8 for faster moving crack. What is most remarkable, however, is that the "universal" roughness exponent characterizing fracture surfaces seems to be valid for a very wide range of length scales from hundreds of meters for earthquakes to the nanometer length scale as determined in atomistic fracture simulations of graphite (29), ceramics (30), and nanophase materials (31).

We have performed MD simulations of fracture in amorphous silicon nitride films (30). Well-thermalized amorphous films were subjected to uniaxial tensile loads. To investigate crack propagation, we insert a notch in an uniaxially stretched film by removing atoms. Figure 6 (b) shows a snapshot of the amorphous silicon nitride film. We observe the formation of voids in front of the crack tip. These voids grow and form a secondary crack, and eventually the secondary crack and the primary crack coalesce. The resulting crack surface is very rough, and its morphology exhibits a striking similarity to an experimental observation in a very different material (titanium-aluminide-based alloy) as shown in Figure 6 (c). As a comparison, we have also investigated crack propagation in crystalline silicon nitride films. Because of the absence of any pre-existing defects and disorder, crack propagation in crystalline systems is quite different from that in amorphous films. In crystalline films, stress is distributed uniformly before the insertion of a crack. When the crack is inserted, the enhanced stress at the crack tip results in the rupture of silicon-nitrogen bonds. The resulting crack is cleavage-like, see Figure 6 (a).

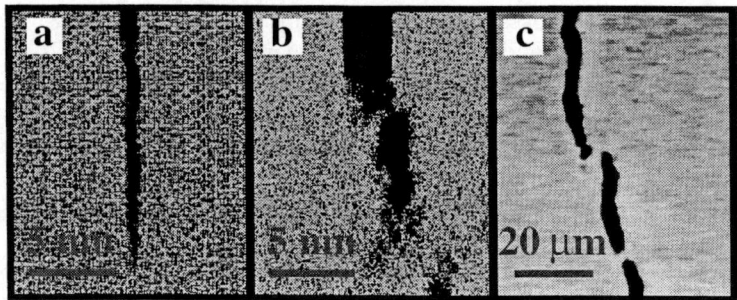

FIGURE 6. Scale invariance of the roughness of fracture surfaces on five orders-of-magnitude in length scales—from 1 nanometer to 100 microns. (a) An MD result on crystalline silicon nitride film exhibits cleavage fracture. (b) An MD result for fracture in amorphous silicon nitride film. Starting from the notch, initially the crack propagates straight. Voids form in front of the crack and grow, finally coalescing and forming a second crack. The same process repeats as the second crack proceeds forming the third crack in the right hand corner of the Figure 1 (b). (c) Experimental fracture surface in titanium-aluminide-based alloy (32).

We have calculated the height-height correlation function $g(r)$ of the crack surface from the height profile, $h(r)$, of the surface (30),

$$g(r) = \left\langle \left[h(r'+r) - h(r') \right]^2 \right\rangle^{1/2},$$

where $\langle f(r') \rangle$ denotes the average of a function $f(r')$ over lateral coordinate, r' (29). To analyze surfaces with branches and overhangs, we also use a return-probability, $P(r)$, that the height profile comes back to the original height after a distance r (29). $P(r)$ is just a pair-correlation function along a line parallel to the surface.

For a self-affine surface, we expect the scaling relation, $g(r) \sim r^\zeta$ and $P(r) \sim r^{-\zeta}$. We found two well-delineated regimes in $g(r)$ (30). In early stage of crack propagation, we observe a crack surface with a smaller roughness exponent. We obtain a roughness exponent of 0.44 ± 0.02 for $r < 2.5$ nm. Beyond 2.5 nm, the surface has a larger roughness exponent, $\zeta = 0.82 \pm 0.02$. Detailed analysis reveals that the smaller exponent corresponds to "slow" crack propagation inside microcracks, while the larger exponent is due to coalescence of microcracks which causes rapid propagation. The MD results support the theoretical model of crack propagation via microcrack formation and coalescence.

Branching instability of crack front

Another challenging problem in fracture is dynamic instabilities and the associated limiting speed of a crack. Recently we have investigated these issues by simulating crack propagation in a graphite sheet using the MD method (29). (These simulations are based on the reactive bond-order potentials for hydrocarbons developed by Brenner (33).) Dynamic fracture is studied in a single 150 nm × 200 nm graphite sheet. Dangling bonds at the boundaries are terminated with hydrogen atoms. Simulations are run with a time step of 1 femtosecond at a temperature of 300 K. While the internal configuration is completely two dimensional, the dynamics of atoms is followed in three dimensions. Typical length of a simulation is 10^5 time steps. We consider two different orientations of the graphite sheet, which exhibit completely different behavior during crack propagation, see Figure 7 (a). Under an applied uniaxial strain of 12% the system $G(1,1)$, in which a fraction of covalent bonds are parallel to the strain, undergoes cleavage-like fracture, see Figure 7 (b). On the contrary, for the $G(1,0)$ orientation at a strain of 12%, the crack front develops multiple secondary branches and overhangs when its speed reaches approximately half the Rayleigh wave speed, see Figure 3 (c). Within the same secondary branch the crack-front profile has a roughness exponent $\zeta = 0.5$, whereas for interbranch fracture surface profiles $\zeta = 0.7$. (The "universal" value of the roughness exponent in a two-dimensional system is ~ 0.7, whereas for three-dimensional systems it is ~ 0.8.) The scenario of multiple local branches and the values of the roughness exponents are in agreement with recent experimental investigations of fracture. Branching is strain-dependent. At a larger strain of 16 % for the $G(1,0)$ orientation, the crack front develops much more branching than at a strain of 12%.

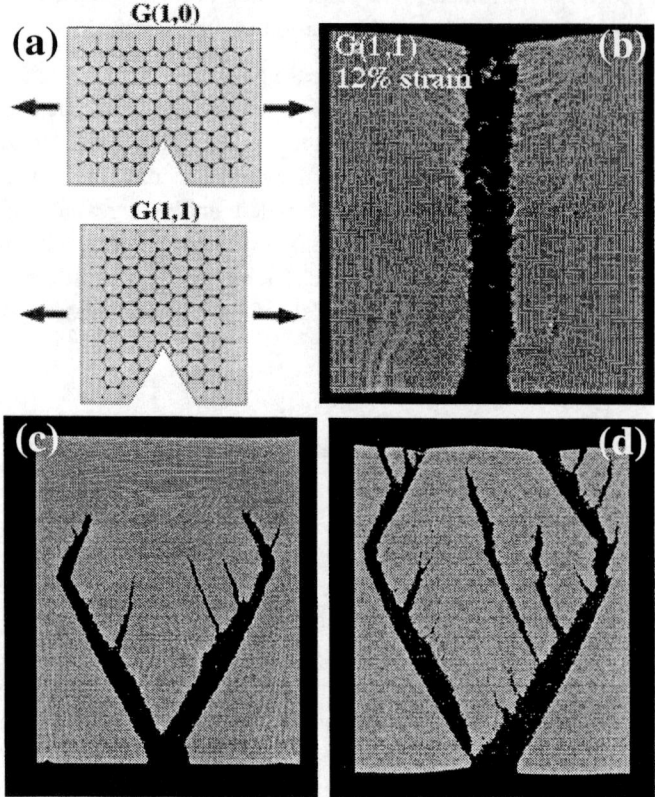

FIGURE 7. Molecular-dynamics simulation of dynamic fracture in graphite to examine the effect of orientation. 1.1 million atoms are used in the simulation. (a) Two different orientations, $G(1,0)$ and $G(1,1)$, used for stretching a graphite sheet. In the $G(1,1)$ orientation, applied strain is parallel to some of the carbon-carbon bonds and in the $G(1,0)$ orientation the strain is perpendicular to some of the bonds. (b) For the G(1,1) orientation, cleavage fracture is observed in a graphite sheet with 1.1 million atoms at applied strain of 12%. (c) For the G(1,0) orientation, multiple branches are observed at applied strain of 12%. (d) More branches are observed at a higher strain of 16% in the G(1,0) orientation.

Fracture toughness of gallium arsenide

Designing materials with high fracture toughness is one of the central issues in materials research. Recently, a reliable method has been developed to calculate the fracture toughness using a strip geometry (34).

Recently we have applied this method to investigate dynamic fracture in crystalline GaAs at various temperatures using large-scale MD simulations (see Fig. 8). The simulation sample is a gallium arsenide strip of 100 million atoms. After inserting a notch, the system is heated gradually to 300 K and then a constant strain is applied in the x direction by displacing atoms in the y-z plane that are within 0.5 nanometer from the boundaries of the system. For the strip geometry, the mechanical energy release rate, G, of the system can be calculated from the knowledge of the applied strain, ε, and the value of the stress, σ, far ahead of the crack tip: $G = W\sigma\varepsilon/2$, where W is the width of the strip (34). In addition to the mechanical energy release rate, we monitor the crack-tip velocities and local stress distributions at various temperatures. At low temperatures, the critical energy release rate at which crack propagation initiates is $G_c = 1.7$ J/m^2 for the (110) surface, which agrees well with experimental values (1.52 - 1.72 J/m^2) (35).

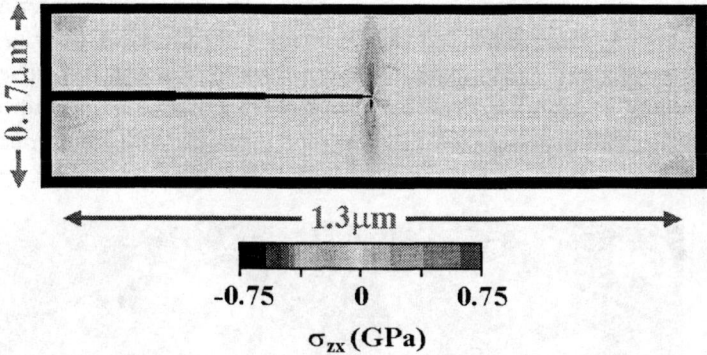

FIGURE 8. 100-million-atom MD simulation of fracture in gallium arsenide. Shear stress distribution near the crack tip in a thin-film strip at low temperature.

Nanophase ceramics with high fracture toughness

High temperature ceramics such as silicon nitride, silicon carbide, and alumina are highly desirable materials for applications requiring extreme operating conditions (36). Light weight, elevated melting temperatures, high strengths, and wear and corrosion resistance make them very attractive for applications in various energy technologies.

Unlike metals, these ceramic materials preserve their hardness at high temperatures. The only serious drawback of ceramics is that they are brittle.

In recent years a great deal of progress has been made in the synthesis of ceramics that are much more ductile than conventional coarse-grained materials (37). These so called nanophase materials are fabricated by in-situ consolidation of nanometer size clusters. Despite a great deal of research, many perplexing questions concerning nanophase ceramics remain unanswered. A realistic simulation of a nanophase solid requires 10^5-10^6 time steps for processing and at least ~ 10^6 atoms since each nanocluster itself consists of 10^3-10^4 atoms.

Molecular-dynamics simulations have been performed to investigate the structure and mechanical behavior of nanophase silicon nitride, silicon carbide, and silica (8-10,31). MD simulations of nanophase silicon nitride involve 108 nanoclusters each containing 10,052 atoms, see Fig. 9 (9). The clusters are cut out of bulk crystal and then relaxed with the conjugate-gradient approach. In the initial nanophase configuration the clusters are placed randomly in a cubic MD cell and then the system is consolidated at 2,000K under pressure ranging from 1 to 15 GPa using the constant-pressure MD approach. Subsequently each sintered system is cooled to room temperature and then the pressure is reduced to zero. In these systems, pair-distribution functions and bond-angle distributions reveal that interfacial regions between clusters are amorphous and they contain a large number of undercoordinated silicon atoms. Systems consolidated at low pressures (~ 1 GPa) have percolating pores whose surface roughness exponents are in excellent agreement with experiments. We also find that the dependence of elastic moduli on porosity and grain size can be understood in terms of a three-phase model for heterogeneous materials.

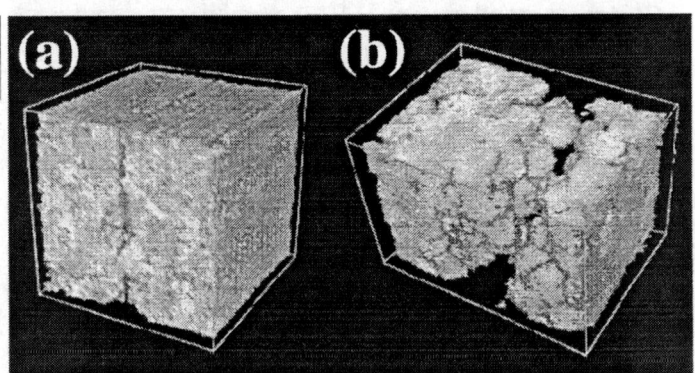

FIGURE 9. (a) Cleavage fracture in crystalline silicon nitride. (b) Nanophase silicon nitride just before it fractures.

Understanding the role of microstructures in fracture is one of the most challenging problems in materials science (36). Nanophase materials are ideal systems to examine this issue at the atomistic level since microstructures in these materials are only a few nanometers in dimension. We have investigated the morphological and dynamic aspects of fracture in nanophase silicon nitride using 1.08 million particle MD simulations (31).

Interfacial fracture, dislocation formation and its motion

Fracture at interfaces has been a subject of numerous experimental and theoretical studies. Cracking patterns range from surface cracks and channeling in the film to substrate damage, spalling and debonding of the interface. For silicon nitride as a dielectric material and passivation layer in silicon and gallium arsenide devices it is therefore important to understand crack initiation, propagation, dislocation emission and fracture on an atomistic level. Molecular dynamics is a powerful method to investigate mechanical behavior of these interface systems. A way to simulate crack initiation and its propagation is to apply uniaxial strain parallel to the interface and study time evolution of the system to analyze its failure resistance.

To account for all the structural correlations for silicon, silicon nitride and the silicon(111)/silicon-nitride(0001) interface, the system is modeled using eight components (38). These consist of: Si^{4+} and N^{3-} in the bulk silicon nitride; Si^{3+}, N^{2-}, and N^{3-} at the silicon nitride side of the interface; 3-fold coordinated silicon at silicon(111) interface, its 4-fold coordinated neighboring silicon in the plane; and bulk silicon.

A schematic of the geometry of the interface system is shown in Fig. 10 (a). After thermalizing the system at 300K, the system is stretched parallel to the interface, i.e. in the [2100] direction for silicon nitride and in the [$\bar{2}$11] direction for silicon until it failed. This is due to the fact that a crack started to form at the top surface of the silicon nitride layer and it propagated through the whole silicon nitride layer. It was found that the crack does not propagate into silicon, but instead emits dislocations which correlates well with an additional drop in σ_{xx}. We have examined the structure of silicon at the interface to determine the nature of defects created by the crack arriving from silicon nitride. In Fig. 10 (b) the extra line of atoms in a Si(111) plane parallel to the interface - an edge dislocation - can be clearly seen.

Time evolution of the emitted dislocation is given in Fig. 10 (c) and (d). Only those silicon atoms whose energies are higher than the average silicon energy by +0.35 eV are shown. In Fig. 10 (c), we see the formation of a dislocation loop at the interfacial plane. The dislocation loop lies on a ($\bar{1}\bar{1}$1) plane denoted with dashed lines in Fig. 10 (c). As time proceeds the dislocation loop grows (see Fig. 10 (d)) till it reaches the silicon surface at the bottom after 13 ps. From our simulation data we estimate the speed of the dislocation motion to be 500 ($\pm$100) m/sec.

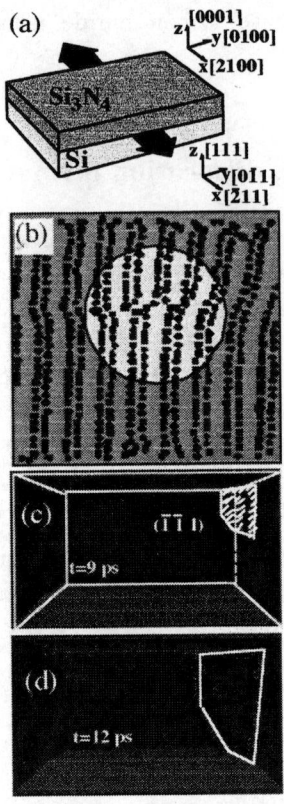

FIGURE 10. (a) Schematics of fracture geometry for the silicon/silicon-nitride interface. (b) Atomic positions in a slice of silicon (111) parallel to the interface showing an edge dislocation. The extra double layer of atoms is clearly seen. (c) and (d) Time evolution of dislocation motion. Only silicon atoms with energies larger than the average silicon atom energy by +0.35 electronvolts are plotted. At 9 ps, formation of a dislocation loop at the interfacial plane and the right-hand silicon surface (surface atoms belonging to the vertical planes have been removed from the plot to make the dislocation loop visible). The dislocation loop lies on a ($1\bar{1}1$) plane denoted by dashed lines.

Nanoindentation of silicon nitride

Nanoindentation testing is a unique local probe of mechanical properties of materials at surfaces and multilayered structures (39). This technique is especially useful in testing surfaces and thin films in microelectronics and other industries. Using multimillion atom MD simulations, we are investigating nanoindentation in silicon nitride (39).

The nanoindentation simulation is performed on a 60 nm × 60 nm × 30 nm crystalline silicon nitride slab consisting of 10 million atoms (see Fig. 11). The sample is indented using a pyramid indentor with a load ~ 10 µN and indentation depth ~ 10 nm. These values are within the resolution of high-quality nanoindentation experiments. Having established the methodology for atomistic simulation of nanoindentation, this virtual nanohardness approach is being used to probe high-temperature properties where actual indentation measurements may not be possible. Our simulations reveal significant plastic deformation and pressure-induced amorphization under the indentor.

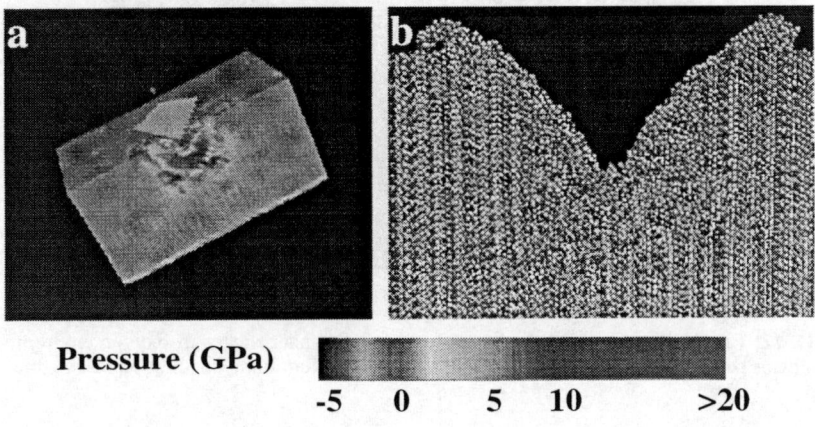

FIGURE 11. (a) A half-slice view of pressure in a silicon nitride sample during nanoindentation. (b) An atomic view showing amorphization under the indentor and material pileup at the edges of the indentor. These figures demonstrate the relationship between residual stresses and atomistic-level amorphization phenomena.

Environmental effects on fracture

In real applications, reactive processes such as oxidation have deleterious effects on the physical and mechanical properties of materials, including premature failure of otherwise properly designed systems. Using the MD approach, we are studying effects of oxidation on fracture in aluminum. This requires models of interatomic potentials that are much more refined than those used in conventional MD simulations.

We use a variable-charge interatomic potential because it can handle bond formation and bond breakage (40).

Recently, we have tested such a scheme by performing the first MD simulation of oxidation of an aluminum nanocluster (Fig. 12) (41). Structural and dynamic correlations in the oxide region and the evolution of various quantities including charge, surface oxide thickness, diffusivities of atoms, and local stresses have been calculated. Structural analysis reveals that a 4 nm thick amorphous oxide scale consisting of mixed octahedral, $Al(O_{1/6})_6$, and tetrahedral, $Al(O_{1/4})_4$, configurations is formed during 466 ps of simulation time. (The canonical ensemble has been used for the simulation). This is in excellent agreement with experimental results on aluminum nanoclusters. The average mass density in the oxide scale is 75% of the crystalline alumina density.

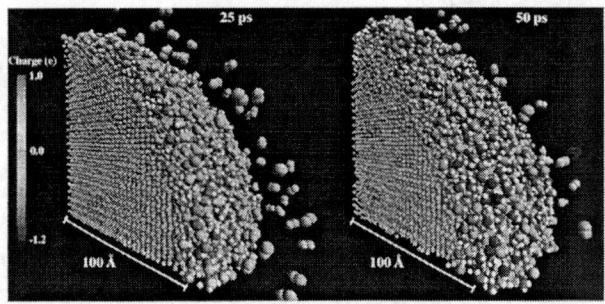

FIGURE 12. Snapshots of a small wedge of an aluminum nanocluster in oxygen environment. The larger spheres correspond to oxygen and smaller spheres to aluminum; shading represents the charge on an atom.

Within ten years, we will be using petaflop computers to perform trillion-atom MD simulations to include the effects of microstructures spanning diverse length scales up to the mesoscale regime above micron. Within the same timeframe, interatomic potential models used in the MD simulations will be refined with inputs from quantum-mechanical calculations of electronic structures. These atomistic simulations will further be combined with continuum schemes based on finite-element methods to model truly macroscopic dynamic fractures at all length scales in a seamless manner (42,43).

ACKNOWLEDGMENTS

Research presented in this article was carried out in collaboration with Martina Bachlechner, Timothy Campbell, Alok Chatterjee, Ingvar Ebbsjö, Hideaki Kikuchi,

Wei Li, Shuji Ogata, Andrey Omeltchenko, Kenji Tsuruta, and Phillip Walsh. This work is partially supported by DOE, NSF, AFOSR, USC-LSU MURI, NASA, and ARO. 1-2 million-atom simulations were performed using in-house parallel computers at the CCLMS at LSU. 10-100 million-atom simulations were performed using parallel computers at Department of Defense's Major Shared Resource Centers under a DoD Challenge Applications Award. We would like to acknowledge continued support and encouragement from Lt. Colonel Dr. Larry P. Davis who has made these very large-scale simulations possible on DoD machines.

REFERENCES

1. de Leeuw, S.W., Perram, J. W., and Smith, E. R., *Proc. R. Soc. London*, **A373**, 27 (1980).
2. Kalia, R. K., de Leeuw, S. W., Nakano, A., and Vashishta, P., *Comput. Phys. Commun.* **74**, 316 (1993).
3. Appel, A. W., *SIAM J. Sci. Stat. Comput.* **6**, 85 (1985).
4. Barnes, J., and Hut, P., *Nature* **324**, 446 (1986).
5. Greengard, L., and Rokhlin, V., *J. Comput. Phys.* **73**, 523 (1987).
6. Allen, M. P., and Tildesley, D. J., *Computer Simulation of Liquids*, Oxford: Oxford Univ. Press, 1987.
7. Skeel, R. D., Zhang, G., and Schlick, T., *SIAM J. Sci. Comput.* **18**, 203 (1997).
8. Vashishta, P., Kalia, R. K., Rino, J. P., and Ebbsjö, I., *Phys. Rev. B* **41**, 12197 (1990); Campbell, T., Kalia, R. K., Nakano, A., Shimojo, F., Tsuruta, K., and Vashishta, P., *Phys. Rev. Lett.*, **82**, 4018 (1999).
9. Vashishta, P., Kalia, R. K., and Ebbsjö, I., *Phys. Rev. Lett.* **75**, 858 (1995); Kalia, R. K., Nakano, A., Tsuruta, K., and Vashishta, P., *Phys. Rev. Lett.* **78**, 689 (1997).
10. Chatterjee, A., Campbell, T., Kalia, R. K., Nakano, A., Omeltchenko, A., Tsuruta, K., and Vashishta, P., *J. Euro. Ceram. Soc.*, in press.
11. Kodiyalam, S., Chatterjee, A., Ebbsjö, I., Kalia, R. K., Kikuchi, H., Nakano, A., Rino, J. P., and Vashishta, P., *Mater. Res. Soc. Symp. Proc.* **536**, 545 (1999).
12. Tuckeman, M. E., Berne, B. J., and Martyna, G. J., *J. Chem. Phys.* **97**, 1990 (1992).
13. Nakano, A., *Int. J. High Performance Comput. Appl.* **13**, 154 (1999).
14. Nakano, A., Kalia, R. K., and Vashishta, P., *Comput. Phys. Commun.* **83**, 197 (1994).
15. Vashishta, P., and Rahman, A., *Phys. Rev. Lett.* **40**, 1337 (1978).
16. Vashishta, P., and Rahman, A., in *Fast Ion Transport in Solids*, 1979, Eds. Vashishta, P., Mundy, J. N., and Shenoy, G.K., Elsevier, p. 527.
17. Jin, W., Kalia, R. K., Vashishta, P., and Rino, J. P., *Phys. Rev. Lett.* **71**, 3146 (1993).
18. Jin, W., Kalia, R. K., Vashishta, P., and Rino, J. P., *Phys. Rev. B* **48**, 9359 (1993).
19. Nakano, A., Bi, L., Kalia, R. K., and Vashishta, P., *Phys. Rev. Lett.* **71**, 85 (1993).
20. Vashishta, P., Kalia, R. K., Antonio, G. A., and Ebbsjö, I., *Phys. Rev. Lett.* **62**, 1651 (1989).

21. Antonio, G. A., Kalia, R. K., Nakano, A. and Vashishta, P., *Phys. Rev. B* **45**, 7455 (1992).
22. Li, W., Kalia, R. K., and Vashishta, P., *Phys. Rev. Lett.* **77** 2241, (1996).
23. Moss, S. C., and Price, D. L., in *Physics of Disordered Materials*, Eds. Adler, D., Fritzsche, H., and Ovshinsky, S. R., New York: Plenum, 1985, p. 77.
24. Nakano, A., Kalia, R. K., and Vashishta, P., *J. Non-Cryst. Solids* **171**, 157 (1994).
25. Wright, A. C., *J. Non-Cryst. Solids* **159**, 264 (1993).
26. Susman, S., Volin, K. J., Montague, D. G., and Price, D. L., *Phys. Rev. B* **43** 11076 (1991).
27. Rino, J. P., Ebbsjö, I., Kalia, R. K., Nakano, A., and Vashishta, P., *Phys. Rev. B* **47**, 3053 (1993).
28. Rino, J. P., Nakano, A., Kalia, R. K., and Vashishta, P., in *Computer-Aided Design of High-Temperature Materials*, Eds. Pechenik, A., Kalia, R. K., and Vashishta, P., Oxford: Oxford Univ. Press, 1999.
29. Omeltchenko, A., Yu, J., Kalia, R. K., and Vashishta, P., *Phys. Rev. Lett.* **78**, 2148 (1997).
30. Nakano, A., Kalia, R. K., and Vashishta, P., *Phys. Rev. Lett.* **75**, 3138 (1995).
31. Kalia, R. K., Nakano, A., Tsuruta, K., and Vashishta, P., *Phys. Rev. Lett.* **78**, 2144 (1997); Tsuruta, K., Nakano, A., Kalia, R. K., and Vashishta, P., *J. Am. Ceram. Soc.* **81**, 433 (1998).
32. Bouchaud, E., *J. Phys.: Condens. Matter* **9**, 4319 (1997).
33. Brenner, D. W., *Phys. Rev. B* **42**, 9458 (1990).
34. Freund, L. B., *Dynamic Fracture Mechanics*, Cambridge: Cambridge Univ. Press, 1990.
35. Messmer, C., and Bilello, J. C., *J. Appl. Phys.* **52**, 4623 (1981); Michot, G., George, A., Chabli-Brkmac, A., and Molva, K., *Scr. Metall.* **22**, 1043 (1988).
36. Pechenik, A., Kalia, R. K., and Vashishta, P., *Computer-Aided Design of High-Temperature Materials*, Oxford: Oxford Univ. Press, 1999.
37. Siegel, R. W., *Scientific American*, Dec. 1996, p. 74.
38. Bachlechner, M. E., Omeltchenko, A., Nakano, A., Kalia, R. K., Vashishta, P., Ebbsjö, I., Madhukar, A., and Messina, P., *Appl. Phys. Lett.* **72**, 1969 (1998).
39. Walsh, P., Omeltchenko, A., Kikuchi, H., Kalia, R. K., Nakano, A., and Vashishta, P., *Mater. Res. Soc. Symp. Proc.* **539**, 119, (1999).
40. Streitz F. H., and Mintmire, J. W., *Phys. Rev. B* **50**, 11996 (1994).
41. Campbell, T., Kalia, R. K., Nakano, A., Vashishta, P., Ogata, S., and Rodgers, S., *Phys. Rev. Lett.* **82**, 4866 (1999).
42. Tadmor, E. B., Phillips, R., and Ortiz, M., *Langmuir* **12**, 4529 (1996).
43. Abraham, F. F., Broughton, J. Q., and Kaxiras, E., *Comput. Phys.* **12**(6), 538 (1998).

Scattering of plane-wave atomic vibrations in disordered structures

S.N. Taraskin and S.R. Elliott

Department of Chemistry, University of Cambridge, Lensfield Road, Cambridge CB2 1EW, UK

Abstract. A theoretical analysis of the scattering of plane-wave atomic excitations in disordered solids has been made in terms of the spectral densities. Hybridization between transverse and longitudinal waves of approximately the same frequency is demonstrated. The analytic results agree well with the results obtained from computer simulation for a toy linear zig-zag chain model and a model of vitreous silica constructed by molecular dynamics.

I INTRODUCTION

Propagation of classical plane waves in random scattering media has attracted a lot of theoretical and experimental attention in recent years [1,2]. The phenomena of wide interest include weak localization [1,2], Anderson localization [3], phonon localization [4–6] and related behaviour of plane waves in the Ioffe-Regel crossover region [7] separating weakly and strongly scattered waves [8,9].

Vibrational plane waves propagate well in disordered structures in the long-wavelength limit, $ka \ll 1$, with k the wavevector and a a measure of the microscopic scale of the structure (being of the order of interatomic distances), when the atomic structure behaves as an elastic continuum and disorder on the microscopic level is not of great importance. The situation is changed with decreasing wavelength and, on the microscopic scale, disorder becomes important and the wavevector is no longer a good quantum value [10–17].

In the investigation of the propagation of the plane-wave vibrational excitations in disordered structures, different decay channels have been suggested to explain their attenuation: (i) disorder-induced channels [18–22], (ii) anharmonic channels [20] and (iii) channels involving two-level systems [23–26]. The anharmonic channels are strongly enhanced with increasing temperature, particularly at temperatures comparable with the glass-transition temperature, $T_g \sim 10^3$K. In contrast, scattering by two-level systems can be important at low temperatures [25], $T \ll T_{TLS} \sim 10 - 100$K [24,27]. In the intermediate temperature range, $T_{TLS} \leq T \ll T_g$, which is considered below, the scattering processes involving two-level systems are suppressed and the atomic dynamics are usually harmonic [21,28], meaning that disorder-induced channels play the most important role in the decay mechanism of plane-wave excitations.

If the harmonic approximation is valid for describing the atomic dynamics, then a normal-mode analysis can be used for the problem under consideration. The normal modes can be found either analytically or numerically. A general theory of atomic vibrations in disordered structures, which in principle should result in normal modes, has been mainly developed for particular simple model structures [18,29–34], which can hardly describe quantitatively the situation in real structures. Therefore, a numerical approach could be very useful in the calculation of normal modes, e.g. by direct diagonalization of the dynamical matrix which can be available, e.g. from molecular dynamics simulations.

The main questions we address in this paper are: how are plane-wave vibrational excitations scattered by disorder and what are the characteristics of the final state after scattering?

Our approach to the problem is based on combining analytical and numerical techniques. First, we create a few structural models of a disordered atomic material: (i) realistic models of v-SiO$_2$ using molecular dynamics and (ii) toy models of a linear zig-zag chain. Then all eigenmodes and eigenfrequencies (in the harmonic approximation) are found numerically. These characteristics fully determine the dynamical response of the system (and final state at $t \to \infty$) to any external excitations, including the plane-wave excitations of present interest. The final state after scattering was investigated then in momentum space (analytically and numerically).

II FORMALISM

The time evolution of any vibrational excitation is fully determined in the harmonic approximation by the eigenmodes and eigenfrequencies of the system. Indeed, the initial excitation can be expanded in the eigenmodes, the time dependence of which is known. The coefficients in such an expansion are defined by the shape of the initial vibrational excitation. Here we consider only plane-wave external initial excitations, mainly because exactly such excitations are generated in a system by inelastic neutron, light and electron scattering [35].

In amorphous materials, because of disorder, the eigenmodes are not plane waves even in the long-wavelength limit. Therefore, an initial plane wave, when expanded over eigenmodes, contains different eigenmodes characterized by different weights in this expansion. The eigenmodes participating in the expansion evolve differently with time, so that the propagating excitation becomes different in shape compared with the initial one. If we expand the vibrational state in plane waves after a certain time, then this expansion contains not only the initial plane-wave component but also other plane waves characterized by different wavevectors. This means that the initial plane wave is scattered by the structure into a different final state. Our aim here is to study the properties of the final state after decay, for different wavevectors of an initial plane wave.

Let us consider an external wave excitation introduced in the system, $\mathbf{u}(t)$, which at the initial moment of time is an ideal plane wave, $\mathbf{w_k}$, characterized by the wavevector $\mathbf{k}$, unit polarization vector $\hat{\mathbf{n}}$ and initial phase ϕ_0:

$$\mathbf{u}(t=0) = \mathbf{w_k} \equiv A\hat{\mathbf{n}}\cos[\mathbf{k}\cdot\mathbf{r}+\phi_0] , \qquad (1)$$

where $\mathbf{u}$ is a $3N$-dimensional displacement vector, the i-th component of which describes the displacement of atom i from its equilibrium position at $\mathbf{r_i}$, A is the normalization constant defined below and the wavevector index $\mathbf{k}$ includes also the polarization index $\hat{\mathbf{n}}$. In our analytical treatment, we assume that eigenmodes and eigenfrequencies are known, e.g. from numerical simulations. The initial displacement vector, Eq. (1), each atomic component of which is multiplied by the mass factor $m_i = M_i N / \sum_i M_i$ (M_i stands for the mass of atom i) can be expanded in eigenmodes as:

$$\mathbf{u}(0) = \sum_{j=1}^{3N} \overline{\alpha}_\mathbf{k}^j \mathbf{e^j}/\sqrt{m} , \qquad (2)$$

where the symbolic script $\mathbf{e^j}/\sqrt{m}$ means that each i-th component of vector $\mathbf{e^j}$ is divided by the factor $\sqrt{m_i}$. The coefficients $\overline{\alpha}_\mathbf{k}^j$ in expansion (2), the squares of which are the spectral densities of the system, are defined by the following equation:

$$\overline{\alpha}_\mathbf{k}^j = \langle \mathbf{e^j} \cdot \sqrt{m}\mathbf{u_k} \rangle \equiv \sum_{i=1}^{N} \sqrt{m_i} \mathbf{e}_i^j \cdot \mathbf{w_{k,i}} . \qquad (3)$$

The spectral-density coefficients, Eq. (3), fully determine the dynamical response of the system to plane-wave excitation. Indeed, at any moment of time t, the displacement vector of the propagating excitation can be represented via eigenmodes developing in time as:

$$\mathbf{u}(t) = \sum_{1}^{3N} \overline{\alpha}_\mathbf{k}^j \frac{\mathbf{e^j}}{\sqrt{m}} \cos\omega_j t . \qquad (4)$$

For the sake of simplicity and without loss of generality (as shown below), we consider the initial excitation to be a standing wave, i.e. $\dot{\mathbf{u}}(0) = 0$, leading to the absence of terms proportional to $\sin\omega_j t$ in expression (4). It is convenient for the initial vector $\sqrt{m}\mathbf{u_k}(0)$ to be normalized to unity, so that

$$\sum_{1}^{3N} |\overline{\alpha}_\mathbf{k}^j|^2 = 1 , \qquad (5)$$

and the normalization constant in Eq. (1) is $A^2 = [\sum_i m_i |\mathbf{u_k}(0)|^2]^{-1}$.

An ideal initial plane wave (1) scatters with time to different plane waves. In order to calculate the weights of different plane-wave components in the propagating excitation, we expand the displacement vector $\mathbf{u}(t)$ in plane waves:

$$\mathbf{u}(t) = \sum_{\mathbf{k}'} \mathbf{u}_{\mathbf{k}'}(t), \qquad (6)$$

where the sum is taken over all wavevectors $\mathbf{k}'$ (allowed by the periodic simulation box in the case of a finite model) and all polarizations (two transverse and one longitudinal for each wavevector), and the waves $\mathbf{u}_{\mathbf{k}'}(t)$ are defined as:

$$\mathbf{u}_{\mathbf{k}'}(t) = a_{\mathbf{k}'}(t) A \hat{\mathbf{n}}' \cos(\mathbf{k}' \cdot \mathbf{r} + \phi_{\mathbf{k}'}(t)). \qquad (7)$$

The same normalization as in Eq. (1) is used here. In order to find the time dependence of the amplitude $a_{\mathbf{k}'}(t)$ and phase $\phi_{\mathbf{k}'}(t)$, it is convenient to rewrite Eq. (7) in the following form:

$$\mathbf{u}_{\mathbf{k}'}(t) = a_{\mathbf{k}',c}(t) \mathbf{w}_{\mathbf{k}',c} + a_{\mathbf{k}',s}(t) \mathbf{w}_{\mathbf{k}',s}, \qquad (8)$$

where

$$\mathbf{w}_{\mathbf{k}',c} = A \hat{\mathbf{n}}' \cos \mathbf{k}' \cdot \mathbf{r} \quad \text{and} \quad \mathbf{w}_{\mathbf{k}',s} = A \hat{\mathbf{n}}' \sin \mathbf{k}' \cdot \mathbf{r}, \qquad (9)$$

so that

$$a_{\mathbf{k}'}(t) = (a_{\mathbf{k}',c}^2(t) + a_{\mathbf{k}',s}^2(t))^{1/2}, \qquad (10)$$

$$\phi_{\mathbf{k}'}(t) = \operatorname{Arctan}[a_{\mathbf{k}',s}(t)/a_{\mathbf{k}',c}(t)]. \qquad (11)$$

The coefficients $a_{\mathbf{k}',c(s)}(t)$ before the cos- (sin-) like components in Eq. (8) can be found by multiplying both sides of this equation by $\mathbf{w}_{\mathbf{k}',c(s)}$ and using Eq. (3), so that

$$a_{\mathbf{k}',c(s)}(t) = \sum_j \overline{\alpha}_{\mathbf{k}}^j \underline{\alpha}_{\mathbf{k}',c(s)}^j \cos \omega_j t / \langle \mathbf{w}_{\mathbf{k}',c(s)}^2 \rangle. \qquad (12)$$

where

$$\langle \mathbf{w}_{\mathbf{k}',c(s)}^2 \rangle \equiv \sum_{i=1}^N |(\mathbf{w}_{\mathbf{k}',c(s)})_i|^2. \qquad (13)$$

The spectral-density coefficients $\underline{\alpha}_{\mathbf{k}',c(s)}^j$ for plane waves of cos- and sin-like type are defined by the following equation:

$$\underline{\alpha}_{\mathbf{k},c(s)}^j = \langle \mathbf{e}^j \cdot \frac{1}{\sqrt{m}} \mathbf{w}_{\mathbf{k},c(s)} \rangle \equiv \sum_{i=1}^N \frac{1}{\sqrt{m_i}} \mathbf{e}_i^j \cdot (\mathbf{w}_{\mathbf{k},c(s)})_i. \qquad (14)$$

The spectral-density coefficients $\underline{\alpha}_{\mathbf{k},c(s)}^j$ in a multicomponent system, in contrast to $\overline{\alpha}_{\mathbf{k}}^j$, instead of being normalized to unity are normalized by the following value:

$$\sum_{j=1}^{3N}|\alpha_{\mathbf{k},c(s)}^{j}|^{2}=\frac{\sum_{i}(m_{i})^{-1}|\mathbf{w}_{\mathbf{k},c(s)}|^{2}}{\sum_{i}m_{i}|\mathbf{w}_{\mathbf{k},c(s)}|^{2}},\qquad(15)$$

having the value $\simeq 0.7$ in the case of vitreous silica.

Eqs. (10) - (15) fully determine the time evolution of different $\mathbf{k}'$ plane-wave components in the propagating vibrational excitation via the spectral densities and the vibrational spectrum itself. Another useful characteristic often used to characterize the decay of an initial external excitation is the time correlation function [36],

$$\frac{\langle \mathbf{u}(t)\cdot\mathbf{u}(0)\rangle}{\langle \mathbf{u}(0)\cdot\mathbf{u}(0)\rangle}=\frac{\sum_{i}\mathbf{u}_{i}(t)\mathbf{u}_{i}(0)}{\sum_{i}\mathbf{w}_{\mathbf{k},i}\mathbf{w}_{\mathbf{k},i}}=\frac{\sum \overline{\alpha}_{\mathbf{k}}^{j}\alpha_{\mathbf{k}}^{j}\cos\omega_{j}t}{\langle \mathbf{w}_{\mathbf{k}}^{2}\rangle},\qquad(16)$$

where the spectral-density coefficient $\alpha_{\mathbf{k}}^{j}$ is defined by Eq. (14) with $\mathbf{w}_{\mathbf{k},c(s)}$ replaced by $\mathbf{w}_{\mathbf{k}}$, and is related to the spectral-density coefficient $\overline{\alpha}_{\mathbf{k}}^{j}$ according to the following equation:

$$\alpha_{\mathbf{k}}^{j}=\sum_{j'}\overline{\alpha}_{\mathbf{k}}^{j'}\langle e^{j}m^{-1}e^{j'}\rangle.\qquad(17)$$

In the case of a one-component system, the spectral-density coefficients $\alpha_{\mathbf{k}}^{j}$ and $\overline{\alpha}_{\mathbf{k}}^{j}$ are obviously identical.

The decay of the external plane-wave excitation can also be characterized via the properties of the final state after decay, averaged over time as $t\to\infty$. An initial plane-wave excitation characterized by the wavevector $\mathbf{k}$ and polarization $\hat{\mathbf{n}}$ is scattered to different plane-wave components characterized by the wavevectors $\mathbf{k}'$ and polarizations $\hat{\mathbf{n}}'$. The distribution, $\rho(\mathbf{k}',\hat{\mathbf{n}}'|\mathbf{k},\hat{\mathbf{n}})$, of the weights of different plane-wave components averaged over time in the final state, $\overline{a_{\mathbf{k}',\hat{\mathbf{n}}'}^{2}(t)}$, is of particular interest. Such a distribution is defined by the following equation:

$$\rho(\mathbf{k}',\hat{\mathbf{n}}'|\mathbf{k},\hat{\mathbf{n}})=\overline{a_{\mathbf{k}'}^{2}(t)}=\frac{1}{2}\left\{\frac{\sum_{j}|\overline{\alpha}_{\mathbf{k}}^{j}|^{2}|\alpha_{\mathbf{k}',s}^{j}|^{2}}{(\langle \mathbf{w}_{\mathbf{k}',s}^{2}\rangle)^{2}}+\frac{\sum_{j}|\overline{\alpha}_{\mathbf{k}}^{j}|^{2}|\alpha_{\mathbf{k}',c}^{j}|^{2}}{(\langle \mathbf{w}_{\mathbf{k}',c}^{2}\rangle)^{2}}\right\},\qquad(18)$$

where, as in Eq. (1) and subsequently, the wavevector index $\mathbf{k}'$ also includes the polarization index $\hat{\mathbf{n}}'$. Eq. (18) is obtained by averaging Eq. (10) over time and using Eq. (12). Bearing in mind that $\langle \mathbf{w}_{\mathbf{k}',s}^{2}\rangle\simeq\langle \mathbf{w}_{\mathbf{k}',c}^{2}\rangle\simeq\langle \mathbf{w}_{\mathbf{k}'}^{2}\rangle$ and that the sum $|\alpha_{\mathbf{k}',s}^{j}|^{2}+|\alpha_{\mathbf{k}',c}^{j}|^{2}$ is independent of the phase of cos- and sin-like components defined by Eq. (9), expression (18) can be transformed to:

$$\rho(\mathbf{k}',\hat{\mathbf{n}}'|\mathbf{k},\hat{\mathbf{n}})=\frac{\sum_{j}|\overline{\alpha}_{\mathbf{k}}^{j}|^{2}|\alpha_{\mathbf{k}'}^{j}|^{2}}{(\langle \mathbf{w}_{\mathbf{k}'}^{2}\rangle)^{2}}.\qquad(19)$$

The distribution (19) averaged over all polarizations $\hat{\mathbf{n}}'$ in the final state is

$$\rho_{\mathrm{av}}(\mathbf{k}'|\mathbf{k},\hat{\mathbf{n}})\simeq 2\rho(\mathbf{k}',\hat{\mathbf{n}}_{\mathrm{t}}'|\mathbf{k},\hat{\mathbf{n}})+\rho(\mathbf{k}',\hat{\mathbf{n}}_{\mathrm{l}}'|\mathbf{k},\hat{\mathbf{n}}),\qquad(20)$$

where the unit vector $\hat{\mathbf{n}}'_t$ stands for transverse polarization in the final state while $\hat{\mathbf{n}}'_l$ refers to longitudinal polarization, and the factor 2 takes into account the existence of two independent and, in glasses, equivalent transverse polarizations. Glasses are isotropic, and an averaging of Eqs. (18)-(20) over the directions of both initial and final wavevectors (including averaging over transverse polarizations in the initial wave) can be made, resulting in:

$$\rho_{av,t(l)}(k'|k) = \langle \rho_{av}(\mathbf{k}'|\mathbf{k}, \hat{\mathbf{n}}_{t(l)}) \rangle_{\Omega_{\mathbf{k},\mathbf{k}'}} . \qquad (21)$$

If we are interested in the contribution of the same plane-wave **k**-component as in the initial excitation, then the wavevector **k**′ should be replaced by **k** in Eqs. (18)-(20) and averaging only over **k**-directions should be made in Eq. (21).

III SPECTRAL DENSITIES

As follows from Sec. II, the coefficients $\overline{\alpha}^j_{\mathbf{k}}$, $\alpha^j_{\mathbf{k}}$ in the expansion of different **k**-plane-waves over the eigenmodes, i.e. projections of plane waves onto eigenvectors, and related spectral densities, $|\alpha^j_{\mathbf{k}}|^2$, $|\overline{\alpha}^j_{\mathbf{k}}|^2$ and $\overline{\alpha}^j_{\mathbf{k}} \alpha^j_{\mathbf{k}}$, fully determine the dynamical response of the system to the initial plane-wave excitation. We calculate below the spectral densities for two models of disordered structures: (i) a model of v-SiO$_2$ constructed by molecular dynamics and (ii) a disordered zig-zag chain.

A Vitreous silica

The models of v-SiO$_2$ have been constructed by NPT-molecular-dynamics simulations, using the potential of van Beest [37]. The van Beest potential has been modified for small interatomic distances according to Guissani and Guillot [38]. At large interatomic distances, we have used a cutoff for short-range interactions, multiplying the modified van Beest potential by a Fermi-like step function. The step function is characterized by the step position at $R_{\text{cut}} = 5.5\text{Å}$ and the step width $\delta R_{\text{cut}} = 0.5\text{Å}$ for all atomic species. The latter cutoff has been used to obtain a density of the glassy structure (at zero pressure), of 2.38g/cm^3, reasonably close to the experimental value of 2.2g/cm^3 (see the discussion of the densification problem in Ref. [38]). Note that a similar cutoff ($R_{cut} = 5.0\text{Å}$ and $\delta R_{cut} = 0$) has been used by Vollmayr et al. [39].

All glassy models have been created by quenching from the melt ($T = 6000$K) to the well-relaxed glassy state ($T \sim 10^{-4}$K) at an average quench rate of ~ 1K/ps. No coordination defects have been found in the models. The fully dense dynamical matrices for the relaxed systems were diagonalized directly, resulting in eigenvectors $\{\mathbf{e}^j\}$ and eigenvalues (ω_j), thus allowing us to perform a complete harmonic vibrational analysis. Structural characteristics and vibrational properties of the models are very similar to those described in Ref. [40].

The models of v-SiO$_2$ were of two types: a cubic model containing $N = 1650$ atoms and of box length $L \simeq 28.4\text{Å}$, and a bar configuration containing $N = 1500$ atoms of size $85.6\text{Å} \times 15.6\text{Å} \times 15.6\text{Å}$ (a bar-shaped model of B$_2$O$_3$ has been also used in Ref. [41]). The bar-shaped models were constructed to allow access to much lower values of k ($\geq 0.07\text{Å}^{-1}$) for modes propagating along the bar than can be obtained for the cubic models ($k \geq 0.22\text{Å}^{-1}$).

Silica is a multicomponent system so that the three spectral densities differ from each other due to different contributions in the coefficients $\overline{\alpha}_k^j$ and α_k^j of the mass factor (see Eqs. (3) and (14)). In the case of vitreous silica, the masses of the atomic species are quite comparable and the mass factor is of the order of unity, so that the different types of the spectral densities differ only slightly from each other and the following approximate relationships can be used: $|\alpha_k^j|^2 \simeq A_1|\overline{\alpha}_k^j|^2$ and $\overline{\alpha}_k^j \alpha_k^j \simeq A_2|\overline{\alpha}_k^j|^2$, with the normalization constants for the corresponding spectral densities being $A_1 = \sum_j |\alpha_k^j|^2$ and $A_2 = \sum_j \overline{\alpha}_k^j \alpha_k^j$, which, in the case of vitreous silica, give $A_1 \simeq 0.7$ and $A_2 \simeq 0.8$.

The shape of the spectral density depends on the characteristics of the plane wave (wavevector and polarization) and on the atomic structure itself. In disor-

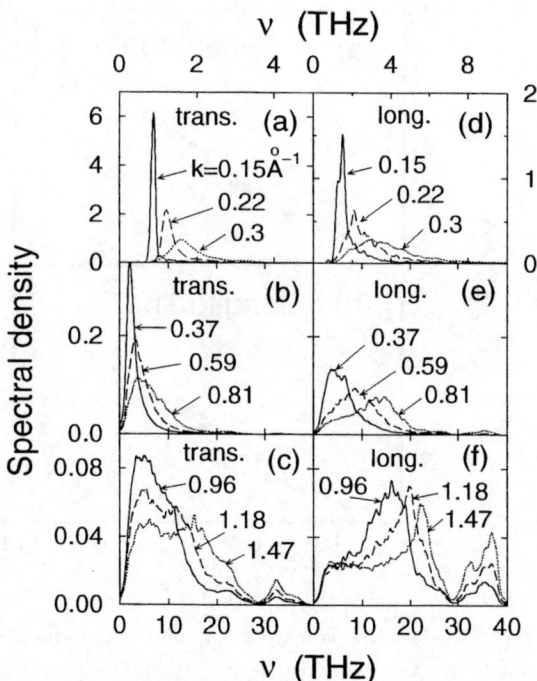

FIGURE 1. The spectral densities $|\overline{\alpha}_k^j|^2$ for transverse ((a), (b) and (c)) and longitudinal ((d), (e) and (f)) initial polarizations for different magnitudes, k, of the initial wavevector as shown in the figure.

dered structures and for small values of the wavevector magnitude ($ka \ll 1$), the spectral density both for longitudinal and transverse polarizations has the shape of a single pronounced peak (see Figs. 1(a) and 1(d)). The positions of these low-frequency peaks at $\nu_{t,l}$ are related to the wavevector magnitude according to the linear dispersion relation (see Fig. 2),

$$\nu_{t,l} \simeq c_{t,l} k/2\pi \ . \tag{22}$$

As seen from Fig. 2, the calculated dots in the low-frequency range lie on the straight lines plotted using experimental sound velocities ($c_t \simeq 37.5 \text{Å/ps}$ and $c_l \simeq 59 \text{Å/ps}$ [42]).

With an increase of the magnitude k of the wavevector, the peak-shaped spectral density shifts to higher frequencies and its width increases (see Figs. 1(b) and 1(e)). At large enough $k \geq 1 \text{Å}^{-1}$, the spectral density no longer consists of a single peak but rather resembles the vibrational density of states (VDOS) (see Figs. 1(c) and 1(f)), clearly showing the two frequency bands found in v-SiO$_2$ [40].

If the spectral densities are peak-shaped, two of their characteristics, the peak position and width, are normally used in order to describe the propagation of ex-

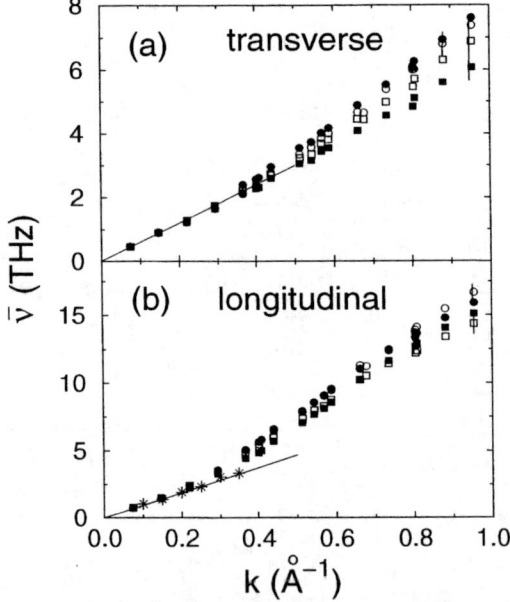

FIGURE 2. The dispersion laws for transverse (a) and longitudinal (b) polarizations of the initial plane-wave excitation. The solid circles and squares were obtained from the fit of the spectral densities by Lorentzians and the DHO model, respectively. The stars in (b) correspond to IXS data by Benassi et al [42]. The solid lines represent the long-wavelength limit characterized by the experimental sound velocities.

ternal plane-wave excitations [19,21,42,43]. The peak position is associated with the average frequency of the propagating excitation, while the peak width is associated with the decay time of the excitation. Indeed, if we look at the **k**-plane-wave component in the propagating excitation, $\mathbf{u}(t)$, its evolution with time is described by relation (7) at $\mathbf{k}' = \mathbf{k}$. The weight (amplitude) of this component, $a_\mathbf{k}(t)$, decays with time according to Eqs. (10) and (12). A rough estimate of the time dependence of $a_\mathbf{k}(t)$ can be obtained if we assume that $a_\mathbf{k}(t) \propto a_{\mathbf{k},c}(t)$, i.e. $a_\mathbf{k}(t)$ is approximately the back cosine Fourier transformation of the spectral density $\overline{\alpha}_\mathbf{k}^j \alpha_\mathbf{k}^j$ (see Eq. (12)). If the spectral density has the shape of a well-defined peak which can be fitted, say, by a Lorentzian, i.e.

$$f_L = \frac{1}{\pi} \frac{(\Gamma_\omega/2)}{(\omega - \overline{\omega}_k)^2 + (\Gamma_\omega/2)^2} , \qquad (23)$$

where the Lorentzian position $\overline{\omega}_k$ and full-width at half-maximum (FWHM) Γ_ω are the fitting parameters, then the back cosine Fourier transform of the function (23) is

$$a_k(t) \simeq A_2 \cos \overline{\omega}_k t \exp\{-\Gamma_\omega t/2\} , \qquad (24)$$

where we have actually used Eq. (23) to fit the spectral density $|\overline{\alpha}_\mathbf{k}^j|^2$ normalized to unity and then took into account the factor A_2 (see the beginning of the section). As clearly seen from Eq. (24), the decay of the **k**-plane-wave component can be characterized by the average radial frequency $\overline{\omega}_k$ and the inverse decay time

$$\tau_k^{-1} \simeq \Gamma_\omega(k)/2 = \pi\Gamma_\nu(k) , \qquad (25)$$

with $\Gamma_\nu(\text{THz}) = \Gamma_\omega/2\pi$.

In Refs. [42,21], the damped harmonic oscillator (DHO) model has been used to fit spectral densities, which gives similar values for the average frequency and width, if $(\Gamma_\omega)^2 \ll \overline{\omega}_k^2$. This inequality holds true in the region $k \leq 1 \text{Å}^{-1}$ and in particular in the IR regime around $k \sim 0.1 \text{Å}^{-1}$ (see below), where the spectral densities have a well-defined peak shape and fitting of the spectral densities by Lorentzian and/or DHO curves makes sense.

We have used fits both by the Lorentzian and DHO models to obtain the average frequency and decay time (not shown, see Ref. [46] for more detail) of the propagating plane-wave excitation as a function of the initial wavevector. The dependence of $\overline{\nu}_k = \overline{\omega}_k/2\pi$ vs. k shown in Fig. 2 can be associated with some sort of "dispersion law". Of course, the propagating plane-wave excitation cannot be characterized by only one wavevector (and single frequency) and instead consists of a packet of plane waves (see Eq. (6)) with different wavevectors (packet of eigenmodes characterized by different frequencies). We chose from the $\mathbf{k}'$-packet only one component characterized by the same wavevector as that of the initial plane wave and followed its time evolution. In that case, the dependencies $\overline{\nu}_k$ presented in Fig. 2 can be regarded as the dispersion laws for a single plane-wave component.

The experimental data for longitudinal external plane-wave excitations from IXS experiments [42], obtained by fitting the experimental curves with the DHO model, are shown by the stars in Fig. 2(b) and they agree well with our results.

Note that the dispersion laws for both branches are practically linear in the low-frequency (long-wavelength) regime for $\nu \leq 3\text{THz}$. Above this frequency, a sort of "fast-sound" behaviour is observed. The increase in the slope of $\bar{\nu}_k$ is related to the changes in the shape of the spectral densities. A shoulder on the high-frequency side of the spectral-density peak for the longitudinal branch starts to appear at $k \geq 0.3\text{Å}^{-1}$ ($\nu \geq 0.3\text{THz}$ - see Figs. 1(d),(e)). A similar transformation happens with the peak for the transverse branch at $k \geq 0.5\text{Å}^{-1}$.

B Disordered zig-zag chain

The other structural model we consider here is a toy model, namely a linear zig-zag chain on the plane. Toy models are very useful in studying atomic dynamics. Usually, scalar models are investigated because of their simplicity [34,44,45]. However, important effects related to mixing of modes of different polarizations [46] are missed in such models. That is why we consider below one of the simplest vectorial models, namely a zig-zag chain in the $x - y$ plane (see Fig. 3).

Consider a chain with 2 atoms in the unit cell: $\mathbf{r}_1^{(0)}\{x_1, y_1\}$ and $\mathbf{r}_2^{(0)}\{x_2, y_2\}$, so that the positions of the other atoms can be found as $\mathbf{r}_{i,1}^{(0)} = \mathbf{r}_1^{(0)} + i\mathbf{a}$ and $\mathbf{r}_{i,2}^{(0)} = \mathbf{r}_2^{(0)} + i\mathbf{a}$ with $i = 0, \pm 1, \ldots$, where $\mathbf{a} = \{a, 0\}$ is a unit-cell vector, so that the chain is directed along the x axis (see Fig. 3).

The nearest neighbours of different types (atoms 1 and atoms 2) are connected by springs with spring constants $\kappa_{1,i}^{(1)}$ in the same unit cell i and with $\kappa_{2,i}^{(1)}$ in different

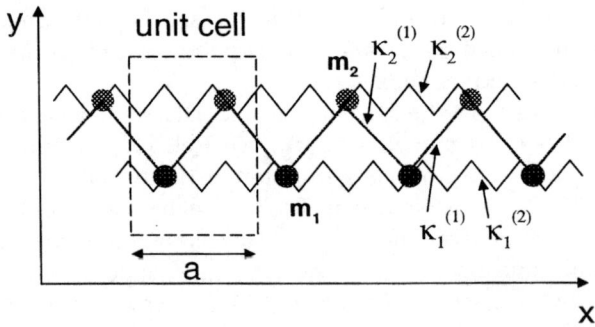

FIGURE 3. Zig-zag chain with two atoms in the unit cell.

unit cells (i and $i+1$), while the nearest neighbours of the same type are linked by springs with constants $\kappa_{1,i}^{(2)}$ for atoms of the first type (the lower row) and $\kappa_{2,i}^{(2)}$ for the atoms of the second type (upper row).

The potential energy then has the following expression:

$$V(\mathbf{r}_{1,1}, \mathbf{r}_{1,2}, \ldots) = \frac{1}{2} \sum_{i}^{N_{uc}} \{\kappa_{1,i}^{(1)} \left(|\mathbf{r}_{i,2} - \mathbf{r}_{i,1}| - |\mathbf{r}_{i,2}^{(0)} - \mathbf{r}_{i,1}^{(0)}|\right)^2$$
$$+ \kappa_{2,i}^{(1)} \left(|\mathbf{r}_{i+1,1} - \mathbf{r}_{i,2}| - |\mathbf{r}_{i+1,1}^{(0)} - \mathbf{r}_{i,2}^{(0)}|\right)^2$$
$$+ \sum_{j=1,2} \kappa_{j,i}^{(2)} \left(|\mathbf{r}_{i+1,j} - \mathbf{r}_{i,j}| - |\mathbf{r}_{i+1,j}^{(0)} - \mathbf{r}_{i,j}^{(0)}|\right)^2\}, \qquad (26)$$

with i numbering all unit cells.

Four branches occur in the dispersion relation: two acoustic (longitudinal and transverse) and two optic branches (see Fig. 4a). The analytical expression for the dispersion is available for the symmetric case ($m_1 = m_2 = m$; $\kappa_{1,i}^{(1)} = \kappa_{2,i}^{(1)} = \kappa^{(1)}$; $\kappa_{1,i}^{(2)} = \kappa_{2,i}^{(2)} = \kappa^{(2)}$; $x_1/a = y_1/a = 0$; $x_2/a = 0.5$):

$$\frac{(\omega(k))^2}{(\kappa^{(1)}/m)} = 1 + \frac{A}{2} \pm \left[1 + \frac{A^2}{4} \pm A\cos(ka/2)\right]^{1/2}, \qquad (27)$$

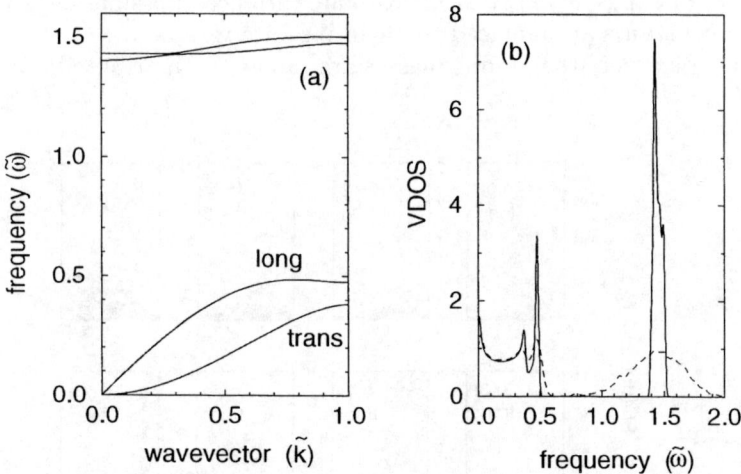

FIGURE 4. (a): Dispersion curves ($\tilde{\omega} = \omega\sqrt{m/\kappa^{(1)}}$ vs $\tilde{k} = k/(\pi/a)$) for the zig-zag chain model characterized by the following parameters: the equilibrium coordinate of atoms in the unit cell, $x_1/a = 0; y_1/a = 0, x_2/a = 0.4; y_2/a = 0.5$, ratio of force constants $\kappa^{(1)}/\kappa^{(2)} = 0.1$ ($\kappa_1^{(1,2)} = \kappa_2^{(1,2)}$), masses $m_1 = m_2 = m$ and the total number of atoms $N = 2000$. (b): The VDOS of the zig-zag chain model with the same set of parameters as in (a) (solid line) together with that for a disordered chain with fluctuations in force constants $\delta\kappa^{(1)}/\kappa^{(1)} = 0.3$, $\delta\kappa^{(2)}/\kappa^{(2)} = 0.3$ (dashed line).

with $A = 2(1 - \cos(ka))(\kappa^{(2)}/\kappa^{(1)})$. From this expression it is not difficult to obtain that in the low-frequency limit ($\omega \to 0$) for the longitudinal acoustic branch not surprisingly $\omega \propto k$, while for the transverse acoustic branch $\omega \propto k^2$. Such a k^2-dependence is typical for transverse vibrations of a linear chain and is related to the much smaller restoring force in the y-direction (compared to that in the x-direction) for long-wavelength vibrations because of the absence of the spring continuum in that direction.

The vibrational density of states $g(\omega)$ contains two bands characterized by typical van Hove singularities around the band boundaries (see Fig. 4b). The k^2-dependence of the transverse acoustic branch results in an $\omega^{-1/2}$ singularity of the VDOS as the frequency approaches zero, $\omega \to 0$.

We are interested mainly in disordered structures. Several different types of disorder can be introduced in the system:

(i) Positional disorder which is due to random positional vectors $\mathbf{r}_i^{(0)}$.

(ii) Force-constant disorder: κ_i are randomly distributed around their mean value, $\overline{\kappa}$,

$$\kappa_i = \overline{\kappa} + \delta\kappa \cdot \delta_{i,\mathrm{ran}}, \tag{28}$$

where $\delta\kappa$ is a typical width of the distribution and $\delta_{i,\mathrm{ran}}$ are random numbers distributed around zero with the density distribution, $\rho(\delta)$, e.g. the normal distribution, $\rho(\delta) = \exp\{-\delta^2/2\}/\sqrt{2\pi}$, with unit variance. Possible negative values of the spring constants are replaced by their absolute values.

(iii) Mass disorder: the atomic masses are randomly distributed, $m_i = \overline{m} + \delta m \cdot \delta_{i,\mathrm{ran}}$.

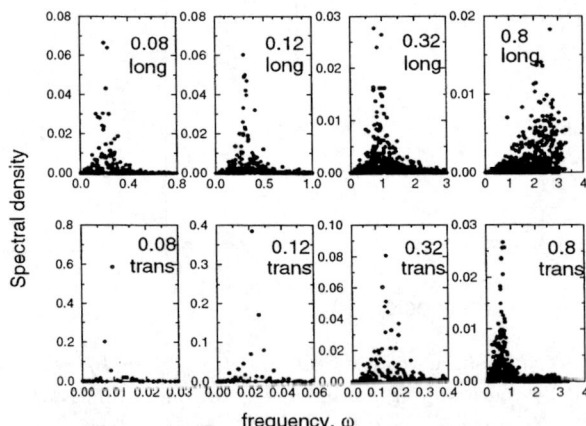

FIGURE 5. Spectral densities for a disordered zig-zag chain described by the following parameters: $m_1 = m_2 = 1, \delta m_1 = \delta m_2 = 0, \kappa^{(1)} = \kappa^{(2)} = 1, \delta\kappa^{(1)} = \delta\kappa^{(2)} = 1.0$, $x_1/a = y_1/a = 0, x_2/a = y_2/a = 0.5, a = 1$ and $N = 1000$. The wavevector magnitude, k, and polarizations (longitudinal or transverse) of the initial plane wave are marked in the figure.

In the case of force-constant and mass disorder (which are considered in what follows), the equilibrium positions of the atoms are not changed. This is quite convenient and an ideal (crystalline) linear chain can be treated as the crystalline counterpart of the disordered chain.

The VDOS of a disordered zig-zag chain is similar to that for the crystalline chain (see Fig. 4b) except all sharp features (excluding the region around zero) are washed out (band tails appear). We should also note that extra states (relative to the Debye spectrum) appear in the low-frequency regime ($\omega \leq c_{t,l}(\pi/a)$, with $c_{t,l}$ being the transverse or longitudinal sound velocity. These extra states, characterized by the change in the VDOS, $\Delta g(\omega) = g(\omega) - g_{\text{cryst}}(\omega)$, are called the boson peak [40].

The spectral densities for the disordered zig-zag chain, being of particular interest, are calculated according to Eq. (3) and presented in Fig. 5.

As expected in the long-wavelength limit, the spectral density has the shape of a peak. The position of the peak can be approximately found from the dispersion for acoustic branches (see below). The width of the peak increases with increasing magnitude of the wavevector of the initial plane wave. This corresponds to more intense scattering of the plane waves by disorder.

IV ANALYSIS OF THE FINAL STATE IN MOMENTUM SPACE

As is well known, a scattering process can be investigated by analysis of the final state. First, we consider the final state of a single **k**-plane-wave component characterized by the same wavevector as the initial one. The phase of this wave has a random value and is not an informative characteristic. The important quantity is the amplitude of the wave, or more precisely its squared average value, defined by Eq. (18) with $\mathbf{k}' = \mathbf{k}$,

$$\overline{a_{\mathbf{k}}^2} = \frac{1}{2} \left\{ \frac{\sum_j |\overline{\alpha_{\mathbf{k}}^j}|^2 |\alpha_{\mathbf{k},s}^j|^2}{(\langle \mathbf{w}_{\mathbf{k},s}^2 \rangle)^2} + \frac{\sum_j |\overline{\alpha_{\mathbf{k}}^j}|^2 |\alpha_{\mathbf{k},c}^j|^2}{(\langle \mathbf{w}_{\mathbf{k},c}^2 \rangle)^2} \right\} \simeq \frac{\sum_j |\overline{\alpha_{\mathbf{k}}^j}|^2 |\alpha_{\mathbf{k}}^j|^2}{(\langle \mathbf{w}_{\mathbf{k}}^2 \rangle)^2}. \quad (29)$$

This value can be easily estimated for a peak-shaped spectral density of width Γ. Indeed, the number of eigenmodes contributing to an initial plane wave is of order $3N \cdot (\Gamma/D)$, where D is the width of the whole vibrational spectrum ($\simeq 40$THz in the case of vitreous silica). Then we can easily evaluate from the normalization conditions Eqs. (5), (17) for the spectral densities the average value of the spectral density in the peak region, $|\overline{\alpha_{\mathbf{k}}^j}|^2 \sim |\alpha_{\mathbf{k},s}^j|^2 \sim (D/\Gamma) \cdot (1/3N)$, and obtain the following estimate for $\overline{a_{\mathbf{k}}^2}$,

$$\overline{a_{\mathbf{k}}^2} \sim \frac{D}{\Gamma} \cdot \frac{1}{3N}, \quad (30)$$

where we have taken into account that $\langle \mathbf{w}_{\mathbf{k}}^2 \rangle \sim 1$ according to Eq. (5). The factor D/Γ in relation (30) shows that the averaged squared amplitude is inversely proportional to the number of initially excited modes and not to all the modes. The factor D/Γ in the Ioffe-Regel (IR) region is much larger than unity; $D/\Gamma \sim 10^2$ for $\Gamma \sim \nu_{\rm IR}/\pi \sim 0.3$THz. Therefore we can say that the **k**-plane-wave component around and below the IR regime is not damped (the squared average amplitude is not of order $1/3N$) but rather is attenuated (scattered), weakly (strongly) below (beyond) the IR limit as discussed below. The **k**-plane-wave component is damped at $k \geq k_* \simeq 1\mathring{A}^{-1}$ when the peak width becomes comparable to the full spectral width, $\Gamma \sim D$.

The analysis of all **k**$'$-plane-wave components in the final state, in particular the distribution (18) of their weights, allows the scattering mechanism to be clarified. Let us consider an initial plane wave characterized by the wavevector **k** and polarization $\hat{\bf n}$. This wave is scattered with time into different plane waves characterized by wavevectors **k**$'$ and polarizations $\hat{\bf n}'$, which do not necessarily coincide with the initial polarization. We would like to know the weights of all plane-wave components in the final state as a function of the wavevector magnitude k'. The distributions of the transverse and longitudinal plane waves, $\rho({\bf k}', \hat{\bf n}'_{\rm t}|{\bf k}, \hat{\bf n})$ and $\rho({\bf k}', \hat{\bf n}'_{\rm l}|{\bf k}, \hat{\bf n})$ (see Eq. (19)), and the distribution averaged over polarizations, $\rho_{\rm av}({\bf k}'|{\bf k}, \hat{\bf n})$ (see Eq. (20)), for both transverse $\hat{\bf n}_{\rm t}$ and longitudinal $\hat{\bf n}_{\rm l}$ polarizations of the initial plane-wave excitation are of particular interest. These distributions depend only on the spectral densities $|\overline{\alpha}_{\bf k}^j|^2$, $|\underline{\alpha}_{\bf k'}^j|^2$ and the vibrational spectrum itself, and can be easily calculated numerically for different k.

A Vitreous silica

The results of such calculations for vitreous silica are presented in Fig. 6. The upper (lower) row describes the scattering of initial longitudinal (transverse) plane waves, characterized by different wavevector magnitudes, into transverse and longitudinal plane waves and also the distribution of the weights averaged over the polarization in the final state.

First, we consider scattering of a longitudinal initial wave (the upper row in Fig. 6). The weight distributions, $\rho({\bf k}', \hat{\bf n}'_{\rm l}|{\bf k}, \hat{\bf n}_{\rm l})$ and $\rho({\bf k}', \hat{\bf n}'_{\rm t}|{\bf k}, \hat{\bf n}_{\rm l})$, characterize the scattering of the longitudinal wave to a longitudinal wave, the $\{l \to l\}$ channel, and of the longitudinal wave to a transverse wave, the $\{l \to t\}$ channel, respectively. As follows from Fig. 6, these distributions are peak shaped but the positions of the peaks are different. The distribution for the $\{l \to l\}$ channel has a maximum around $k'_{\rm ll} \simeq k_{\rm l} \equiv k$ (or maybe a bit below the initial wavevector), while the distribution for the $\{l \to t\}$ channel is mainly concentrated at a higher wavevector value, $k'_{\rm lt} > k_{\rm l}$. The distribution, $\rho_{\rm av}({\bf k}'|{\bf k}, \hat{\bf n}_{\rm l})$, averaged over polarizations of the final state is a sum of double the distribution for the $\{l \to t\}$ channel plus the distribution for the $\{l \to l\}$ channel. If the peaks related to the individual channels and constituting the average distribution are narrow enough, then the distribution function $\rho_{\rm av}({\bf k}'|{\bf k}, \hat{\bf n}_{\rm l})$ is doubly peaked (not clearly seen in Fig. 6). If the peaks are too wide, then $\rho_{\rm av}({\bf k}'|{\bf k}, \hat{\bf n}_{\rm l})$ looks like a single wide peak (see Fig. 6) with a maximum position $k'_{\rm l,av}$ close to $k'_{\rm lt}$.

Such a shape of the distributions of the weights of plane waves in the final state can be qualitatively understood in the following way. The distribution function $\rho(\mathbf{k}', \hat{\mathbf{n}}'_t | \mathbf{k}, \hat{\mathbf{n}}_l)$ of the transverse waves is an integral (sum in the case of a finite-size model) of the product of two spectral densities, $|\overline{\alpha}^j_{\mathbf{k},\hat{\mathbf{n}}_l}|^2$ for longitudinal and $|\alpha^j_{\mathbf{k}',\hat{\mathbf{n}}'_t}|^2$ for transverse polarization. Around the IR region and below it, these peak-shaped spectral densities have maxima at $\nu_l \simeq c_l k/2\pi$ and $\nu'_t \simeq c_t k'/2\pi$, respectively, which generally do not coincide with each other. Therefore, the distribution $\rho(\mathbf{k}', \hat{\mathbf{n}}'_t | \mathbf{k}, \hat{\mathbf{n}}_l)$ has a maximum around k'_{lt} satisfying the equation, $\nu_l \simeq c_l k/2\pi \simeq c_t k'_{lt}/2\pi \simeq \nu'_t$, i.e.

$$k'_{lt} \simeq c_l k/c_t , \qquad (31)$$

which is obviously greater than the wavevector of the initial longitudinal wave.

The distribution of longitudinal waves for the $\{l \rightarrow l\}$ channel can be analysed in a similar manner. The main difference from the $\{l \rightarrow t\}$ channel is that the spectral density of the longitudinal plane wave in the final state coincides with the spectral density of the initial longitudinal plane wave at approximately the same wavevector magnitude as for the initial wave,

$$k'_{ll} \simeq k . \qquad (32)$$

Actually, the value k'_{ll} should be slightly shifted to lower values, because the height of the peak for the spectral density $|\alpha^j_{\mathbf{k}'}|^2$ increases with decreasing k' and the

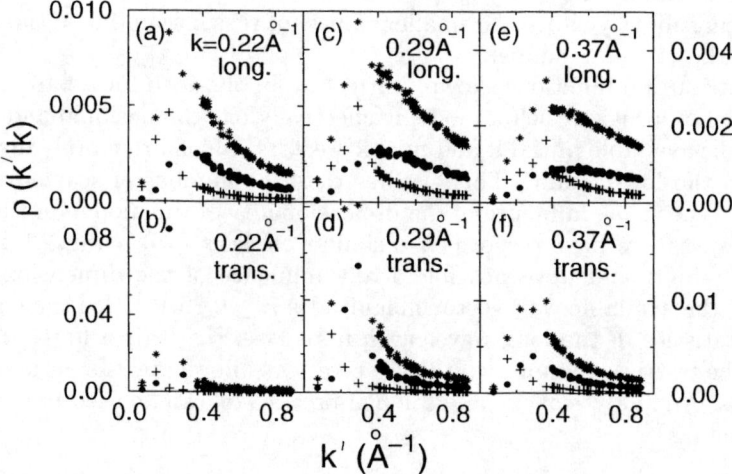

FIGURE 6. The distribution functions $\rho(\mathbf{k}', \hat{\mathbf{n}}'_t | \mathbf{k}, \hat{\mathbf{n}})$ (circles), $\rho(\mathbf{k}', \hat{\mathbf{n}}'_l | \mathbf{k}, \hat{\mathbf{n}})$ (pluses) and $\rho_{av}(\mathbf{k}' | \mathbf{k}, \hat{\mathbf{n}})$ (stars) for longitudinal ((a), (c) and (e)) and transverse ((b), (d) and (f)) initial polarizations of plane waves characterized by different initial wavevector magnitudes k for a structural model of v-SiO$_2$.

maximum of the product of the spectral densities is reached in the low-frequency tail of the spectral density for the initial plane wave.

The scattering of an initially transverse plane wave occurs similarly. In particular, the conclusion that the average frequency, ν', of the majority of the plane-wave components comprising the final state coincides with the average frequency, ν, of the initial plane wave,

$$\nu' \simeq \nu , \qquad (33)$$

holds true independently of the polarization of the initial plane-wave excitation. Therefore, we can roughly say that the disorder-induced scattering of the plane wave is approximately "elastic" (on average). This is not an absolutely precise conclusion because, first, the plane-wave components are distributed in frequency (composed of eigenmodes having different frequencies) in the initial and final states and, second, even the maximum of the distribution in the final state is slightly shifted to lower frequencies as compared to the initial one, as discussed above.

In the case of the scattering of an initial transverse plane wave, two channels are available: $\{t \to l\}$ and $\{t \to t\}$. The distribution functions, $\rho(\mathbf{k}', \hat{\mathbf{n}}'_l | \mathbf{k}, \hat{\mathbf{n}}_t)$ and $\rho(\mathbf{k}', \hat{\mathbf{n}}'_t | \mathbf{k}, \hat{\mathbf{n}}_t)$, of the weights of plane waves in the final state for these channels have peaks located around the following values:

$$k'_{tl} \simeq c_t k / c_l \quad \text{and} \quad k'_{tt} \simeq k . \qquad (34)$$

As follows from Eq. (34) and Figs. 6(b),(d),(f), the peak for longitudinal waves lies below the initial k, while for transverse waves the peak approximately coincides with k, being slightly shifted to smaller values for reasons similar to those discussed above for the $\{l \to l\}$ channel.

The distribution functions shown in Fig. 6 were obtained for a bar-shaped structural model of v-SiO$_2$. Such a model is effectively one-dimensional and has restrictions for the available initial $\mathbf{k}$ and final $\mathbf{k}'$ vectors, which are mainly directed along the bar in the low-k limit. This also restricts the number of scattering channels. In order to check the influence of the dimensionality of the model on the scattering of plane waves, we have performed a similar analysis for a cubic (3-dimensional) model of v-SiO$_2$ and have not found any influence of the dimensionality of the model for the available wavevector magnitudes $k \geq 0.22 \text{Å}^{-1}$ (for the cubic model).

Poor statistics in the long-wavelength limit (see Fig. 6) is a finite-size effect related to the restricted number of the wave-vectors allowed by the periodic boundary conditions. An analytical extrapolation approach [46] can be used to overcome such a shortcoming.

B Zig-zag chain

Another possible way to overcome the disadvantages of finite-size 3-D numerical models is to analyse low-dimensional models. Much lower wavevectors

$k \geq k_{\min}^{(d)} = 2\pi/N^{1/d}a$ are available, for example, in one-dimensional ($d = 1$) models as compared to the 3-D case, and the acoustic spectrum appears to be be much more dense. In order to check and support the analytical and numerical approaches presented above for the 3-D case, we have performed numerical experiments for a disordered 1-D zig-zag chain (see Sec. III B) and calculated the distribution function $\rho(\mathbf{k}', \hat{\mathbf{n}}'|\mathbf{k}, \hat{\mathbf{n}})$ for it.

Our main purpose here is to calculate the distribution function $\rho(\mathbf{k}', \hat{\mathbf{n}}'|\mathbf{k}, \hat{\mathbf{n}})$ characterizing the scattering of a plane-wave excitation. First, we have calculated this distribution function for the crystalline counterpart ($\delta \kappa_i = 0$) and not surprisingly we found for $k \leq \pi/a$ only $\{t \to t\}$ and $\{l \to l\}$ channels (see the lower row in Fig. 7 marking the peaks at $k'_{tt} = k$ and $k'_{ll} = k$ for the $\{t \to t\}$ and $\{l \to l\}$ channels, respectively). Disorder changes the situation dramatically and gives rise to the occurrence of $\{t \to l\}$ and $\{l \to t\}$ channels (see the upper row in Fig. 7), in complete agreement with the results of the k-analysis given in Fig. 6 for the case of v-SiO$_2$. The positions of the additional peaks, at k'_{tl} ($\{t \to l\}$ channel), and k'_{lt} ($\{l \to t\}$ channel) can be obtained from the dispersion laws for the crystalline zig-zag chain by solving the equations:

$$\omega_t(k) = \omega_l(k'_{tl}) , \qquad (35)$$

and

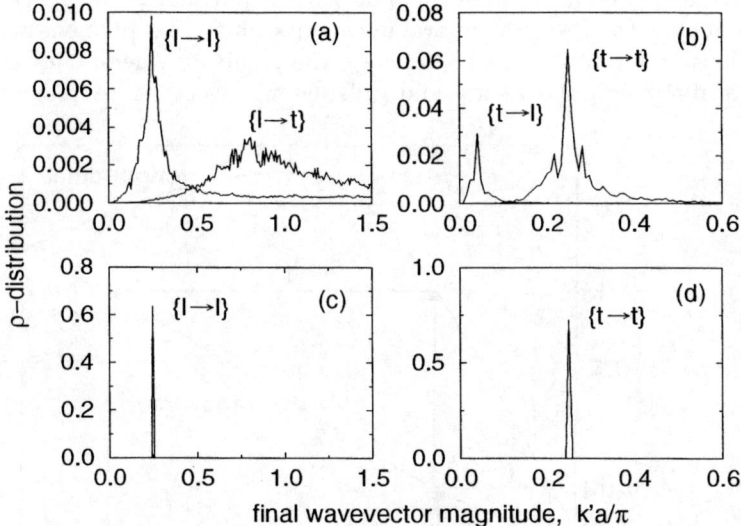

FIGURE 7. The distribution function $\rho(\mathbf{k}', \hat{\mathbf{n}}'|\mathbf{k}, \hat{\mathbf{n}})$ for different scattering channels (as marked in the figure) for a disordered ((a) and (b)), $\delta\kappa^{(1)} = \delta\kappa^{(2)} = 1$, $\delta m_1 = \delta m_2 = 0$) and an ordered ((c) and (d)) zig-zag chain characterized by the following parameters: $m_1 = m_2 = 1$, $\kappa^{(1)} = \kappa^{(2)} = 1$, $x_1/a = y_1/a = 0$, $x_2/a = y_2/a = 0.5$. The initial wavevector magnitude $ka/\pi = 0.25$.

$$\omega_l(k) = \omega_t(k'_{lt}) , \qquad (36)$$

respectively (see arrows in Fig. 8). The width of the peaks increases with increasing disorder. We have also found a similar shape of the distribution function ρ (for four channels) for all wavevectors $k \leq \pi/a$ with the corresponding ω_t and ω_l lying in the range of the dense spectrum.

V SCATTERING MECHANISM

From numerical calculations for both the 1-D zig-zag chain model and the 3-D model of v-SiO$_2$, we have found that plane waves scatter not only to modes of approximately the same wavelength (and polarization) but also to modes of different wavelength (and polarization) but of similar frequency. The reason for such scattering is a natural question.

In our simulations on zig-zag chains, we have found the $\{t \to l\}$ and $\{l \to t\}$ scattering channels even for models not showing an appreciable increase of the VDOS in the low-frequency regime. Hence, the appearance of the $\{t \to l\}$ and $\{l \to t\}$ channels should be explained in terms of existing transverse and longitudinal acoustic waves. Indeed, in the crystal, transverse and longitudinal acoustic phonons are orthogonal to each other, and hence transverse (longitudinal) plane waves cannot be scattered into longitudinal (transverse) plane waves (as we checked numerically; see the lower row in Fig. 7). Disorder leads to changes in the interaction energy, with the result that an acoustic phonon with a particular energy can couple to other phonons with closely comparable energies, including phonons with different polarizations and wavevectors. Therefore, the resulting eigenmodes contain components of different polarization and different wavevectors. A plane wave is not

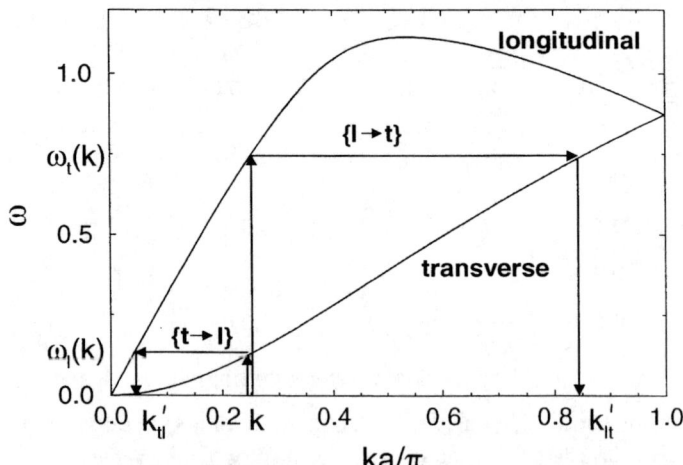

FIGURE 8. The acoustic branches for an ordered zig-zag chain characterized by the same parameters as in Fig. 7. The construction used to estimate values of k'_{tl} and k'_{lt} is shown.

an eigenmode in the disordered stucture and it is scattered into eigenmodes of approximately the same energy as that of the plane wave. These eigenmodes contain both transverse and longitudinal components and therefore an original plane wave, independent of its polarization, is scattered into both transverese and longitudinal plane waves. This gives a qualitative explanation of the existence of $\{t \to l\}$ and $\{l \to t\}$ scattering channels.

In the 3-D case, the situation can be more complicated. Apart from the scattering mechanism due to the disorder-induced mixing of transverse and longitudinal plane waves discussed above, extra states (comprising the Boson peak) relative to the Debye spectrum (e.g. optic modes pushed down by disorder (see [46,47])) could participate in the hybridization between plane waves with different polarization.

REFERENCES

1. *Scattering and Localization of Classical Waves in Random Media*, edited by Sheng P. (World Scientific, Singapore, 1990).
2. Sheng P., *Introduction to Wave Scattering, Localization, and Mesoscopic Phenomena* (Academic Press, San Diego, 1995).
3. Mott N.F. and Davis E.A., *Electronic Processes in Non-Crystalline Materials*, 2nd ed. (Clarendon Press, Oxford, 1979).
4. John S., Sompolinsky H., and Stephen M.J., Phys.Rev. B **27**, 5592 (1983).
5. Aharony A., Alexander S., Entin-Wohlman O., and Orbach R., Phys. Rev. Lett. **58**, 132 (1987).
6. Alexander S., Phys.Rev. B **40**, 7953 (1989).
7. Ioffe A.F. and Regel A.R., Prog. Semicond. **4**, 237 (1960).
8. Sheng P., Zhou M., and Zhang Z.-Q., Phys. Rev. Lett. **72**, 234 (1994).
9. Anglaret E., Hasmy A., Courtens E., Pelous J., and Vacher R., Europhys. Lett. **28**, 591 (1994).
10. Grest G.S., Nagel S.R., and Rahman A., Phys. Rev. Lett. **49**, 1271 (1982).
11. Hafner J. and Crajci M., J.Phys.: Cond.Matter,**6**, 4631 (1994).
12. Schober H.R. and Laird B.B., Phys. Rev. B **44**, 6746 (1991).
13. Gurevich V.L., Parshin D.A, Pelous J., and Schober H.R., Phys.Rev., B **48**, 16318 (1993).
14. Mazzacurati V., Ruocco G., and Sampoli M., Europhys.Lett., **34**, 681 (1996).
15. Sampoli M., Benassi P., Dell'Anna R., Mazzacurati V. and Ruocco G., Phil.Mag., B **77**, 473 (1998).
16. Taraskin S.N. and Elliott S.R., Europhys. Lett., **39**, 37 (1997).
17. Ribeiro M.C.C., Wilson M., and Madden P.A., J.Chem.Phys., **108**, 9027 (1998).
18. Allen P.B. and Feldman J.L., Phys. Rev. B **48**, 12581 (1993),
19. Feldman J.L., Kluge M.D., Allen P.B., and Wooten F., Phys. Rev. B **48**, 12589 (1993).
20. Fabian J. and Allen P.B., Phys. Rev. Lett., **77**, 3839 (1996).
21. Dell'Anna R., Ruocco G., Sampoli M., and Viliani G., Phys. Rev. Lett. **80**, 1236 (1998).

22. Feldman J.L., Allen P.B., and Bickham S.R., Phys. Rev. B **59**, 3551 (1999).
23. Wischnewski A., Buchenau U., Dianoux A.J., Kamitakahara W.A., and Zarestky J.L., Phys. Rev. B **57**, 2663 (1998).
24. Sheng P. and Zhou M., Science **253**, 539 (1991).
25. Buchenau U., Galperin Yu.M., Gurevich V.L., Parshin D.A., Ramos M.A. and Schober H.R., Phys.Rev., B **46**, 2798 (1992).
26. Gaganidze E., Konig R., Esquinazi P., Zimmer K., and Burin A., Phys. Rev. Lett. **79**, 5038 (1997).
27. Vacher R., Pelous J., and Courtens E., Phys. Rev. B **56**, R481 (1997).
28. Horbach J., Kob W. and Binder K., J.Non-Cryst.Sol., **235**, 320 (1998).
29. Maradudin A.A., Montroll E.W., Weiss G.H., and Ipatova I.P., *Theory of Lattice Dynamics in the Harmonic Approximation* (Acad. Press, N.Y., 1971).
30. Leibfried G. and Breuer N., *Point Defects in Metals I. Introduction to the theory* (Springer, Berlin, 1978).
31. Dederichs P.H. and Zeller R., in *Point Defects in Metals II. Dynamical Properties and Diffusion Controlled Reactions* (Springer, Berlin, 1980).
32. Klinger M.I., Phys.Rep., **165**, 275 (1988).
33. Galperin Yu.M., Karpov V.G., and Kozub V.I., Adv.Phys. **38**, 669 (1989).
34. Schirmacher W., Diezemann G., and Ganter C., Phys. Rev. Lett. **81**, 136 (1998).
35. Elliott S.R., *Physics of Amorphous Materials* 2nd Edn. (Longman, N.Y., 1990).
36. Hansen J.-P. and McDonald I.R., *Theory of Simple Liquids* 2-nd Ed. (Academic Press, London, 1990).
37. van Beest B.W.H., Kramer G.J., and van Santen R.A., Phys.Rev.Lett., **64**, 1955 (1990).
38. Guissani Y. and Guillot B., J.Chem.Phys., **104**, 7633 (1995).
39. Vollmayr K., Kob W., and Binder K., Phys.Rev., **B54**, 15808 (1996).
40. Taraskin S.N. and Elliott S.R., Phys.Rev., B56, 8605 (1997).
41. Fernandez-Perera R., Bermejo F.J., and Enciso E., Phys. Rev., B **53**, 6215 (1996).
42. Benassi P., Krisch M., Masciovecchio C., Mazzacurati V., Monaco G., Ruocco G., Sette F., and Verbeni R., Phys. Rev. Lett. **77**, 3835 (1996).
43. Foret M., Courtens E., Vacher R., and Suck J.-B., Phys. Rev. Lett. **77**, 3831 (1996).
44. Nakayama T., Phys. Rev. Lett. **80**, 1244 (1998).
45. Parshin D.A. and Schober H.R., Phys. Rev., B **57**, 10232 (1998).
46. Taraskin S.N. and Elliott S.R., J.Phys.: Cond.Matter, **11**, A219 (1999).
47. Taraskin S.N. and Elliott S.R., Phys.Rev., B, **59**, 8572 (1999) .

Molecular dynamics in amorphous solids and liquids

H.R. Schober

Institut für Festkörperforschung, Forschungszentrum Jülich, D-52425 Jülich, Germany

Abstract. Three different types of glasses, a mono-atomic soft sphere glass, amorphous selenium and amorphous CuZr, have been studied by molecular dynamics simulations. The vibrational spectrum at low temperatures shows excess vibrations which form the so called boson peak. These can be understood as resonant modes (low frequency quasi localized modes). These modes interact strongly with the sound waves and at the boson peak frequency these become over-damped and the different modes are completely intermixed. In a basis of sound waves and low frequency local modes, the latter can be described by the soft potential model. These modes are collective motions of chains of around 20 atoms. The same chain structures are observed in the relaxations in the glass. With increasing temperature both the total jump length and the number of participating atoms increases. In the melt of CuZr, additionally to this collective motion, jumps of Cu atoms into nearest neighbor sites are observed.

I INTRODUCTION

Glasses and amorphous solids which we will not distinguish here differ from crystalline solids by the absence of long-range order, whereas they show short-range order. In many aspects glasses are similar to their crystalline counterparts. Usually their specific weight is somewhat lower and their elastic constants somewhat smaller than in a crystal. Also their vibrational spectrum is similar to the one of the ordered phase. However, taking a closer look differences of the vibrational properties are noted. At low temperatures one finds well-defined sound waves in both states. At higher frequencies the vibrations in glasses can no longer be described by extended phonons. Because of the strong scattering by disorder, the mean free path would be shorter than the wave length. The vibration spectra of the glasses are smeared but resemble the ones of the crystals.

The most obvious differences between the dynamics of crystals and glasses are observed at low temperatures [1,2]. In an electrically non-conducting crystal, such as SiO_2, the specific heat increases with temperature as T^3. In a glass this Debye contribution from the sound waves is slightly higher due to the reduced elastic constants (sound velocities) and, more important, additional contributions are observed. At the lowest temperatures one finds an additional contribution $\propto T$ to the specific heat, which is attributed to tunneling systems. Above approximately 2K another contribution, $\propto T^5$, dominates, which stems from low frequency vibrations. These additional vibrational modes become particularly obvious when one plots the low-frequency part of the spectrum divided by the squared frequency. In this representation the contribution from the sound waves, corresponding to the Debye spectrum, is a constant. The contribution $\propto T^5$ to the specific heat corresponds to an increase $\propto \nu^2$ of $g(\nu)/\nu^2$. Typical for glasses is the maximum in $g(\nu)/\nu^2$, the boson peak.

Additionally to sound waves, tunneling and soft vibrations, one observes in glasses aperiodic motions, relaxations. All these excitations are found in all types of glasses: metallic glasses, network glasses (e.g. SiO_2) and polymer glasses. There are indications that the different excitations have a common origin. They are localized to several atoms or molecules. Whereas the low temperature properties of the glasses are nearly universal, at higher frequencies the structural and chemical differences become important.

During the last years several models have been developed to explain the low temperature properties of glasses. The behavior below 1K is described by the tunneling model [3,4]. It was extended by the soft potential model in order to describe also local vibrations and relaxations [5]. Fitting the parameters of the model to the experimental data one finds that 20-100 atoms/molecules participate in the local excitations [6]. Such a large number of atoms, which participate in a quantum mechanical motion (tunneling), is of course unusual.

However, models cannot show the microscopic origin of the excitations. Also the experimental information is not sufficiently detailed. Therefore one tries to simulate glasses. Nature offers one way in colloidal suspensions which one can understand as a kind of glass with large atoms (100Å-1000Å). A different way is computer simulation which can show its strength in this respect.

II COMPUTER SIMULATION

An introduction into the techniques of computer simulation is not possible here. I refer to the literature, e.g. [7,8]. We restrict ourselves here to the classical molecular dynamics method. There one solves for a system of N particles (atoms) the classical equation of motion

$$m_j \ddot{\mathbf{R}}_j = -\nabla_j V(\{\mathbf{R}\}) = \mathbf{F}_j. \tag{1}$$

$\mathbf{R}_j(t)$ is the position of the particle j, m_j is its mass, and V the total potential energy of the system. This equation of motion is solved numerically as function of time t. This means one discretizes time into intervals of lengths Δt. The time-step of the simulation should not exceed about 1/20 of the maximal frequency. The typical timestep is, therefore, of order of a few fs. In one million time-steps one describes a time-span of just a few ns. If one quenches in that time from 1000 K to $T = 0$ one gets a quench rates $\Delta T \approx 10^{12}$ K/s. Such rates are faster by orders of magnitude than can be achieved in usual experiments, except special radiation experiments. It is therefore difficult to investigate with the computer differences of glasses caused by different production techniques. On the other hand, one can produce glasses on the computer which, due to their fast tendency to crystallize, cannot easily be produced experimentally, e.g. mono-atomic metallic glasses.

For short-range interactions computer time increases linearly with the product of particle number and time. Therefore one tries to optimize $N \times t$ for each problem. Typical values are $N \approx 1000$. This relatively small number of atoms means that nearly all atoms are near the surface. To reduce this effect one usually takes periodic boundary conditions, i.e. one simulates a crystal with a unit cell of N atoms. Pressure and temperature oscillate and cannot be determined exactly. A further problem of all simulations is the huge amount of data.

Here we deal mainly with simulations of typical glass properties. These have been investigated by a number of groups for different substances [9–11]. In the following we will mainly concentrate on work done in Jülich [12–18] and references therin.

Before one starts a computer simulation, one should think whether the investigated quantity can be determined with sufficient accuracy with the available computer power. For the problem at hand, a direct search for tunneling systems does not look promising. According to experiment there are $10^{-6} - 10^{-5}$ tunnel systems/atom. That means one would have to simulate about a million atoms to find about ten tunneling systems, and this would not allow to obtain statistically relevant distributions. A different type of simulation by Heuer and Silbey [19] gives results on distributions of tunneling systems by extrapolation. A direct simulation of soft local vibrations at low temperatures on the other hand is possible. From experiment one expects $10^{-3} - 10^{-2}$ modes/atom. A further advantage is that vibrations can, to a good approximation, be treated classically. Also relaxations at finite temperatures and diffusion in the liquid should be observable.

Essential to any MD simulation is the choice of interaction potential. If one wants to calculate "typical" properties one chooses simple potentials such as Lennard Jones or hard and soft spheres. If one wants to calculate material specific properties or do a quantitative comparison with experiment much more elaborate interaction potentials are needed. The form of these potentials depends on the class of material, e.g. metal or covalently bonded network. The parameters are then fixed to as many as possible available experimental data and in some cases to results of electronic calculations.

The results presented in the following are derived with three different interaction potentials, each corresponding to a different class of material. As a simple model system we studied a soft sphere glass (SSG), described by the potential

$$V(r) = \epsilon \left(\frac{\sigma}{r}\right)^6 + A\left(\frac{r}{\sigma}\right)^4 + B. \tag{2}$$

To simplify the simulation the potential is cut off at $r/\sigma = 3.0$ and shifted by a polynomial with $A = 2.54 \times 10^{-5}\epsilon$ and $B = -3.43 \times 10^{-3}\epsilon$. The interaction is purely repulsive and the calculations are done with a fixed atomic density, $\rho/\sigma^3 = 1$ and periodic boundary conditions. The configurations were obtained by a quench from the liquid to $T = 0$ K. From the pair correlation one finds a nearest neighbor distance of around 1.1σ. For more details see [12,15]. Eq. 2 gives basically a one parameter potential and we set $m = \epsilon = \sigma = 1$. This interaction is much softer than the $1/r^{12}$ of the Lennard Jones potential and corresponds roughly to alkali metals. Due to the strong tendency to crystallization this model system can be used for studies at low temperatures or in the liquid but not in the temperature range of the glass transition.

As an example of a good mono-atomic glass former we treated Se. The interatomic interaction was of the Stillinger-Weber type including short range three-body terms [20]. Se predominantly forms covalent bonds to two neighbors. The glass is formed from chains and rings and the coordination number is close to two. System sizes of up to 1470 atoms were used. The configurations were monitored for a few ns. Hopping processes were investigated for temperatures below up to T_g.

As an example for a binary metallic glass we investigate Cu-Zr described by a modified embedded atom model [21]. The configurations consist of 667 Zr and 333 Cu atoms, were quenched from the melt with a rate of 10^{12} K/s and subsequently aged.

III VIBRATIONS IN GLASSES

The vibrations in a glass at low temperatures can be most easily studied in the harmonic approximation. For this we quench samples of up to 5488 atoms of SSG (Eq. 2) to $T = 0$ K. The atomic configuration will then correspond to a minimum of the potential energy and we can expand the energy in terms of the displacements from this minimum energy. The quadratic term defines a dynamical matrix:

$$D^{ij}_{\alpha\beta} = \frac{1}{\sqrt{m_i m_j}} \frac{\partial^2 E_{pot}(\{\mathbf{R}\})}{\partial R_{i,\alpha}\, \partial R_{j,\beta}}. \qquad (3)$$

Here we denote by i,j the atoms and by α,β the Cartesian coordinates. The atomic masses m_i will, for simplicity, set to unity in the following. For glasses of up to a few thousand atoms this dynamical matrix can be diagonalized on larger computer using standard library routines. For larger systems, due to memory limitations, sparse matrix methods have to be employed. Diagonalization gives in harmonic approximation the frequencies of the eigenmodes of vibration and their eigenvectors, i.e. their spatial structure. Fig. 1 shows the spectrum of all modes as well as the Debye spectrum calculated from the elastic constants. The area between the two curves corresponds to the fraction of excess low frequency modes modes, typical for the glassy structure. In a plot of $Z(\nu)/\nu^2$ one finds a maximum around $\nu = 0.1$, the boson peak. The spectrum has the typical shape for a mono-atomic glass. In average it is similar to the one of the fcc crystal. The sharp structures, present in the spectra of crystals, and the van Hove singularities are smeared out. The small indentation at $\nu \approx 1.5$ vanishes for harder glasses.

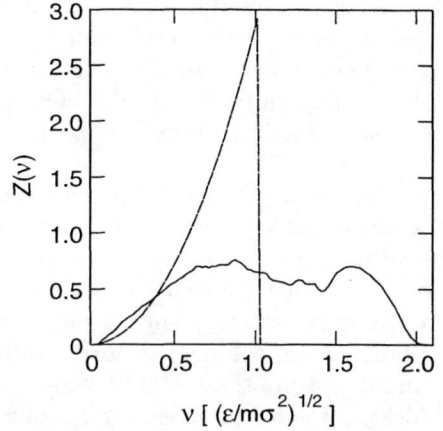

FIGURE 1. Configurationally averaged vibrational density of states of the soft sphere glass (solid line) and Debye spectrum (dashed line) [15].

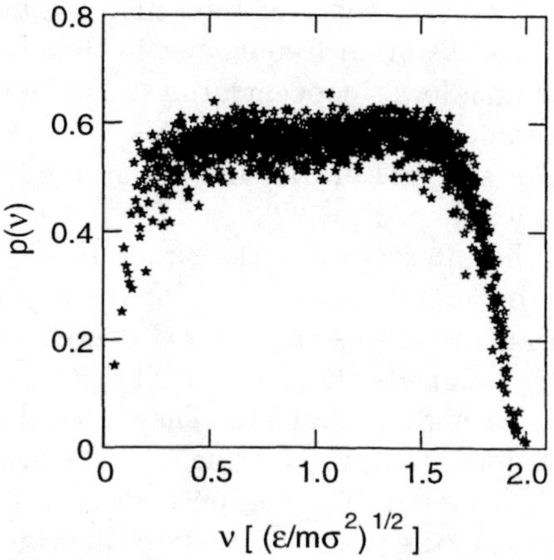

FIGURE 2. Participation ratios p of the individual modes in one configuration of 500 atoms plotted against the mode frequencies.

The eigenvectors can be used as a first indication of localization of the modes. A frequently used measure of localization is the participation ratio

$$p^\sigma = \left\{ N \sum_j^N (\mathbf{e}_j^\sigma \cdot \mathbf{e}_j^\sigma)^2 \right\}^{-1} \quad (4)$$

Here $\mathbf{e}_j^\sigma$ is the component on atom j of the eigenmode σ. For a translation one has $p = 1$, for an undamped sound-wave $p \approx 0.6$ and for a vibration localized on n atoms with all others at rest $p = n/N$. This scaling with $1/N$ should hold for all localized modes. Fig. 2 shows the participation ratios for a single configuration of 500 atoms. Obviously the vast majority of the modes are extended over the entire system. At the largest frequencies a number of modes with very low p are observed. These are high frequency localized modes, well known from lattice dynamics of defect crystals. There, these are modes with frequencies above the continuum of frequencies of the host lattice. The amplitude of the vibration decays exponentially with distance from the center. Despite the absence of a well defined "maximum host frequency" this notion survives in the glass. This can be seen from $p(\nu) \propto 1/N$ for these modes when N is changed. With smaller ν the decay of the amplitudes becomes slower and the number of modes increases. Both effects cause $p(\nu)$ to increase just as in defect crystals.

Fig. 2 shows low participation ratios also at the lowest frequencies. At the low end one finds again scaling with $1/N$. However the frequency range where this scaling is valid shifts with larger N to lower ν, a clear indication of interaction. This signature of the "low frequency quasi-localized modes" is again well known from the resonant modes in defect lattices. For these modes the amplitudes decay much weaker than for their high frequency counterparts, namely with a power law. Modes of similar frequency will therefore interact strongly over larger distances, i.e. at much lower concentrations. Furthermore these low frequency modes will always interact with the sound waves. They can be observed as individual modes only for finite systems. In an infinite system they are observed as low frequency peaks of the local spectra [22].

The interaction with these quasi-localized modes causes in turn an attenuation of the sound waves. At a frequency near the boson peak the mean free path of the transverse sound waves decreases to the wavelength. The sound waves are over-damped, the Ioffe-Regel limit is reached. For higher frequencies the modes are strongly entangled. They extend over the whole system but can no longer be considered as propagating. See the article by Elloitt in this volume for more details.

To compare the computer results with the model assumptions it is necessary to separate the exact modes into their building blocks, extended phonons and localized modes. Since their interaction operator is not known it is not possible to switch off the interaction to obtain the "naked" (interaction free) states. A few modes can be dissected by inspection. An example is given in Fig. 3. Shown are all atoms whose squared amplitude is at least 20% of the maximal one. The illustrated mode with $\nu = 0.061$ and $p = 0.21$ is clearly recognizable as a superposition of four local modes. The frequency is below the minimal sound wave frequency and, therefore, there is no admixture of sound waves. In general such a separation by inspection is impossible. For comparison with show in Fig. 4 a mode with a strong sound wave contribution. At low frequencies a separation can be achieved by minimizing the participation ratio. A naked mode $\mathbf{e}'(\nu')$ with frequency ν' will interact with other modes of similar frequency. The resulting eigenmodes $\mathbf{e}(\nu)$ are then linear combinations of the naked modes

$$\mathbf{e}(\nu) = \sum_{0.8\nu < \nu' < 1.2\nu} a(\nu, \nu')\mathbf{e}'(\nu') \tag{5}$$

with the expansion coefficients $a(\nu, \nu')$. The restriction of the interaction to $\pm 20\%$ is to stabilize the calculation. Starting from an initial base $\mathbf{e}' = \mathbf{e}$ we search by an orthogonal transformation of all modes with $\nu(m\sigma^2/\epsilon)^{1/2} < 0.2$ for a basis consisting of local and extended modes. To avoid an artificial break-up of modes we restrict ourselves to 1% of all modes. In the resulting basis of naked modes the local modes are decoupled and their phonon parts are separated. Although we optimized only the local naked modes, also the extended phonons are reconstructed.

Fig. 5 shows the participation ratios of the naked modes gained by the deconvulution. One can now distinguish two groups of modes: localized ($p < 0.1$) and extended ($p > 0.4$) ones. The increase near $\nu = 0.2$ is an artifact of the cutoff to higher frequencies. The two most long-wave groups of phonons are formed by the correct number of modes. A comparison of the frequencies gives a small shift of a

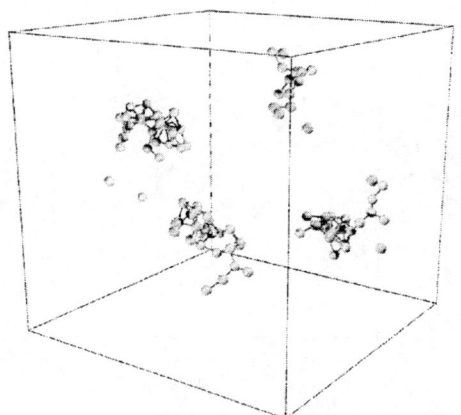 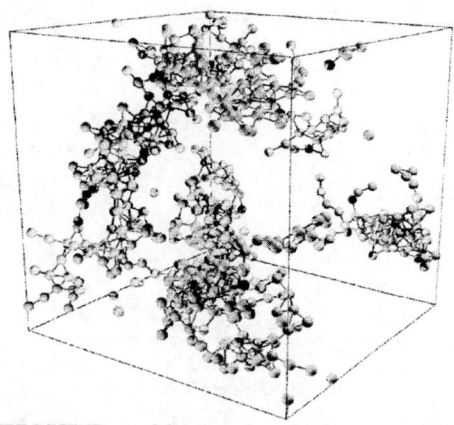

FIGURE 3. Most active atoms in a vibrational mode: $\nu = 0.061$, $p = 0.21$ and $m_{\text{eff}} = 148$. Shown are all atoms whose squared amplitude is at least 20% of the maximal one. Bonds show nearest neighbours. The cubes depict the periodicity volume (N=5488) [15].

FIGURE 4. Most active atoms in a vibrational mode: $\nu = 0.079$, $p = 0.52$ and $m_{\text{eff}} = 545$. Shown are all atoms whose squared amplitude is at least 20% of the maximal one. Bonds show nearest neighbours. The cubes depict the periodicity volume (N=5488) [15].

few percent due to the interaction. For the mode shown in Fig. 3 one finds that it consists to 80% of four localized modes corresponding approximately to the four parts in Fig. 3.

The naked localized modes gained in this way are centered on approximately 20 atoms ($m_{\text{eff}} = 20m$). For the naked modes the participation ratio scales with system-size as $1/N$. The cores of these de-convoluted low frequency localized modes form chainlike structures with side-branching and have a dimension ≈ 1.5. The modes are centered at irregularities of the glass structure. The first neighbour shell is reduced in number but compressed, i.e. the modes are not connected with holes in the glass. The simultaneous participation of individual atoms in low and high frequency "localized" vibrations indicates to local strains as origin [22]. In Se the situation looks very similar [23]. There the chain structure of the modes might be influenced by the Se-chains forming part of the glass.

We can use the interaction free modes, thus obtained, to test the validity of the "soft potential model" [5,6]. One basic assumption of this model is that soft local vibrations and relaxations have a common origin and can be described with a common distribution of parameters. The motion of the central atom of a vibrational or relaxational mode is described by a potential

$$V^{\text{SP}}(x) = \epsilon^{\text{SP}}\left[\eta^{\text{SP}}(x/\sigma)^2 + t^{\text{SP}}(x/\sigma)^3 + (x/\sigma)^4\right]. \qquad (6)$$

Depending on the values of η^{SP} and t^{SP} one has single or double well potentials which describe vibrations and relaxations, respectively. Deviating from the usual length scale we have used here a scaling with σ of the soft sphere potential.

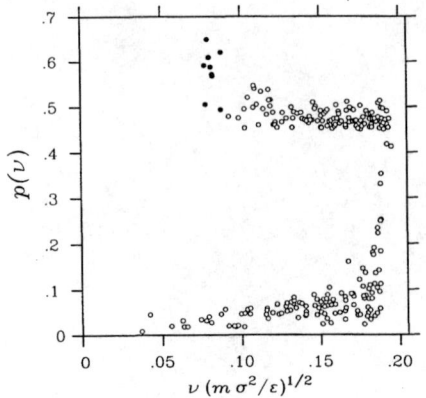 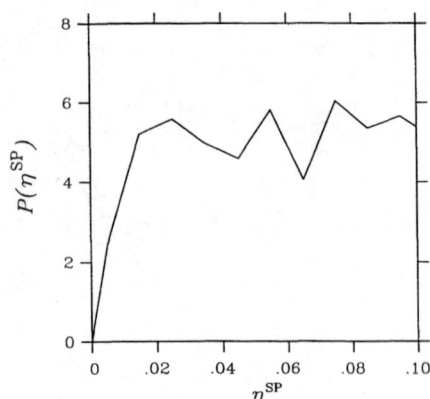

FIGURE 5. Participation ratios of low frequency modes of one soft sphere glass configuration with $N = 5488$ atoms plotted against frequency. Solid circles depict those modes whose overlap with the idealized phonons of longest permitted wavelength exceeds 0.5 [15].

FIGURE 6. Distribution function of the curvature parameter η^{SP} in the soft potential model gained from the de-mixed modes [15].

The scale factor ϵ^{SP} for the energy is in this formulation equal to the value for a single atom times the effective number of atoms participating in the motion, $\epsilon^{SP} = N_s \epsilon_a^{SP}$. In agreement with the assumptions of the model we find for ϵ_a^{SP} a well defined narrow distribution. For the the quadratic term the model predicts for small values of η^{SP} a distribution

$$P(\eta^{SP}) = |\eta^{SP}| P_o, \qquad (7)$$

the so called "sea gull singularity". Fig. 6 shows this for our "interaction free modes". We find the distributions of the harmonic terms and the asymmetry to be independent. At finite temperatures there is a strong correlation between the relaxations and the vibrations. This correlation is, however, weaker than assumed in the model [18].

IV RELAXATIONS IN GLASSES

To study relaxations we heated the two mono-atomic glasses from $T = 0$ K to the desired temperature where they were kept approximately 0.5 ns. (For the soft sphere glass we set $2(m\sigma^2/\epsilon)^{1/2} \approx 10$THz to get a time-scale.) After an initial equilibration time of ≈ 50 ps at each temperature we monitored the total displacement from the starting configuration [14]:

$$\Delta R(t) = \sqrt{\sum_j (\mathbf{R}_j(t) - \mathbf{R}_j(0))^2}. \tag{8}$$

We observe at various times sudden changes which we interpret as hopping transitions to a new minimum. Averaging the atomic positions over three periods of a typical soft vibration we obtain average atomic positions $\mathbf{R}_m^f$ and comparing these with the ones of the previous configuration $\mathbf{R}_m^i$ we define a total jump length by summing over all atoms m

$$(\Delta R)^2 = \sum_m \left(\mathbf{R}_m^i - \mathbf{R}_m^f \right)^2. \tag{9}$$

Suitable cutoffs in residence time and minimum jump length were used to avoid spurious effects. Both reversible and irreversible jumps were observed. We did not see any marked difference in their structure. As a measure of the collective nature we calculated for each jump the participation ratio

$$p_{\Delta R} = \left\{ N \left[\sum_m | \mathbf{R}_m^i - \mathbf{R}_m^f |^4 \right] / \left[(\Delta R)^2 \right]^2 \right\}^{-1}. \tag{10}$$

Because of its instability against crystallization the soft sphere glass was only monitored up to temperatures of about $0.15\, T_g$ whereas in the Se-glass we could observe relaxations up to T_g, below 300 K in our simulation. Qualitatively both systems behave equally. Fig. 7 shows the total jump length, ΔR, of single relaxational jumps in Se as function of temperature. At the lowest temperatures ΔR is about one nearest neighbor distance. Since about 10 atoms participate strongly in the jump, the single atoms move only about one tenth of a nearest neighbor distance. It is a highly collective motion. The total jump length increases according to a law $\Delta R \approx 0.87\sqrt{T}$ (dashed line). Simultaneously with this increase also the number of atoms participating rises strongly (Fig. *). For both ΔR and $p_{\Delta R}$ we find a large spread of values. This behavior is again qualitatively equal in the soft sphere glass. We have observed a merging of jumping units with rising temperature, similar to the merging of localized vibrations with rising frequency. Such a situation was illustrated in [13]. We find a strong correlation between the eigenvectors of the localized vibrations and the displacement vector of the relaxation. However, in general, this correlation is to the set of localized modes and not necessarily to a single one [14].

In the binary amorphous $Cu_{33}Zr_{67}$ similar chain-like relaxations are observed where the maximal displacement of a single atom is much less than the nearest neighbour distance. The relaxations are predominantly along chains of the smaller Cu atoms. At temperatures above 400 K additionally to these collective jumps, a second jump type is observed. Cu atoms hop into nearest neighbour sites. Such hops can form replacement chains or rings, i.e. Cu atoms jump into sites vacated by another Cu atom. So far we have not been able to check quantitatively how far the occurrence of this second jump type is related to aging.

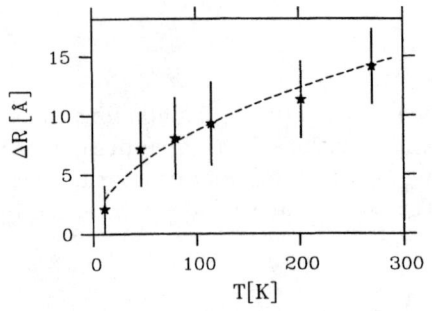
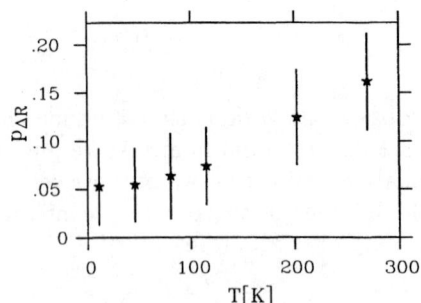

FIGURE 7. Total length of single jump processes in Se against temperature [16]. Dashed line: fit see text.

FIGURE 8. Participation ratio of single jump processes in Se against temperature, calculated for $N = 1470$ atoms [16].

V DIFFUSION IN THE MELT

In the liquid state it is no longer possible to separate single jump processes or relaxational steps. We studied the binary metallic system $Cu_{33}Zr_{67}$ both below and above its glass transition, $T_g \approx 950$ K. To get some insight into the structure of the relaxations of the melt we calculated atomic displacements as differences between the time averaged positions. The atoms with the largest displacements are depicted in Fig. 9. Again we observe the familiar chain structures.

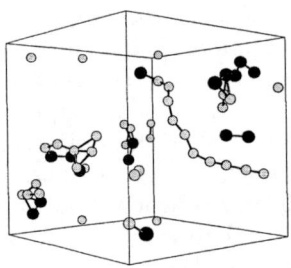
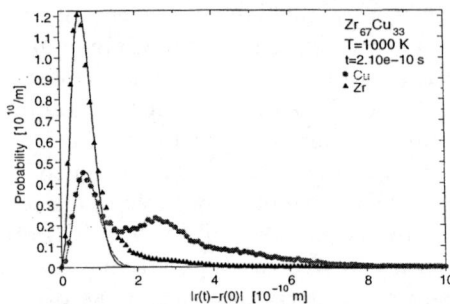

FIGURE 9. Atoms in a melt of $Cu_{33}Zr_{67}$ at $T = 1200$ K with the largest displacements between time averaged configurations separated by 6.5 ps. Shown are Cu (light spheres) and Zr (dark spheres) atoms displaced by more than 1.6Å and 1.45Å, respectively [16].

FIGURE 10. Distribution, $P(r,t)$, of the atomic displacements in liquid $Cu_{33}Zr_{67}$ at $T = 1000$ K after 210 ps. The symbols are the results of the MD simulation and the lines are Gaussian fits [21].

To get some insight into the atomic motions contributing to diffusion we analyze the radial pair self-correlation function (SCF)

$$P(r,t) = \langle \delta\left(r - |\mathbf{R}_j i(t'+t) - \mathbf{R}_j(t')|\right)\rangle_{i,t'} \tag{11}$$

as function of time and temperature. At high temperatures $P(r,t)$ is, for both Zr and Cu a near perfect Gaussian (apart from the geometrical weighting factor $4\pi r^2$) broadening with time. This signifies flow not involving an elementary length scale. Below $T = 1500$ tails to larger distances appear, the SCF becomes for intermediate times non-Gaussian. At $T < 1200$ K, additional structure evolves. For the Cu-atoms a hump grows at the NN-distance and the tail increases in intensity. As example Fig. 10 shows the distribution of the atomic displacements of $Zr_{67}Cu_{33}$ at $T = 1000$ K after a time $t = 210$ ps i.e. during the early time of the so called α-relaxation. This indicates that the Cu-atoms perform, apart from the vibrations, two types of motion: they frequently jump by a NN-distance and are additionally subject to a more flow-like motion as evidenced by the broadening of the first peak. [17]. This latter motion reminds one of the collective jumps in the glassy state discussed in the introduction. The relative importance of the two types of motion is still investigated.

The diffusion constants and the intermediate self-scattering function can be fitted by mode coupling theory [24]. This theory has been treated extensively in other lectures of this school.

VI CONCLUSION

The dynamics of glasses was studied by molecular dynamics simulations of soft spheres, amorphous Se and $Cu_{33}Zr_{67}$ in the temperature range from $T = 0$ K to well above the glass transition. At low temperatures different types of vibrational modes, high and low frequency localized and extended modes are observed. There is a strong interaction between the different low frequency modes resulting in a delocalisation of the exact modes. The pure, i.e. non-interacting, modes can be described in the framework of the soft potential model. These modes are collective motions of 10 and more atoms, typically forming chain-like structures. They are correlated with relaxations by jumps in the glasses at finite temperatures. At the lowest temperature the total jump lengths are of order of the nearest neighbour distance with a single atom moving only a fraction of this. Increasing temperature both the total jump lengths and the number of atoms participating in a jump increases. In the binary CuZr at higher temperatures for the smaller component (Cu) a second type of jump is observed where Cu-atoms jump to nearest neighbour sites. Diffusion seems to be mediated by both types of motion.

VII ACKNOWLEDGMENTS

Part of this work was supported by the Deutsche Forschungsgemeinschaft in the Schwerpunkt "Unterkühlte Metallschmelzen: Phasenselektion und Glasbildung".

REFERENCES

1. W.A. Phillips (ed.), *Amorphous Solids: Low Temperature Properties*, (Berlin: Springer 1981).
2. H.-J. Günterodt and H. Beck (eds.) *Glassy Metals I*, (Berlin: Springer 1981).
3. P.W. Anderson, B.I. Halperin and C.M. Varma, Phil. Mag. **25**, 1 (1972).
4. W.A. Phillips, J. Low Temp. Phys. **7**, 351 (1972).
5. V.G. Karpov, M.I. Klinger and F.N. Ignatiev, Sov. Phys. JETP **57**, 439 (1987); M.A. Il'in, V.G. Karpov and D.A. Parshin, Sov. Phys. JETP **65**, 165 (1987).
6. U. Buchenau, Yu.M. Galperin, V.L. Gurevich and H.R. Schober, Phys. Rev. B **43**, 5039 (1991).
7. D. W. Heermann, *Computer Simulation Methods*, Springer Verlag, Berlin 1990.
8. D. C. Rappaport, *The Art of Molecular Dynamics Simulation*, Cambridge University Press, Cambridge (UK) 1996.
9. L.J. Lewis and M.L. Klein in: *Dynamical Properties of Solids* (G.K. Horton and A.A. Maradudin eds.) p.383, (New York: Elsevier 1990).
10. P.H. Gaskell in: *Materials Science and Technology* **9**, *Glasses and Amorphous Materials* (R.W. Cahn, P. Haasen and E.J. Kramer eds.) p.177, (Weinheim: Chemie Verlag 1991).
11. J.N. Roux, J.L. Barat and J.P. Hansen, J. Phys. Condens. Mat. **1**, 7171 (1989); H. Miyagawa, Y. Hiwatari, B. Bernu and J.P. Hansen, J. Chem. Phys. **88**, 3879 (1988).
12. B.B. Laird and H.R. Schober, Phys. Rev. Lett. **66**, 636 (1991); H.R. Schober and B.B. Laird, Phys. Rev. B. **44**, 6746 (1991).
13. H.R. Schober, C. Oligschleger and B.B. Laird, J. Noncryst. Sol. **156-158**, 965 (1993).
14. C. Oligschleger and H. R. Schober, Solid State Commun. **93**, 1031 (1995).
15. H.R. Schober and C. Oligschleger, Phys. Rev. B **53**, 11469 (1996).
16. H.R. Schober, C. Gaukel and C. Oligschleger, Prog. Theor. Phys. Suppl. **126**, 67 (1997).
17. C. Gaukel and H. R. Schober, Solid State Commun. **107**, 1 (1998).
18. C. Oligschleger and H. R. Schober, Phys. Rev. B **59**, 811 (1999).
19. A. Heuer and J. R. Silbey, Phys. Rev. Lett. **70**, 3911 (1993).
20. C. Oligschleger, R. O. Jones, S. M. Reimann and H. R. Schober, Phys. Rev. B **53**, 6165 (1996).
21. C. Gaukel, *Dynamics of Glasses and Undercooled Melts of Zr-Cu*, Berichte des Forschungszentrums Jülich **3556**, Jülich 1998.
22. P.H. Dederichs and R. Zeller in: *Point Defects in Metals II*, Springer Tracts in Modern Physics **87**, (Berlin: Springer 1980).
23. C. Oligschleger, *Dynamik struktureller Gläser*, Berichte des Forschungszentrums Jülich **2968**, Jülich 1994; C. Oligschleger and H. R. Schober, Physica A **201**, 391 (1993).
24. U. Bengtzelius, W. Götze and A. Sölander, J. Phys. C **17**,5915, (1984).

The Liquid-Glass Transition in the $Ni_{0.5}Zr_{0.5}$ System: Molecular Dynamics Studies

H. Teichler

Institut f. Materialphysik, Univ. Göttingen, D-37073 Göttingen, Germany

Abstract. Results from MD simulations are discussed about the liquid-glass transition in a model system adapted to $Ni_{0.5}Zr_{0.5}$. The calculations rely on a semi-phenomenological approximation to the interatomic couplings of the Hausleitner / Hafner hybridized nearly-free-electron–tight-binding-bond model of binary amorphous transition metal systems. The investigations provide results on the caloric glass transition temperature T_G, the equilibrium melting temperature T_m, entropy and free enthalpy differences between undercooled melt and crystal, the isentropic Kauzmann temperature T_K, the change of the atomic dynamics through the glass transition, diffusion coefficients in the liquid and the arrested state, the dynamic glass transition temperature T_c, the intermediate scattering functions, and the parameter $g_m(T)$ which gives a relative measures how close the undercooled liquid has approached the idealized structural arrest of the glassy state. Finally, results are considered from simulations for thin $Ni_{0.5}Zr_{0.5}$ films, which show an enhancement of the atomic dynamics by two orders of magnitude in a sheet of nm-thickness below the surface at temperatures above and below T_G and T_c. Relations of this phenomenon to dynamical heterogeneity are stressed.

INTRODUCTION

The liquid-glass transition is one of the unsolved challenging problems in actual solid state and materials sciences. In contrast to usual phase transformations, the glass transition seems to be primarily dynamic in origin, and its heating rate dependence signals significant non-equilibrium features. Therefore new theoretical concepts have to be developed for the description of this transition. At present we seem far away from a complete understanding of all the underlying phenomena.

Since long molecular dynamics (MD) simulations are applied to gain knowledge of the atomic processes that take place during the glass transition. The greatly increased computer capacities from the last years and the deepened understanding of the atomic interactions in solids, in particular in metals, now make it possible to carry out long-time MD simulations of undercooled melts and glasses with realistic atomic couplings for 0.1 to 1 µs of nominal time. By this one can create well relaxed, dense amorphous structures with sufficient high viscosity to model the vitrification of the liquid.

The glass transition is a general phenomenon observed for rather different types of materials, e.g., in the melts of the Silica- and Oxidglasses, of the Chalcogenides, of organic molecule systems, polymeres, and metallic alloy systems. Below, we consider one example from the metallic glasses, being aware that the glass transition may develop rather special features for each class of these materials.- It is worth mentioning that in recent years metallic glasses have attracted much interest due to presumed innovative technical applications related to their properties like, e.g., hardness, elasticity, plasticity, corrosion resistance, casting, and soft magnetic behavior.

The present contribution describes a number of results which we deduced from MD simulations [3-7] for a $Ni_{0.5}Zr_{0.5}$ model, and the conclusions on the glass transition we draw therefrom. Similar results are available for other compositions [6, 8] but for conciseness we shall limit ourselves to the mid-concentration regime. The NiZr-system here stands for the broad class of binary, glass forming late and early transition metal alloys. Glass formation in this system is well studied by experiments (e.g., [9-12]) and there are in the literature a number of further MD investigations on this system (e.g. [13-16]). As the latter rely on other atomic potentials than ours and since they in most cases consider much less relaxed structures, we shall not discuss them in details.

In the following we shortly characterize our model. We then provide results on the caloric glass transition as observed in the enthalpy data of the system, including an estimate of the Kauzmann temperature T_K. The further results concern the dynamical glass transition, that means transition from viscous high temperature dynamics in the melt to low temperature atomic motions in the solid-like glass, the related diffusion coefficients, and the characterization of the dynamic transition in terms of the critical temperature T_c introduced by the mode coupling theory [1] from analyzing the structural fluctuations in the system. Finally, the question of dynamical heterogeneity is addressed in the context of inhomogeneous dynamics of undercooled thin films.

MODEL AND METHOD

The MD calculations are carried out as isothermal-isobaric (N,T,p) simulations for a binary A_xB_{1-x} alloy. (Technical details on MD simulations can be found in various textbooks, e.g., the book by Alder and Tildesley [17].) In our studies, the equations of motion of N particles at temperature T and (zero) external pressure p are numerically integrated by a fifth-order predictor-corrector algorithm with time steps typically 2.5×10^{-15} s. Temperature is introduced as the kinetic temperature determined from time averages of the kinetic energy. In our calculations for amorphous $Ni_{0.5}Zr_{0.5}$, an ensemble of N=648 atoms is considered, in the case of crystalline samples 960 atoms. For the amorphous systems, the rather small number of atoms was chosen in order to allow long-time computer runs with up to 4×10^8 integration steps, which corresponds to simulations of the time evolution for nominally up to 1 μs.

The interatomic couplings are modeled by short-ranged pair potentials, aimed at taking care of the electronic d-state interactions, and a volume dependent s,p-state part [3-7]. For the pair potentials a short range expression of the Stillinger-Weber form [18] is used. The potentials are adapted to the interactions deduced by Hausleitner and Hafner [13] within the hybridized nearly-free-electron--tight-binding-bond model for binary amorphous transition metal alloys. Adopting the volume dependent term from [19] with parameters for NiZr completes our model.

CALORIC GLASS TRANSITION AND MELTING TEMPERATURE

Fig.1 is taken from [7]. It provides a general view of our results about the caloric glass transition temperature T_G and the melting temperature T_m. Enthalpy values H from MD simulations are shown as function of temperature. The trivial vibration part $3k_BT$ is subtracted to make more obvious the effects of structural changes. Fig. 1 includes data from simulations starting with the crystalline configuration (triangles), those from liquid structures (open circles), and those from sandwich structures of alternating layers of crystalline and molten material (filled circles). Under isothermal annealing the latter turn into a fully molten state for temperatures $T_a > T_m$. They totally crystallize for $T_a < T_m$. This procedure allows a precise estimation of the model melting temperature and gives $T_m = 1952 \pm 2$ K [7], while the experimental value is 21% lower, at 1533 K.

The lower curve are the enthalpy values for crystalline $Ni_{0.5}Zr_{0.5}$ in the orthorhombic B33 structure, the experimental ground state of the system. Our model predicts for this structure a cohesion energy of 6.046 eV/atom [7] which agrees well with the experimental value of 5.88 eV/atom. Estimation of the formation energies of simpler structures like the B1 (NaCl) or B2 (CsF)-configuration, indeed, yields energies 0.141 eV/atom resp. 0.167 eV/atom above the B33-value [7], which indicates that the B33 arrangement is an energetically rather favorable configuration. Well above T_m the crystal enthalpy curve terminates when the samples in the computer become unstable against homogeneous melting.

The upper curve are the MD calculated values for the liquid phase, where the values by open circles are from [4] and the triangles indicate data from melting the crystal model [7]. The enthalpy of the liquid structure decreases - more or less - linearly with temperature up to the caloric $T_G = 1050$ K [4]. Here the slope reduces drastically. Be-

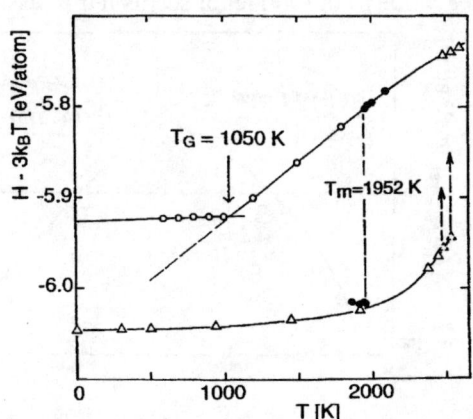

FIGURE 1. MD simulated reduced enthalpy values of liquid and crystalline $Ni_{0.5}Zr_{0.5}$ [7].

low T_G the amorphous system is no longer able to further reduce its enthalpy by suitable structure changes on the time scale of the applied cooling rate. The present T_G has to be ascribed to a rate of 2×10^9 K/s [7]. One should note that the enthalpy of the glassy structure below T_G already is significantly below the enthalpy of the crystalline B1 and B2 structures, which thus cannot energetically compete with the amorphous state.

At the estimated T_m the free enthalpy of crystalline and liquid phase coincide. Hence there is an entropy difference

$$\Delta S(T_m) = \Delta H(T_m)/T_m = 1.32 \; k_B/\text{atom} \tag{1}$$

between liquid and the crystalline state. At arbitrary temperature $\Delta S(T)$ and therefrom the corresponding difference of the free enthalpies $\Delta G(T)$ can be calculated by use of $\Delta S(T_m)$ and of the isobaric specific heat from the enthalpy data in Fig.1. However, evaluation of $\Delta S(T)$ below T_G demands some additional remarks. The liquid state below T_m may be considered as a metastable phase. The transition of the liquid structure into the glassy configuration, however, depends on the cooling rate and means capturing of the system below T_G in a non-equilibrium configuration. For evaluating $\Delta S(T)$ and $\Delta G(T)$ it therefore seems appropriate to substitute H_{liq} below 1100 K by a linear extrapolation from higher temperatures. Fig. 2 displays $\Delta S(T)$ and $\Delta G(T)$ calculated this way [7].

The condition

$$\Delta S(T_K) = 0 \tag{2}$$

determines the isentropic 'Kauzmann' temperature and leads to $T_K = 750$ K for the present model. There are discussions in the literature [20, 21] that T_K means the lower limit of the caloric glass temperature accessible at lowest cooling rates. With regard to the fact that the experimental melting temperature for $Ni_{0.5}Zr_{0.5}$ is about 21% below the nominally calculated value of our model, it seems fair to assume that the calculated T_K

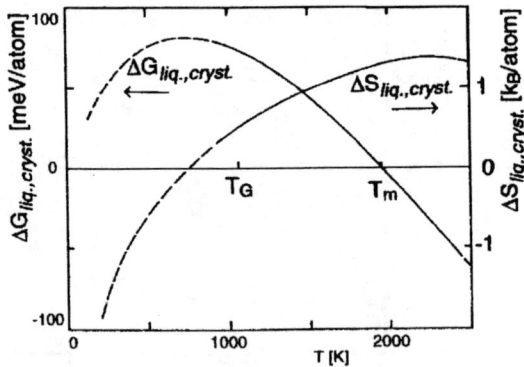

FIGURE 2. Entropy and free enthalpy difference between the undercooled liquid and crystalline state of MD simulated $Ni_{0.5}Zr_{0.5}$ [7].

overestimates the true value by the same percentage. This gives a corrected $T_{K,corr} \approx 600$ K. Similar arguments may be applied to the caloric glass transition, yielding $T_{G,corr} \approx 830$ K. The difference of about 230 K between $T_{G,corr}$ and $T_{K,corr}$ found here seems plausible regarding the rather large cooling rate of 2×10^9 K/s around T_c used in the simulations.

DIFFUSION COEFFICIENTS AND VAN HOVE CORRELATION FUNCTIONS

According to our present understanding of the glass transition in metallic systems, basically guided by the results of the mode coupling theory (MCT) [1], there takes place the fundamental 'dynamic' glass transition around a critical temperature T_c above T_G. In the undercooled melt around T_c the system changes its dynamics from viscous flow in the liquid state to atomic hopping characteristic for a solid.

This change of the dynamics is visible in the self part of the van Hove correlation function

$$P_A(u,t) = <\delta(|x_i(t_v+t) - x_i(t_v)| - u)>_{i \in A} \qquad (3)$$

and may be used according to Roux, Barrat, and Hansen [22] as a fingerprint of the glass transition in metals. Fig. 3 displays representative results for $P_A(u,t)$ (A=Ni, Zr) from our MD modeling. The upper row shows the behavior in the viscous regime ($Ni_{0.5}Zr_{0.5}$ at 1200 K), the lower row that in the Ni hopping regime ($Ni_{0.8}Zr_{0.2}$ at 860 K). In the viscous regime the distribution $P_A(u,t)$ broadens with time. In the hopping regime the broadening with time is negligible but there raise additional peaks in $P_A(u,t)$

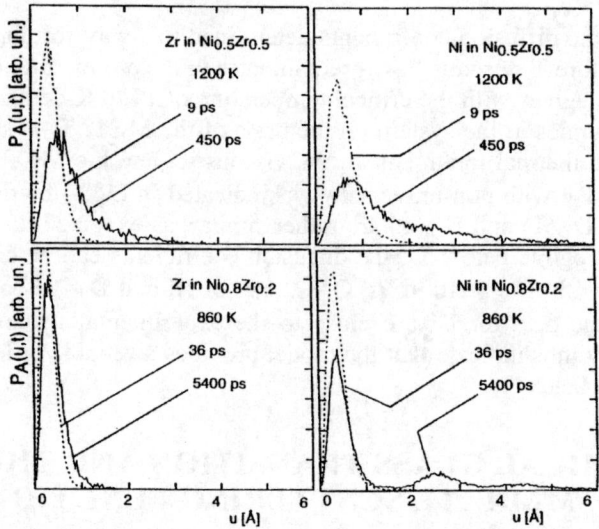

FIGURE 3. van Hove correlation functions $P_A(u,t)$ (A=Ni, Zr) in the regimes of viscous flow (upper row) and Ni hopping motion (lower row) [6].

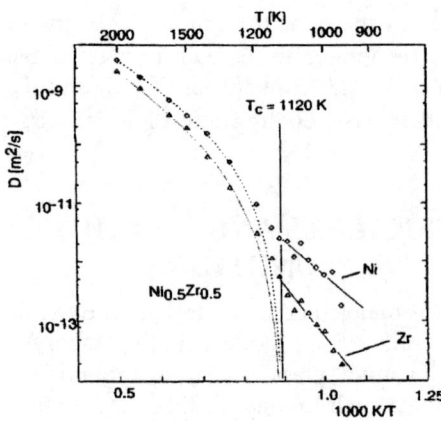

FIGURE 4. Arrhenius plot of the diffusion coefficients from MD simulations for molten and glassy $Ni_{0.5}Zr_{0.5}$ (from [6]).

around distances u of nearest and next nearest neighbor jumps of the atoms. These peaks indicate that the Ni atoms adjust their positions to the structure of the less mobile (Zr-) matrix. This phenomenon we shall call 'hopping motion' below.

The diffusion coefficients are the time derivatives of the atoms mean square displacements (MSD) and can be evaluated from the $P_A(u,t)$ or directly from the slope of the MSD in the linear regime,

$$D_A(t) = 1/6 \lim_{t \to \infty} < \delta(|x_i(t_v+t)-x_i(t_v)|^2) >_{i \in A} . \qquad (4)$$

Fig.4 shows the diffusion coefficients determined this way for $Ni_{0.5}Zr_{0.5}$. Above the critical temperature T_c viscous flow predominates in the atomic motion, below T_c hopping motion. T_c agrees with the critical temperature of 1120 K deduced in [4] from the fluctuation dynamics in the system on the basis of the MCT. T_c is slightly larger than T_G for the same thermal treatment. In the viscous regime the MCT predicts a critical behavior $D \sim |T-T_c|^\gamma$ with non-universal γ, as indicated in Fig. 4 by dashed lines. In the viscous regime $D_{Zr}(T)$ and $D_{Ni}(T)$ are rather similar, as expected for viscous behavior. In the hopping regime below T_c, the diffusion coefficients can be characterized by an Arrhenius law with $D_0 = 5 \cdot 10^{-7}$ m²/s, Q=1.2 eV for Ni and $D_0 = 10^{-4}$ m²/s, Q=2.0 eV for Zr. Note that the data for Ni are close to the experimental values $D_0 = 1.7 \cdot 10^{-7}$ m²/s, Q=1.3 eV. They thus indicate that the model provides a reliable order of magnitude of atomic hopping below T_G.

DYNAMICAL GLASS TRANSITION AND THE INTERMEDIATE SCATTERING FUNCTION

The details of the dynamic glass transition predicted within the MCT are visible in the structural fluctuation dynamics as reflected by the intermediate scattering function (ISF). The central object of our related studies is the self-part of the ISF,

$$\Phi(q,t) = <\exp\{i\mathbf{q}(\mathbf{x}_i(t_v+t)-\mathbf{x}_i(t_v))\}> \qquad (5)$$

Fig.5 presents ISF values [5] obtained from our MD data in a time window between 10^{-14} and 10^{-8} s. It displays by wiggled lines the short time data while the symbols give the long time results. The curves are extrapolated by assuming for the asymptotic regime a stretched exponential law $\phi_0 \exp\{-(t/\tau)^\beta\}$ with common value $\beta=0.65(4)$ for all temperatures. For completeness Fig. 5 also shows the susceptibilities $\omega\Phi_c(\omega)$ ($\Phi_c(\omega)$: cos-Fourier transform).

The initial decay of the ISF below 0.2 ps in Fig.5 is due to the atomic vibrations. Above about 0.2 ps the so-called β-regime starts, which according to our results in [3] reflects configuration fluctuations in the glass, that means breaking of nearest neighbor pairs and formation of new pairs. In the β-regime reconstruction of broken pairs yields that the configuration remains similar to the initial atomic arrangement [3] while in the final α-regime the total decay of the configuration takes place. Above T_c this α-decay is carried by viscous flow of the structure. Below T_c it is carried by thermally activated atomic hopping and thus shifts rapidly towards longer, macroscopic relaxation times with decreasing temperature.

According to the MCT the time dependence of correlation functions like the ISF follows a damped harmonic oscillator equation with retarded damping. The latter is described by a memory kernel $F(q,t)$. In [5] we demonstrated that $F(q,t)$ can be evaluated when $\Phi(q,t)$ is known. A decisive role is played by the auxiliary function introduced in [4]

$$G(\Phi) = F(\Phi) \cdot (\Phi^{-1} - 1) \qquad (6)$$

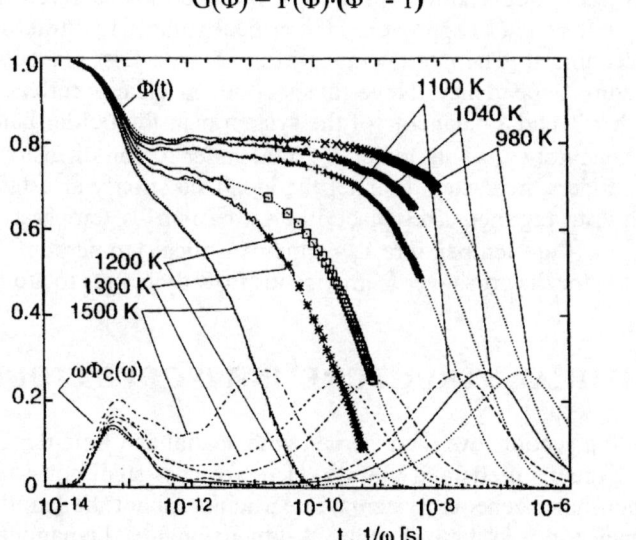

FIGURE 5. Structural correlation functions (intermediate scattering function) $\Phi(q,t)$ and susceptibility $\omega\Phi_c(\omega)$ from MD simulations [5] for $Ni_{0.5}Zr_{0.5}$.

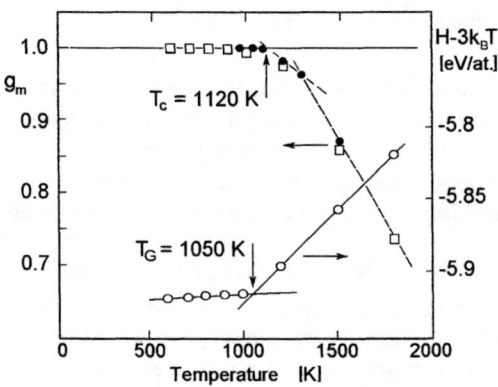

FIGURE 6. The parameter g_m which measures the approach of the liquid towards the arrested state, and the enthalpy of melt and glass for $Ni_{0.5}Zr_{0.5}$ from MD modeling [4,5].

with $\Phi \in [0,1]$. If $G(\Phi)<1$ for all $\Phi \in [0,1]$ then solutions of the damped harmonic oscillator equation decay to zero with time and correlations between structure fluctuations vanish at large times. Hence $\Phi(t)$ describes an ergodic liquid. If, on the other hand, $g_m := \text{Max}\{G(\Phi) | \Phi \in [0,1)\} > 1$ a structural arrest may take place where structural fluctuations and the actual configuration of the atoms are effectively frozen. This characterizes the non-ergodic low-temperature situation of an idealized glass.

As proposed in [4] the g_m-values may be used as a relative measure to judge how close the undercooled liquid has approached the idealized structural arrest. Fig. 6 provides data for $g_m(T)$ deduced in [4] for $Ni_{0.5}Zr_{0.5}$ and those derived from [5]. With decreasing temperature $g_m(T)$ approaches the critical value 1 for structural arrest without, however, exceeding it. The $\Phi(q,t)$ curves in Fig.5 on a first glance show no break in their temperature dependence. Nevertheless, the g_m clearly reflect ergodic behavior ($g_m<1$) at higher T and a balancing of the system near the border between ergodic and non-ergodic behavior ($g_m \approx 1$) at lower T. In the latter region diffusion-like atomic hopping motion hinders the system to enter the idealized strictly arrested state. The crossover between both regimes takes place in a rather narrow temperature window. From extrapolating the high temperature g_m-values a critical temperature $T_c \approx 1120$ K was estimated in [4] for the crossover from viscous flow dynamics to atomic hopping.

DYNAMICAL STRUCTURE IN VISCOUS LIQUID FILMS

The preceding sections were concerned with spatial and time-dependent fluctuations in else homogeneous melts and glasses. Here we now shall consider some of our results about non-homogeneous systems, in particular, about the length scale of dynamical heterogeneity in a system of reduced dimensionality. Dynamical heterogeneity is an actual objective of present discussions on the glass transition mechanism. It is related to the picture that, e.g., in the undercooled melt the glass transition is signaled by the occurrence of long living domains of reduced dynamics embedded in the liquid phase. The length scale of the domains in this heterogeneous state and of the embedding liquid and their temperature variation around the glass transition are questions of concern in this context as they characterize the structure of this state.

In our MD studies [23] we considered a free standing thin film of an undercooled, highly viscous melt or amorphous solid and looked for changes of the dynamics in the film as function of the depth. The MD modeling uses a simulation box with two free surfaces in z-direction and periodic repetition of the cell in the x- and y-directions. The surfaces artificially break the homogeneity of the system. They are regions of increased mobility. The mobility decreases to its bulk value with increasing distance from the surface. The decay of the mobility enhancement as function of depth in the film provides information on the length scale of the dynamical heterogeneity.

In our investigation, the local atomic dynamics are measured by two different quantities, both amenable to computational extraction from MD simulation data. One is the nearest–neighbor – atom-pair decay time τ_{nn} used in our earlier studies [3] to describe the configuration dynamics in viscous liquids. τ_{nn} means the averaged time pairs of nearest neighbor atoms survive when they have been nearest neighbors at time zero. Analyzing τ_{nn} vs. z (pair position below the surface) yields the desired information on the spatial variation of the atomic dynamics. Similar information is obtained from depth dependent chemically averaged diffusion coefficients $D(z) = c_{Zr}D_{Zr} + c_{Ni}D_{Ni}$, determined in accordance with Eq.4 from local mean squares displacements. Details are given in [23, 24] where [24], in particular, describes the way care is taken in the numerical analysis of drift motions of the particles across the film during the investigation time.

For brevity, only results on D(z) shall be given here. Figure 8 displays the variation of D(z) across the film for a simulated $Ni_{0.5}Zr_{0.5}$-sample of 4.2 nm thickness in the temperature range around the glass transition, including the regime of T_c. D(z) takes on

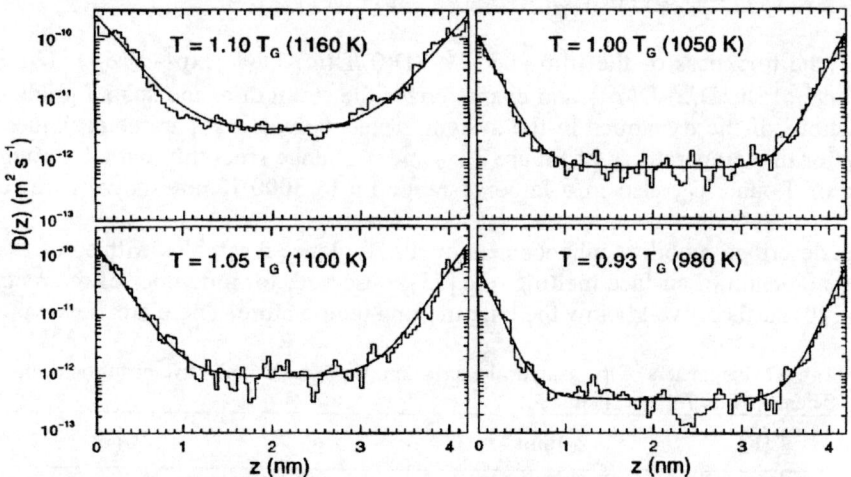

FIGURE 7. Profiles of effective diffusion coefficients across a film of 4.2 nm thickness in MD simulated $Ni_{0.5}Zr_{0.5}$ for temperatures near the glass transition [23,24]. (Lines: analytical fit to expression eq.7.)

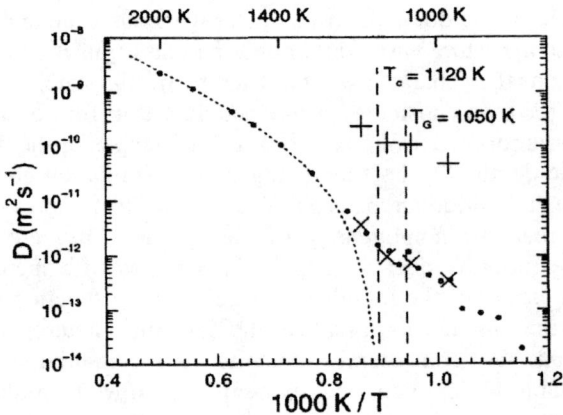

FIGURE 8. Arrhemius plot of the composition averaged diffusion coefficients from MD simulations for $Ni_{0.5}Zr_{0.5}$ in the bulk material (dots), in the interior of the film ($D(\infty)$, symbol ×) and at the film surface ($D(0)$, symbol +) [23, 24]

its bulk value, $D(\infty)$, in the interior of the film. The surface value, $D(0)$, shows a relative enhancement of about two orders of magnitude while the depth of the decay regime is of the order of 1 nm. Detailed analysis indicates that there are neither density fluctuations nor chemical decompositions of significance on the scale of this depth. Moreover, comparable variations of the atomic mobility regarding magnitude and length scale are found by using a different, EAM-type of atomic interactions.

The shape of $D(z)$ is well described by the analytical expression

$$D(z) = D(\infty)\,[1+q'(\exp\{-z/z_0\} + \exp\{-(d_0-z)/z_0)\})] \tag{7}$$

with d_0 the thickness of the film and $q' = (D(0)/D(\infty)-1)/(1+\exp\{-d_0/z_0\})$. $1/z_0$ is the derivative of $\ln\{D(z)-D(\infty)\}$ and characterizes the strength of the spatial gradients of fluctuations of the dynamics in the system. Table 1 gives the parameter values of q' and z_0 for the temperatures of Figure 8. z_0 and q' change smoothly with T through the region of T_c and T_G, also in a larger T-range up to 3000 K not shown here (comp. [24]).

The described mobility enhancement on a first glance resembles with some respects the phenomenon of surface melting (e.g.[25]) observed, for a number of fcc metals on their (110) surfaces well below the bulk melting temperature. There are, however, fun-

Table 1. Parameters of the analytical expression (Eq.7) for the effective diffusion coefficients in Fig. 7. ($D(\infty)$ in $\mu m^2 s^{-1}$)

T [K]	z_0 [nm]	q'	$D(\infty)$
980	0.13	180	0.36
1050	0.15	180	0.72
1100	0.20	170	0.95
1160	0.25	81	3.17

damental differences. Surface melting of crystalline samples is, naturally, limited to the crystalline phase and, according to current interpretations (comp. [25]), is due to the competition of two different thermodynamic phases, the crystalline and the liquid, with different surface enthalpies. In our case, the mobility enhancement seems to be the property of one single phase and is found for the viscous melt at higher temperatures as well as for the arrested, "amorphous" film at lower temperatures.

As mentioned, the slope in the mobility enhancement reflects the length scale on which dynamical heterogeneities are formed. Another relevant quantity is the thickness which the interior part of the film must have in order to exhibit the diffusion properties of the bulk, that means the stability of a glass. The data presented in [23] indicate that there is no lower limit of this thickness: Even for an interior region of an extension of less than 0.5 nm the bulk value of the diffusion coefficients is found. For situations where the film is too thin to reach the bulk value of D, the exponential decay of the surface mobility takes place with the decay parameter z_0 up to the middle of the film.

It has to be emphasized that the mobility enhancement described here was deduced in [23,24] for viscous melts and "amorphous" samples of a particular type of a system, namely a transition metal system with partial covalent interatomic bondings. Although it was shown [23, 24] that the phenomena are found in an EAM-model for NiZr, too, it is an open question to which extend they depend on the involved interactions and on the physics of the atomic dynamics, that means the mechanisms of matter transport, in the liquid and amorphous structure of the bulk material.

REFERENCES

1. Götze, W., and Sjögren, L., *Rep. Prog. Phys.* **55**, 241-376 (1992)
2. Jäckle, J., *Rep. Prog. Phys.* **49**, 171-327 (1986)
3. Teichler, H., *phys. stat. sol. (b)* **172**, 325-335 (1992)
4. Teichler, H., *Phys. Rev. Letters* **76**, 62-65 (1996)
5. Teichler, H., *Phys. Rev. E* **53**, 4287-4290 (1996)
6. Teichler, H., *Defect and Diffusion Forum*, **143-147**, 717-722 (1997)
7. Teichler, H., *Phys. Rev. B* **59**, 8473-8480 (1999)
8. Mutiara, B., *Moden-Kopplungs-Theorie und Glasübergang im MD-simulierten $Ni_{20}Zr_{80}$*, diplomawork, Univ. Göttingen 1996; Spangenberg, M., *MD Simulationen mit EAM-Potentialen zu $Ni_{33}Zr_{67}$- und $Zr_{60}Al_{15}Ni_{25}$-Legierungen*, diplomawork, Univ. Göttingen 1997
9. Altounian, Z., Guo-Hua Tu, and Strom-Ohlsen, J.O., *J. Appl. Phys.* **54**, 3111-3116 (1983)
10. Eckert, J., Schultz, L., and Urban, K., J. Mater. Res. **6**, 1874-1882 (1991)
11. Gachon, J.R., Durand, M., and Hertz, J., J. Less-Comm. Met. **92**, 307-319 (1983)
12. Gärtner, F., *Phasenseparation in amorphen Zirkon-Legierungen*, doctoral Thesis, Univ. Göttingen 1992
13. Hausleitner, Ch. and Hafner, J., *Phys. Rev. B* **45**, 115-127 (1992); **45**, 128-142 (1992)
14. Massobrio, C., Pontikis, V., and Martin, G., *Phys. Rev. B* **41**, 10486-10497 (1990)
15. Aihara, T., Aoki, K., and Masumoto, T., *Scr. Metall.* **28**, 1003-1008 (1993); *Mater. Trans. JIM* **36**, 399-407 (1995); Aihara, T., Kawazoe, Y., and Masumoto, T., *Mater. Trans. JIM* **36**, 835-841 (1995)
16. Devanathan, R., Lam, N.Q., Okamoto, P.R., Sabochick, M.J., and Meshii, M., *J. Alloys and Compounds* **194**, 447-454 (1993)
17. Allen, M.P., and Tildesley, D.J., *Computer Simulation of Liquids*, Oxford: Clarendon, 1987

18. Stillinger, F.H., and Weber, T.A., *Phys. Rev. B* **31**, 5262-5271 (1985)
19. Finnis, M.W., *J. Phys. F* **4**, 1645-1656 (1974)
20. Fecht, H.J., *Mat. Trans. JIM* **36**, 777-793 (1995)
21. Bormann, R. and Zöltzer, K., *phys. stat. sol. (a)* **131**, 691-705 (1992)
22. Roux, J.N., Barrat, J.L., Hansen, J.P., *J. Phys. C* **1**, 7171-7186 (1989)
23. Böddeker, B., and Teichler, H., *Phys. Rev. E* **59**, 1948-1956 (1999)
24. Böddeker, B., *Oberflächennahe Dynamik amorpher, freistehender, metallener $Ni_{0.5}Zr_{0.5}$-Filme in der Molekulardynamik Simulation*, Göttingen, Cuvillier, 1999; and doctoral thesis, Univ. Göttingen, 1999
25. Barnett, R.N., and Landman, U., *Phys. Rev. B* **44**, 3226-3239 (1991)

V CONTRIBUTED TALKS

Dynamical transition in orientationally disordered crystals

F. Affouard and M. Descamps

Laboratoire de Dynamique et Structure des Matériaux Moléculaires,
CNRS ESA 8024, Université Lille I,
59655 Villeneuve d'Ascq Cedex France

Abstract. Results of molecular dynamics numerical simulations are presented for a simple model of an orientationally disordered chloroadamantane crystal. Molecules are considered as rigid, linear and the adamantane part $C_{10}H_{15}$ is modeled by one super atom in order to decrease CPU time and reach the nanosecond regime. We find a plastic-plastic dynamical transition at $T_x \simeq 310$ K interpreted as the rotational analogue of the Goldstein crossing temperature between free diffusion and activated regime. We check the Mode Coupling Theory predictions and find a critical temperature $T_c \simeq 235$ K.

INTRODUCTION

The phenomenology of liquid glassformers [1, 2] has been shown experimentally to be also displayed by some orientationally disordered crystals *i.e glassy crystals* [3, 4]. At high temperature, the average molecular centers of mass of these compounds are ordered in a lattice while the orientations are dynamically disordered. Upon cooling, rotational motions are frozen and glassy crystals present many properties characteristic of conventional molecular liquid glasses both in the way the glass occurs (undercooling) and the thermodynamics signatures. In opposition to the so called *orientational glasses* [5], for example $(KCN)_x(KBr)_{1-x}$, glassy crystals are not frustrated by quenched disorder externally imposed by a dilution.

Substituted adamantane are excellent experimental candidates. Some of them exhibits a glass transition signature [6] as the cyanoadamantane $C_{10}H_{15} - C \equiv N$. A recent MD numerical simulation [7] of a simple model of this compound has shown that the system evolves critically in its plastic crystal from quasi free small-step rotational diffusion to jump-like motion associated to a two-step process observed in the rotational relaxation functions and predicted by the Mode Coupling Theory (MCT) [8]. However, for this compound, owing to the large CN group, the rotational motions are relatively long. This MD simulation has predicted that the MCT dynamical singularity should occur in the real compound at very high temperature close to the melting point which makes it hard to detect experimentally. Here, we present MD calculation results concerning a chloroadamantane crystal in its orientationally disordered phase. This compound belongs to the substituted adamantane family too and shows a plastic phase structure isomorphous to cyanoadamantane but the chloroadamantane molecule $C_{10}H_{15}Cl$ possesses a smaller substitute and a faster dynamics. Chloroadamantane undergoes at $T \simeq 244K$ a first order transition from an ordered monoclinic structure to a fcc rotator phase [9]. In addition to the librational motions, molecules show large tumbling reorientations in the orientationally disordered phase. The most probable molecular orientations are localized along the four-fold crystallographic axes of the cubic cell (see figure 1). The plastic-liquid transition occurs at $T_m \simeq 442$ K.

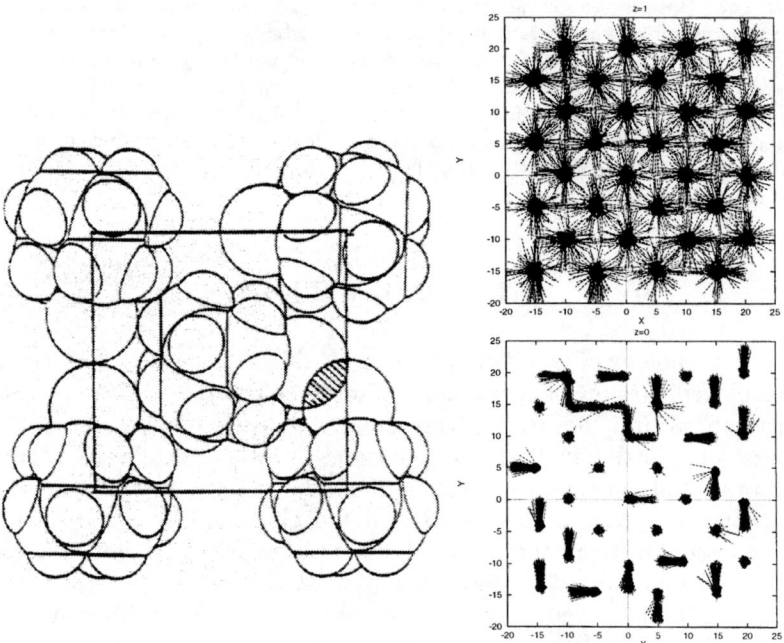

FIGURE 1. On the left part, steric hindrance of one configuration of the chloroadamantane molecules in the rotator cubic phase is displayed. On the right part, snapshots of the $\vec{\mu}$ dipolar moments, in the rotator phase, projected on one xy plane at different instants in a run of 250 ps at two different temperatures. At $T = 220$ K (bottom), jump-like motions are seen. Some orientations are not populated since the system is not completely equilibrated *over this duration*. Collective orientational motions of neighbouring molecules are detected in the snapshot which correspond to the breaking of the orientational cages. At $T = 400$ K (top), quasi free rotational diffusion is observed.

MODEL AND DETAILS OF THE SIMULATION

Real chloroadamantane molecules are displayed in the figure 1. A complete description of the model used previously to simulate cyanoadamantane using MC and MD techniques is given in [7, 10], so we give here only the essential details. The simulated system is composed of rigid linear molecules with two sites : one chlorine atom (noted Cl) and one super atom (noted Adm) that models the adamantane part $C_{10}H_{15}$. Molecular dynamics calculations were performed on a system of $N = 256$ modeled molecules ($4 \times 4 \times 4$ fcc crystalline cells) interacting through a Lennard-Jones short range site-site potential of the form

$$v(r) = 4\epsilon \left((\sigma/r)^p - (\sigma/r)^q\right)$$

where r is the distance between two different sites. Parameters ϵ, σ, p and q are given in table 1.

TABLE 1. Parameters for the two-site chloroadamantane model.

Site-Site	p	q	ϵ (kJ/mol)	σ (Å)
Cl - Cl	12	6	1.441	3.350
Cl - Adm	14	8	3.087	4.786
Adm - Adm	16	11	12.47	6.200

The chloroadamantane molecule possesses a relatively large dipolar moment $\vec{\mu}$ (2.39 Debyes) which is parallel to the molecular axis. The electrostatic interactions were handled by the Ewald method with two partial charges ($q = \pm 0.151$ e) localized on both sites. Newton's equations of motion were solved with a time step of $\Delta t = 5$ fs. We worked in the NPT statistical ensemble with periodic boundaries conditions where the simulating box is allowed to change in size and shape at constant atmospheric pressure and where the average temperature is fixed.

RESULTS AND DISCUSSION

MD simulations were done at 29 different temperatures from $T = 220$ to 500 K for a sample corresponding to the experimental crystalline fcc rotator phase. Owing to the very long MD runs which are required, we succeeded in investigating truly equilibrated states of the system down to 220 K. Below this temperature, the system falls out of equilibrium in the time window of our MD simulation.

For rotational degrees of freedom, if we focus our attention on the motion of the dipole vector $\vec{\mu}$, convenient functions are the first two self angular correlation functions ($l = 1, 2$) [11]

$$C_l(t) = \frac{1}{N} \sum_{i=1,N} \langle P_l\left(\vec{\mu}_i(t).\vec{\mu}_i(0)\right)\rangle$$

where P_l is the l-order Legendre polynomial and $\vec{\mu}_i(t)$ the dipolar moment of the molecule i at the instant t. Experimentally, $C_1(t)$ can be measured in dielectric relaxation and $C_2(t)$ in Raman scattering [12]. Figure 2 shows the $C_2(t)$ time

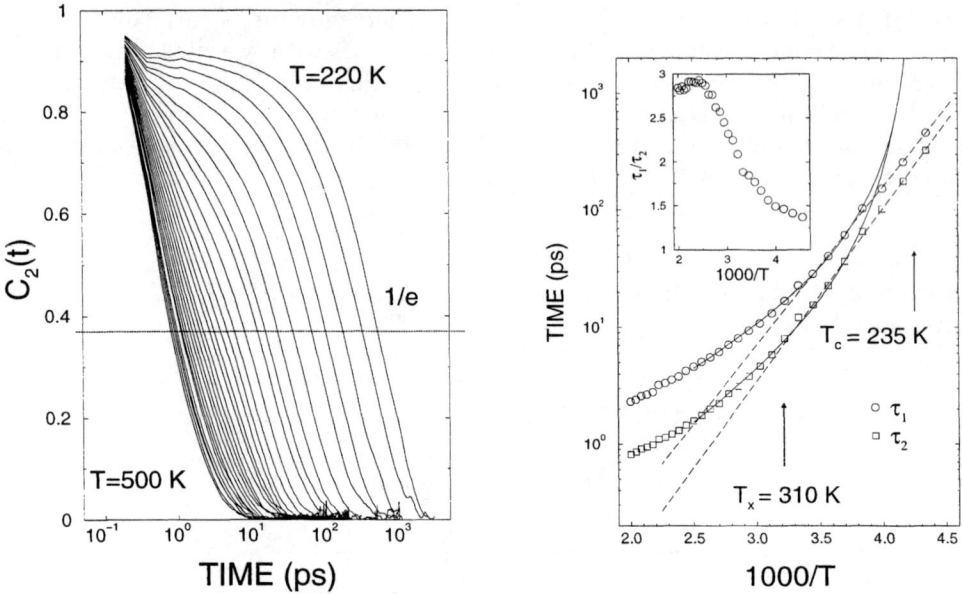

FIGURE 2. Orientational correlation functions $C_2(t)$ as function of time t is displayed at different temperatures from 220 to 500 K (left). Orientational α-relaxation time τ_1 and τ_2 is shown as function of the temperature T (right). The ratio τ_1/τ_2 is given in the inset.

correlation functions [11] for all investigated temperatures above 220 K. Clearly, when lowering the temperature, a two-step relaxation behaviour as predicted by MCT and already observed in supercooled molecular liquids [11, 13, 14] is shown in our simulation. At short times, we observe a fast decay, followed by a plateau-like region, which becomes more pronounced at the lowest temperatures. This feature is identified with the β regime. This plateau domain proves as displayed in the figure 1 the presence of an orientational caging. It means that the tumbling of one molecule is allowed only if orientational rearrangement of its local neighbors occurs. This is the rotational analogue of the translational caging classically observed for liquids. At the end, the last step is a decay to zero corresponding to the α process which can be associated in our system, as checked for cyanoadamantane [7], to large dipolar tumbling motion between four fold crystallographic directions.

The τ_1 and τ_2 relaxation times are computed as the time it takes to the time correlation function to decay e^{-1} of its initial value. They are displayed in the figure 2 in an Arrhenius plot. Clearly, close to $T_x \simeq 310$ K the orientational τ_1 and τ_2 relaxation times deviate from the Arrhenius law when temperature increases. This is a plastic-plastic dynamical transition. T_x can be interpreted as the crossing temperature where over-barrier motions start to play an important role for dynamics. T_x can be also understood as the rotational analogue of the Goldstein's crossover temperature predicted 30 years ago [15] i.e the temperature at which relaxation in liquids is governed by thermally activated crossings of potential energy barriers. This dynamical change is confirmed by the snapshot displayed in the figure 1 and by the ratio τ_1/τ_2. This latter evolves from a value equal to 3 corresponding to small-step rotational diffusion to a value of 1 associated to activated jump-like motion. A fit with a power law $1/(T-T_c)^\gamma$, as predicted by MCT, has been computed and works relatively well. Fitting procedure of the relaxational times τ_1 and τ_2 gives a critical temperature $T_c \simeq 235$ K. The critical exponent γ is about 1.68 for τ_1 and 2.05 for τ_2.

ACKNOWLEDGMENTS

The authors wish to acknowledge the use of the facilities of the IDRIS, Bâtiment 506, Faculté des Sciences d'Orsay F-91403 Orsay, where some of the simulations were carried out.

REFERENCES

1. M. D. Ediger, C. A. Angell, and S. R. Nagel, J. Phys. Chem. **100**, 13200 (1996).
2. A. P. Sokolov, J. Non-Cryst. Solids **235-237**, 190 (1998).
3. H. Suga and S. Seki, J. Non-Cryst. Solids **16**, 171 (1974).
4. M. Descamps and C. Caucheteux, J. Phys. C **20**, 5073 (1987).
5. U. T. Höchli, K. Knorr, and A. Loidl, Adv. Phys. **39**, 405 (1990).
6. M. Descamps, J. Willart, and O. Delcourt, Physica A **201**, 346 (1993).
7. F. Affouard and M. Descamps, Phys. Rev. B **59**, 9011 (1999).
8. W. Götze and L. Sjögren, Rep. Prog. Phys. **55**, 241 (1992).
9. M. Foulon, T. Belgrand, C. Gors, and M. More, Acta Cryst. **B45**, 404 (1989).
10. B. Kuchta, M. Descamps, and F. Affouard, J. Chem. Phys. **109**, 6753 (1998).
11. S. Kämmerer, W. Kob, and R. Schilling, Phys. Rev. E **56**, 5450 (1997).
12. R. M. Lynden-Bell and K. H. Michel, Reviews of Modern Physics **66**, 721 (1994).
13. F. Sciortino, P. Gallo, P. Tartaglia, and S.-H. Chen, Phys. Rev. E **54**, 6331 (1996).
14. L. J. Lewis and G. Wahnström, Phys. Rev. E **50**, 3865 (1994).
15. M. Goldstein, J. Chem. Phys. **51**, 3728 (1969).

Simulation of a silica glass from combined classical and *ab initio* molecular–dynamics.

Magali Benoit, Simona Ispas, Philippe Jund and Rémi Jullien

*Laboratoire des Verres, Université Montpellier II
cc 69, Place E. Bataillon, 34095 Montpellier, France*

Abstract. We present structural and electronic characteristics of a vitreous silica glass obtained from combined classical and Car-Parrinello (CP) molecular–dynamics (MD) simulations. The equilibration of the liquid, quench and relaxation of the glass are performed classically using the van Beest *et al.* (BKS) potential and the resulting configuration is used as input for the CP simulation. A remarkable stability of the CP dynamics is observed justifying this procedure and validating the BKS potential.

Silica is a common material that plays an important role in chemistry and geology as a network forming glass [1]. In the past twenty years, numerical simulations have been extensively used in order to investigate the structural and vibrational properties of amorphous silica [2–4] which are still largely misunderstood. More recently, first-principles studies of disordered silica have become accessible [5] thanks to the development of parallel computing. The explicit treatment of the electronic structure has permitted to improve the description of the structure and to give access to the electronic properties of disordered SiO_2. But *ab initio* simulations of disordered systems can not be carried out without encountering several difficulties: the size of the system that can be treated (box of edge length ≈ 15 Å) and the maximum length of the trajectory that can be calculated (≈ 10 ps) are very limited. In order to partially prevent these difficulties, we have performed an *ab initio* MD simulation of vitreous SiO_2, using the Car-Parrinello technique [6], combined with a classical MD simulation using the BKS interaction potential [7]. In this novel approach, the study of the vitreous SiO_2 system with first-principles is thus accesssible at a much lower CPU time cost since the equilibration of the liquid, the quench and a part of the relaxation are carried out within the framework of classical MD.

A liquid sample of 78 atoms, in a cubix box of edge length 10.558 Å, was equilibrated classically at 5000 K during ≈ 35 ps. A time step of 0.7 fs was used. The system was then cooled to 300 K at a quench rate of 5×10^{13} K/s. Although the glass transition temperature is anormaly high ($T_g \approx 3000$ K) compared to experiments, due to the high cooling rate, the structural [3,8], the dynamical [4] and the thermal [9] properties of the resulting glass fit well with experiments. The obtained glass was relaxed during ≈ 40 ps and then it was used as the initial configuration for the Car-Parrinello simulation performed with the CPMD code developped in Stuttgart [10]. The glass was equilibrated during 0.7 ps and data were accumulated on the following 5.0 ps. A time step of 5 a.u. (0.12 fs) and a fictitious electronic mass of 600 a.u. were used to integrate the equations of motion. The electronic structure was described within the density functional theory in the local density approximation (LDA) and wavefunctions were expanded at the Γ point of the supercell in plane waves up to an energy cutoff of 70 Ry. We adopted a pseudopotential approach using conventional pseudopotentials for silicon [11] and oxygen [12].

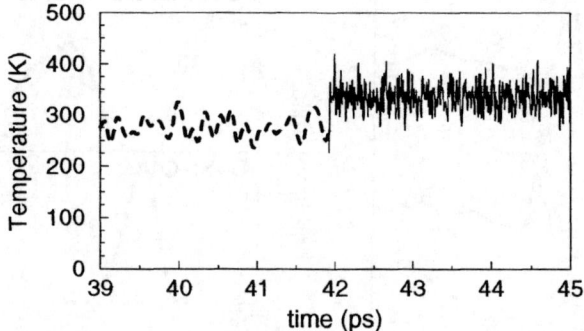

FIGURE 1. Evolution of the temperature of the ions as a function of the simulation time in picoseconds. The end of the classical MD trajectory is shown in dashed line, the beginning of the CP trajectory is shown in solid line.

We report in Fig.1 the evolution of the ionic temperature as a function of the simulation time for the end of the classical MD trajectory and the beginning of the CP trajectory. After a small heating up of the ions when the CP dynamics is switched on, the system remains remarkably stable and the temperature oscillates around 335 K. This temperature jump of approximatly 50 K is unexpectedly small if one considers the strong differences between the two descriptions.

In order to check the structure of the generated glass, we have analyzed several structural characteristics of the sample and compared them to experiments when possible. The pair correlation functions $g_{\alpha\beta}(r)$ ($\alpha, \beta = $ Si,O) of the silica glass sample, averaged over the 5 ps of the CP trajectory, are presented in the left side of Fig.2. The average first neighbour distances are equal to 1.62 Å for Si-O, to 2.66 Å for O-O and to 3.08 Å for Si-Si and are in very good agreement with experimental values [13]. The coordination of the silicon atoms is equal to 4, which reflects a perfect tetrahedral arrangement. These pair correlation functions are overall very similar to those published in recent *ab initio* molecular-dynamics simulations [5].

We have also evaluated the distributions of the intra-tetrahedral O-Si-O angles and the inter-tetrahedral Si-O-Si angles. These distributions, shown in the right side of Fig.2, are in very good agreement with experimental values [14], therefore validating the structural characteristics of our sample. The averaged values are equal to 109° ± 7 for O-Si-O angles and to 145° ± 13 for Si-O-Si angles. The latest is closer to the experimental angle (140-150°) [14] than the one obtained by Pasquarello *et al.* (136°± 14) [5]. This small difference might be due to the use of a different mass density, or to a different cooling rate.

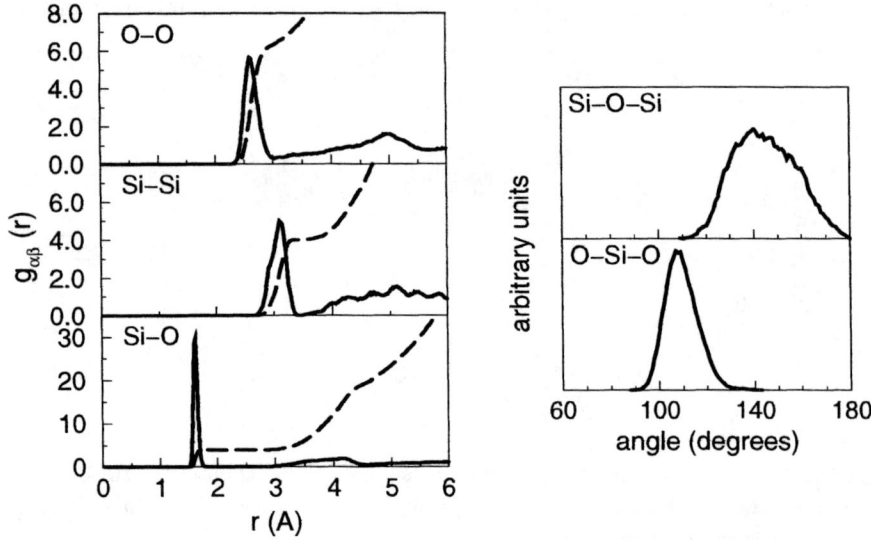

FIGURE 2. Left side: Pair correlation functions $g_{\alpha\beta}(r)$ for O-O (upper graph), Si-Si (middle graph) and Si-O (lower graph). The dashed lines are the integrated coordination numbers. Right side: inter-tetrahedral Si-O-Si angle distribution (upper graph) and intra-tetrahedral O-Si-O angle distribution (lower graph).

We can compare directly the structure to experiments by calculating the static structure factor $S(q)$ as obtained from neutron diffraction. This structure factor, computed by Fourier transforming the pair correlation functions, is depicted in Fig.3 where it is compared to the results of Ref. [15]. One may notice that the agreement between our results and experimental ones is very good for large values of q and still quite good for small q concerning the maxima and minima positions. However the amplitudes of the first and second peaks are slightly too small compared to experiment, which is probably due to our limited system size. The first sharp diffraction peak (FSDP) appears at $q \approx 1.48$ Å^{-1} which is to be compared with the experimental value of 1.52 Å^{-1}.

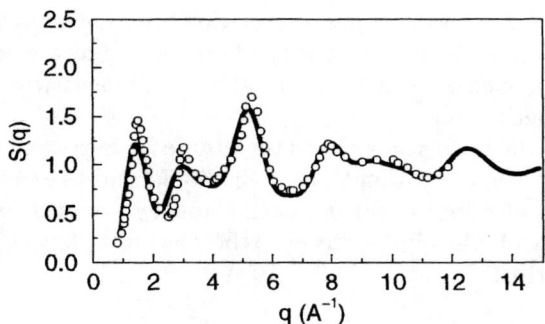

FIGURE 3. Comparison between the simulated $S(q)$ (black line), with scattering lengths of 4.149 fm for Si and 5.803 fm for O, and the experimental neutron $S(q)$ from Ref. [15] (circles).

The electronic properties of the silica glass were analysed in terms of the electronic density of states (see Fig.4). The density agrees well with other calculations [5] and gives a band gap of 5.0 eV (to be compared to 4.8 eV [5]) but underestimates the experimental value ($\sim$ 9 eV). This discrepancy is usually attributed to the use of the LDA in the density functional theory scheme. The different states are overall well reproduced.

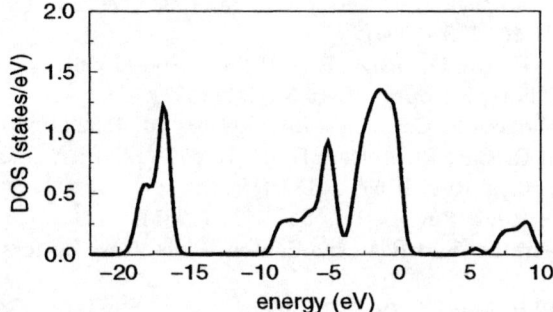

FIGURE 4. Electronic density of states of the glass sample at $\approx$ 5 ps. The curve was smoothed with a gaussian broadening.

In summary, using a combination of classical and *ab initio* molecular–dynamics simulations, we have studied structural and electronic properties of a silica glass. The originality of this work resides in the preparation of the glass sample: using classical MD, a silica glass was generated and taken as the initial configuration of a first-principles MD simulation. The remarkable stability of the Car-Parrinello simulation compared to the classical trajectory (temperature and diffusion) validates this method and our choice of the classical potential. The structural properties of the obtained glass were analyzed in terms of pair correlation functions, angle distributions and structure factor. The electronic density of states was also calculated. Both structural and electronic characteristics are in good agreement with former *ab initio* calculations and experiments.

In conclusion, this method allows to study the glass sample by means of *ab initio* molecular–dynamics saving the equilibration time of the liquid and the time of the quench. This approach could also be extended to other binary systems for which good classical potentials are available, giving access to their accurate local structure and to their electronic properties.

Acknowledgments

M. Benoit would like to thank J. Hutter, D. Marx, M. Parrinello and M. Tuckerman for helpfull discussions and support. Calculations have been performed on the CRAY/T3E at I.D.R.I.S. in Orsay (France) and on the IBM/sp2 at C.N.U.S.C. in Montpellier (France).

REFERENCES

1. *The Physic and Technology of Amorphous SiO_2*, edited by R.A.B. Devine (Plenum Press, New York, 1988).
2. R.G. Della Valle and H.C. Andersen, J. Chem. Phys. **97**, 2682 (1992); A. Nakano, L. Bi, R.K. Kalia, P. Vashishta, Phys. Rev. Letters **71**, 85 (1993); W. Jin, R. K. Kalia, P. Vashishta, Phys. Rev B **50**, 118 (1994).
3. K. Vollmayr, W. Kob and K. Binder, Phys. Rev. B **54**, 15808 (1996).
4. S.N. Taraskin and S.R. Elliott, Europhys. Lett **39**, 37 (1997).
5. J. Sarnthein, A. Pasquarello and R. Car, Phys. Rev. Letters **74**, 4682 (1995); J. Sarnthein, A. Pasquarello and R. Car, Phys. Rev. B **52**, 12 690 (1995); A. Pasquarello, J. Sarnthein and R. Car, Phys. Rev. B **57**, 14133 (1998);
6. R. Car and M. Parrinello, Phys. Rev. Letters **55**, 2471 (1985).
7. B.W.H. van Beest, G.J. Kramer and R.A. van Santen, Phys. Rev. Letters **64**, 1955 (1990).
8. P. Jund and R. Jullien, Phil. Mag. A **79**, 223 (1999).
9. P. Jund and R. Jullien, Phys. Rev. B **59**, 13707 (1999).
10. CPMD Version 3.0, J. Hutter, P. Ballone, M. Bernasconi, P. Focher, E. Fois, St. Goedecker, M. Parrinello, M. Tuckerman, MPI für Festkörperforschung and IBM Research 1990-96.
11. G.B. Bachelet, D.R. Haman and M. Schlüter, Phys. Rev. B **26**, 4199 (1982).
12. N. Trouiller and J.L Martins, Phys. Rev. B **43**, 1993 (1991).
13. P.A.V. Johnson, A.C. Wright and R.N. Sinclair, J. Non-Cryst. Solids **58**, 109 (1983).
14. R. Dupree and R.F. Pettifer, Nature **308**, 523 (1991).
15. S. Susman, K.J. Volin, D.L. Price, M. Grimsditch, J.P. Rino, R.K. Kalia, P. Vashishta, G. Gwanmesia, Y. Wang and R.C. Liebermann, Phys. Rev. B **43**, 1194 (1991).

A Study of Conformational Jumping of Terrylene in *p*-terphenyl by Molecular Modelling

Patrice Bordat and Ross Brown

*Centre de Physique Moléculaire Optique et Hertzienne,
umr 5798 du C.N.R.S. et de l'Université Bordeaux I,
33405 Talence Cedex, France*

Abstract. Terrylene in *p*-terphenyl has been widely studied by optical techniques. Single molecule methods show photo-induced, well defined and reversible frequency jumps of molecules in one of the sites occupied by the guest, constituting light-driven, molecular flip-flops. Empirical molecular modelling is applied to determine the nature of the substitution sites of terrylene in *p*-terphenyl. Simulated annealing shows further that some of the sites may exist in several conformational states. Consideration of the numbers of minima and of spectral lines and of their symmetries leads us to a plausible identification of the nature of the molecular motions involved in the flip-flops.

INTRODUCTION

The anomalous properties of glasses concerning the thermal conductivity, the specific heat, etc ... at very low temperatures, have been described phenomenologically for many years by the standard tunneling model of two level systems (TLS) [1,2], but only in exceptional cases is the microscopic nature of the tunneling systems known. The study of TLS's in crystals offers the opportunity of a better understanding. For example, tunneling relaxation of host molecules at domain walls is thought to be responsible for spectral diffusion of single molecules of pentacene in *p*-terphenyl [3]. Photo-induced spectral diffusion of terrylene in *p*-terphenyl, figure 1, has been observed as well defined, reversible frequency jumps of the guest molecule [4], corresponding to flipping of the environment between two conformations. Understanding such optical flipping offers a path towards tunneling systems and may be relevant to the elaboration of high density optical memories. The purpose of this work is to elucidate the insertion sites of terrylene in *p*-terphenyl and the nature of the flipping, by molecular dynamics.

MODEL

The models of the p-terphenyl molecule and crystal will not be discussed here. The optimization of these models was the subject of an earlier paper [5]. We have optimized the geometry of terrylene by semi-empirical and *ab-initio* calculations. In both case, the molecule is planar. The empirical model used for terrylene is entirely flexible to reproduce the out of plane distorsions which may be important as for the system pentacene/p-terphenyl [6]. To reproduce the out of plane distorsions, we have optimized the intramolecular parameters of the terrylene model to obtain the lower vibrational frequencies in a range from 0 to 300 cm^{-1} with a maximum error of 5% compared to the experimental values [7]. The intramolecular potential is a sum of harmonic bond, angle and torsion potentials. For the intermolecular interactions, we have proceeded in the same way as for the p-terphenyl model with Buckingham potentials [8] and electrostatic interactions with charges on atoms. With these models, we have done room temperature simulations, minimizations and simulated annealing with the DLPOLY package [9]. The crystal has $4 \times 4 \times 2$ cells (256 molecules) centered on the guest with periodic boundary conditions. The system is coupled to a Berendsen heat bath and barostat at atmospheric pressure. The electrostatic interactions are calculated by the shifted potential method (cutoff 10 Å) and the runs range from 10 to 500 ps.

RESULTS

Before commenting the numerical results, we summarize the main experimental results. The single molecule experiments on terrylene/p-terphenyl show four main electronic origins $X_1 - X_4$ [10]. It is natural to think that these four spectroscopic sites can be linked to the four different crystallographic sites in the low temperature phase of the p-terphenyl crystal $M_1 - M_4$. F. Kulzer *et al* [4] show by single molecule spectroscopy that X_1 has photo-induced jumps, X_3 is an unstable site, and X_2 and X_4 are photo-stable during a very long irradiation time. Their experiments show four other spectral positions for X_1 molecules called XY, XY', XY'' and XY'''. Moreover, a Stark effect experiment [11] on single terrylene molecules in the X_1, XY, X_2 and X_4 sites shows that for XY, the Stark effect is linear and for X_1, X_2 and X_4, the Stark effect is quadratic. This means that the environment of the terrylene molecule is centro-symmetric except for XY. Such is the case only if terrylene replaces one p-terphenyl molecule. We have checked by molecular dynamics that the most stable system is indeed that in which terrylene replaces only one p-terphenyl molecule rather than two and that the sites are centro-symmetric. In all subsequent dynamics simulations, terrylene replaces one p-terphenyl molecule. The results of the simulated annealing are summarized in figure 2. Each point is a minimization of the system terrylene/p-terphenyl and between two points, a room temperature simulation is carried out. The axes of the figure are chosen only to distinguish the different minima. We can see that M_2 and M_3 have several minima like the experimental results which have photo-induced jumps for X_1 and at least two sites for the unstable X_3 site. Moreover, M_1 and M_4 have only one minimum like X_2 and X_4 which are stable. So, we can suggest that M_2 and M_3 are X_1 and X_3 in either order. We turn now to the origin of the different minima of M_2 or M_3 to explain the photo-induced jumps. Figure 3 shows the absolute minimum of M_3. We can see that the environment of terrylene is very distorted because the terrylene molecule is bigger than the p-terphenyl molecule, but the matrix is always centro-symmetric. Indeed, we have calculated the Stark effect on terrylene by coupled Hartree-Fock semi-empirical calculations and we have found a mainly quadratic Stark effect. Figure 4 shows the second minimum of M_3. We can see that the conformation is broadly the same as that of the absolute minimum except the highlighted molecule where the central phenyl ring (coloured black) has flipped through the molecular plane. The environment is not centro-symmetric in the second minimum and the numerical Stark effect for terrylene is actually linear like the experimental results. Taking account of the numbers of minima, energies, molecular orientations, Stark effects and so on, we can suggest that the absolute minimum of M_3 corresponds to X_1 and the second minimum is XY, after the flip of the central phenyl ring of the highlighted p-terphenyl molecule. This attribution of the spectral lines to the matrix sites will be further tested by calculations of the spectral shifts and of the polarization of excitation spectra, both of which can be checked experimentally. A full report is in preparation [12].

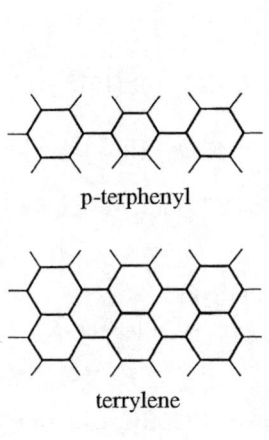

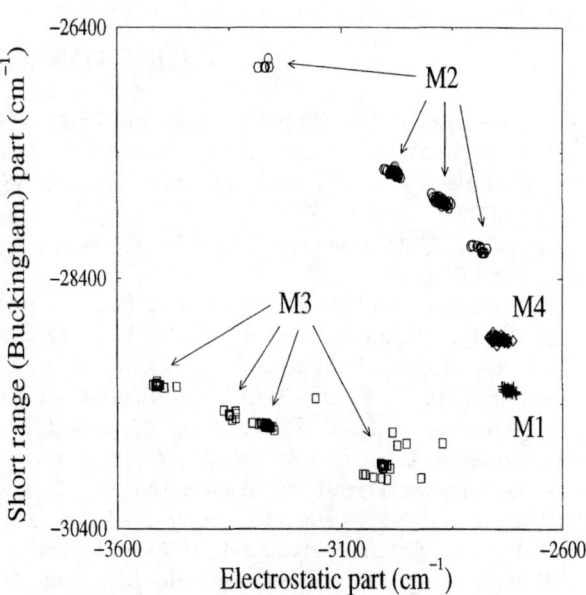

FIGURE 1. Host-guest system.

FIGURE 2. Scatter plot of the terrylene-matrix interactions for terrylene replacing one of the host sites $M_1 - M_4$.

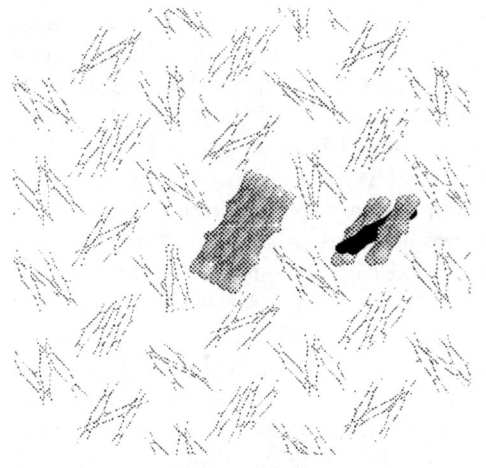
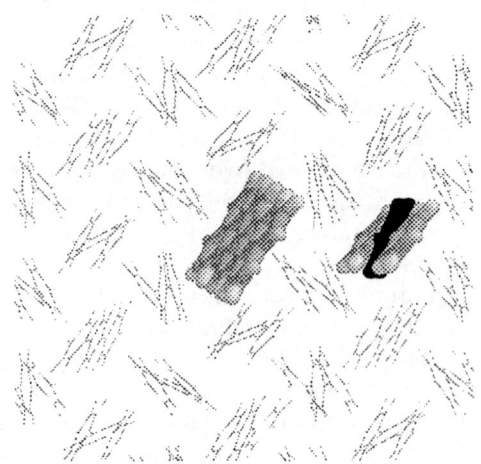

FIGURE 3. View of the model (a,b) plane with terrylene in site M_3: Lowest minimum.

FIGURE 4. Second minimum of M_3.

We thank H. Cailleau, M. H. Lémée-Cailleau and F. Kulzer for helpful discussions. Computations were done with the CRAY T3E computers of the pôle MNI of the Université Bordeaux I and the IDRIS super-computer centre.

REFERENCES

1. Anderson, P. W., Halperin, B. I., and Varma, C. M., *Phil. Mag* **25**, 1-9 (1972).
2. Phillips, W. A., *J. Low Temp. Phys.* **11**, 757-763 (1973).
3. Ambrose, W. P., Basché, Th., and Moerner, W. E., *J. Chem. Phys.* **10**, 7150-7163 (1991).
4. Kulzer, F., Kummer, S., Matzke, R., Braüchle, C., and Basché, Th., *Nature* **387**, 688-691 (1997).
5. Bordat, P., and Brown, R., *Chem. Phys.*, in press.
6. Bordat, P., and Brown, R., *Chem. Phys. Lett.* **291**, 153-160 (1998).
7. Palewska, K., Lipiński, J., Sworakowski, J., Sepiol, J., Gygax, H., Meister, E. C., and Wild, U. P., *J. Chem. Phys* **99**, 16835-16841 (1995).
8. Williams, D. E., *J. Chem. Phys.* **47**, 4680-4684 (1967).
9. Forester, T. R., and Smith, W., *The* DL_POLY 2.0 *User Manual*, CCLRC, Daresbury Laboratory, Daresbury, Warrington WA4 4AD, U. K., (1995).
10. Kummer, S., Kulzer, F., Kettner, R., Basché, Th., Tietz, C., Glowatz, C., and Kryschi, C., *J. Chem. Phys.* **107**, 7673-7684 (1997).
11. Kulzer, F., Matzke, R.; Braüchle, C., and Basché, Th., *J. Phys. Chem. A* **103**, 2408-2411 (1999).
12. Bordat, P., and Brown, R., in preparation.

Molecular Dynamics simulations of crack propagation mode in silica

Laurent Van Brutzel* and Elisabeth Bouchaud[†]

*SRMP CEA Saclay, Gif sur Yvette, France
[†]SRSIM CEA Saclay, Gif sur Yvette, France

Abstract. Using Molecular Dynamics simulations, we investigate the initiation and the propagation of a crack at the atomic scale in β-cristobalite and in amorphous silica. It was found that in the crystal the crack propagates by decohesion whereas in the amorphous the crack propagates by growth and coalescence of cavities like in ductile materials. These different behaviors find their origin in the differences of the two structures.

I INTRODUCTION

Silica glass has been widely studied in the past, but its structure is still somewhat unclear and its mechanical properties at the atomic scale are also unknown.

Ten years ago however, the latter have been studied by Swiler and al. [1,2]. It has been shown that the crack in amorphous silica is not only due to surface defects but strongly depends on the intrinsic structure of the material [3]. Nevertheless, these simulations were not large enough to consider these phenomena on a large scale.

In this paper, we investigate the relationships between the crack propagation mode in the amorphous silica at the atomic scale and the material structure at an intermediate range. In the following section we present the simulation method, in section III we consider the structure of the amorphous silica and in section IV we present the results concerning crack propagation in amorphous silica.

II THE MOLECULAR DYNAMICS METHOD

In this study we use a potential composed of two and three-body terms. The two body interaction between particles i and j is described by :

$$\phi(r_{ij}) = A_{ij} \exp\left[\frac{-r_{ij}}{\rho_{ij}}\right] + \frac{Q_i.Q_j}{r_{ij}} erfc\left(\frac{r_{ij}}{\eta}\right) \qquad (1)$$

Where r_{ij} is the distance between atoms, Q_i and Q_j are the ionic charges and A_{ij}, ρ_{ij} and η are adjustable parameters. The first term is a short range steric Born-Mayer-Huggins repulsion potential effective at distances smaller than the ionic size. The second one, can be viewed as a screened coulombic term. The complementary error function acts on the ionic charge Q_i and Q_j, reducing them to lower values in function of the interaction distance r_{ij}.

The three-body term is given by :

$$\phi(r_{ij}, r_{ik}, \theta_{jik}) = \lambda_i \exp\left[\frac{\gamma_{ij}}{r_{ij} - r_c} + \frac{\gamma_{ik}}{r_{ik} - r_c}\right] \times (cos\theta_{jik} - cos\theta_o)^2 \qquad (2)$$

where θ_o is the equilibrium angle. λ_i, γ_{ij}, γ_{ik} are adjustable parameters. r_c is the cutoff radius. It relates to the covalent Si-O bond. It includes the effects of bond bending and bond stretching.

To solve the Newton motion equations we have used Verlet's algorithm with a time step of 10^{-15} secondes.

The amorphous silica is elaborated by quenching a liquid at 6000 K to 300 K at a cooling rate of 10^{13} K/sec. Temperature was controlled by a rescaling of the atoms velocities.

To study the propagation of the crack the following procedure has been applied. The atoms contained in the bottom and top layers of the sample are frozen. Namely, their dynamics is not computed. To freeze immediatly the upper and the lower atoms prevents a surface reconstruction, and limitates the boundary effects. Furthermore, the periodic boundary conditions along the Z direction are removed. Third a notch is created in the middle of the sample by removing some atoms. The neutral charge of the box is conserved. To simulate crack propagation a displacement on the frozen atoms is imposed every 20 steps. It is equal to impose a constant strain rate. All the simulations presented here are performed with a strain rate of .5/ps.

The simulation box contains 60.000 atoms, and the geometry of the box is 250Å×40Å×80Å. The simulation is performed on a massive parallel machine T3E.

III STRUCTURE OF AMORPHOUS SILICA

The structure of amorphous silica has been fully studied during the past [4–8]. All these simulations find the same short range order, determined by the radial distribution function and the angle distribution function. The first distance Si-O is found to be equal to ∼1.62 Å, Si-Si ∼3.2 Å, O-O ∼2.6 Å. A broad pick centered at 145° is found for the for Si-O-Si angle.

The Molecular Dynamics simulation also gives a picture of the structure of amorphous silica at the atomic scale as being a network of rings of different sizes: 75 % of the structure is composed of 5-7 membered rings. Each ring is composed of

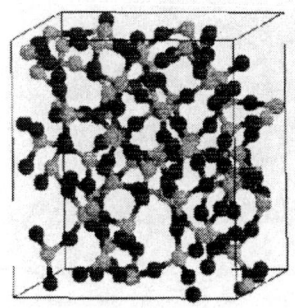 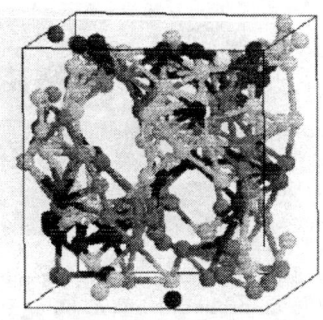

FIGURE 1. Figure a), left picture, is part of 20 Å^3 of a simulated amorphous silica, each point corresponds to an atom. There are no density fluctuations. Figure b), right picture, pictures the structure costituted by the mean positions of rings in figure a). Each point corresponds to a mean position of a ring. We see some regions with great density of points.

tetrahedra, the center of which is occupied by a silicon atom while the tops are occupied by oxygens. These tetrahedra are connected together by so called 'bridging oxygens'.

The structure of amorphous silica at intermediate length scales is still poorly known. This structure is also investigated here in terms of 'ring density' fluctuations. The structure of the network composed of the average positions of the rings is equally studied.

We can see in figure 1b) some regions of greater 'ring density'. Between these regions other regions exist, of lower 'ring density'. The sizes of these regions range between 20 Å and 40 Å. They have no specific geometry. They could be viewed as balls of rings of all sizes which intermingle together. The regions of weak 'ring density' could be viewed as balls combed out. These structures are crucial to determine crack propagation mode.

IV FRACTURE OF AMORPHOUS SILICA

The β-cristobalite, the structure of which is close to a diamond structure, i.e a 6 membered-ring, is first considered.

Under tension, the crack propagates from the notch by successive decohesions, see figure 2. The Si-O bonds are broken near the crack tip, where the stresses are concentrated. The fracture mode is typically brittle. Then the structure relaxes on the new created surfaces. The maximum stress is reached after 25 ps, as shown in Fig.2. The product of this maximum stress by the square root of the length of the notch gives an estimate of the fracture toughness K_{Ic}: $K_{Ic} \simeq 1.5 MPa\sqrt{m}$.

The same experiment on amorphous silica gives rise to a very different propagation mode at the atomic scale. At the beginning, nucleation of small cavities

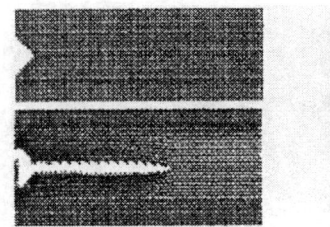 

FIGURE 2. Snapshots of a crack propagating in cristobalite at 0 and 32 picosecondes (2 left pictures) and in amorphous silica at 30 and 43 picosecondes (2 right pictures).

can be observed everywhere in the sample. Some of them subsequently grow and coalescence together to join the macro-crack, which provokes complete failure. This growth and coalescence propagation mode is very close to what is classically observed in the ductile fracture of metallic alloys, or in SiN_3 MD simulations [9]. K_{Ic} is determined in the same way as for the cristobalite and a slightly lower value of the fracture toughness - $1.1 MPa\sqrt{m}$ - is obtained. Note that these estimates of K_{Ic} compare well with the experimental values [10].

The explanation of the difference between the propagation modes in the amorphous and in the crystal resides in the material structure. The amorphous silica network evolves during fracture. A few differences in the radial distribution function are observed during crack propagation, and this indicates that only the ring organisation at intermediate length scales is perturbed by the crack. The average positions of the rings undergo two phases, the growth phase and the coalescence phase. During growth, the rings in the lower 'ring density' region disappear. The great 'ring density' regions behave as hard clusters and remain stable. During coalescence, the greatest rings coalesce together even through some hard clusters regions. Complete fracture is immediately followed by a network relaxation.

V CONCLUSION

We have seen that, at the atomic scale, crack propagations in amorphous silica and cristobalite are very different, due to the differences in structure. In the amorphous silica crack propagation looks like a ductile one. Amorphous silica exhibits regions (of roughly 50 Å) of great 'ring density' which play the role of hard clusters in the silica network. Micro-cavities of an atomic size are nucleated in the weak 'ring density' region, then grow and coalesce. They prevent a localised crack concentration, and lead to a later failure. If the fluctuations of 'rings density' is considered as a 'structural defect', it would be interesting to make a comparison with other types of glasses, including different defects, like alkali glasses.

ACKNOWLEDGEMENTS

We would like to thank Pr. P. Vashishta, Pr. R. Kalia and their group for enlightening discussions about MD simulation in Baton Rouge. We also acknowledge Dr. S. Roux for many fruitful conversations.

REFERENCES

1. Swiler Th.,Simmons J. and Wright C., *J. of non cryst. solids* **182**, 68 (1995).
2. Ochoa R., Swiler Th. and Simmons J., *J. of non cryst. solids* **128**, 57 (1991).
3. PhD Thesis of Guilloteau E., 'La fracture du verre: phénomènes clés à l'échelle nanométrique (1995).
4. Soules Th., *J. of non cryst. solids* **49**, 29 (1982).
5. Woodcock L., Angell C. and Cheesman P., *J. Chem. Phys.* **65**(4), 1565 (1976).
6. Vashishta P., Kalia R. and Rinoo J., *Phys. Rev. B* **41**(17), 12197 (1990).
7. Feuston B. and Garofalini S., *J. Chem. Phys.* **89**(9), 5818 (1988).
8. Vollmayr K., Kob W. and Binder K., *Phys. Rev. B* **54**(22), 15808 (1996).
9. Nakano A., Kalia R. and Vashishta P., *Phys. Rev. Letters* **75**(17), 3138 (1995).
10. Wiederhorn S., *J. Am. Ceram. Soc.* **52**(2), 99 (1969).

Empirical Potential for Si-B-N Ceramics

Marcus Gastreich*, Julian Gale†, and Christel M. Marian‡

*Institut für Physikalische und Theoretische Chemie,
Wegelerstr. 12, 53115 Bonn, Germany

† Department of Chemistry, Imperial College,
South Kensington, London, SW7 2AY, United Kingdom

‡ GMD Forschungszentrum Informationstechnik GmbH,
Schloss Birlinghoven, 53754 St. Augustin, Germany

Abstract. To model amorphous ceramics containing silicon, boron and nitrogen, we have fitted an empirical potential that represents the total energy of a system by means of two- and three-body expressions. The latter are set up to properly describe short- and long-range behavior. We will briefly document the principles behind the fitting procedure, the analytical forms chosen and the early stages of probing the application of the parameter set to ternary compounds which did not form part of the training set.

BOROSILAZANES FROM PRECURSORS

Borosilazane ceramics are high-demand industrial materials consisting of silicon, boron and nitrogen. They are easily synthesised through the so-called precursor-route [1–3] employing molecular compounds reacting with various simple molecules and subsequent heating. More examples are given in [4,5].

The material we are particularly interested in is $Si_3B_3N_7$. It can be obtained from the molecular precursor [(trichlorosilyl)dichloro]aminoborane that is depicted in Fig. 1 by a "one-pot-synthesis". [1] As an intermediate, a polymer is formed,

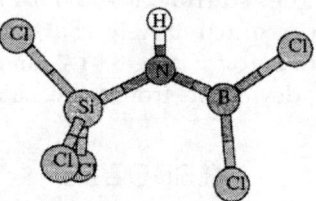

FIGURE 1. [(Trichlorosilyl)dichloro]aminoborane, a molecular precursor to $Si_3B_3N_7$

from which, if desired, fibres may be drawn. Upon subsequent heating, the polymer forms into the ceramic whose stoichiometry has been found to closely approximate

$Si_3B_3N_7$. The resulting material withstands temperatures of about 1700°C and is extremely corrosion resistant (cf. Fig. 2).

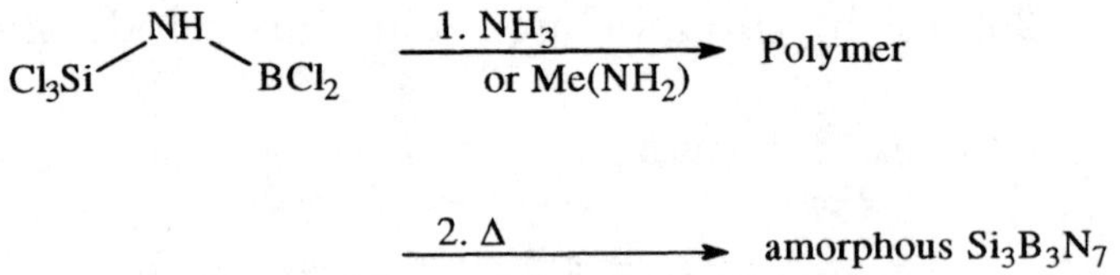

$$Cl_3Si{-}NH{-}BCl_2 \xrightarrow{\text{1. } NH_3 \text{ or } Me(NH_2)} \text{Polymer}$$

$$\xrightarrow{\text{2. } \Delta} \text{amorphous } Si_3B_3N_7$$

FIGURE 2. Synthesis of $Si_3B_3N_7$

As the material is amorphous by design, experimental data is extremely hard to gain. It is therefore of crucial interest to have a good structural model which may be determined by computer simulation, the first stage of which is the parameterization of an interatomic potential.

DERIVATION OF POTENTIAL

For all fitting, we employed the program GULP. [6] Fitting consisted of three stages: (a) parameterization of an appropriate model for hexagonal, rhombohedral, and cubic boron nitride (BN), (b) similarly also for α-Si_3N_4 and β-Si_3N_4, and (c) the merging and refinement having combining the previous sets obtained.

We have utilised an empirical fit based on a many-body description of the potential energy. The observables included were structural data (cell parameters, atomic positions) as well as second-order properties such as elastic constants or phonons close to the Brillouin zone center. It is assumed that all atoms are in mechnical equilibrium at the experimental crystal structures. Thus all energy gradients for the respective structures were fitted to be zero. A detailed description of all analytical formats and parameters shall soon be given in a separate publication.

In principle, we have employed a Coulombic model that is balanced by two- and three-body terms. Correct short- and long-range behavior was ensured by tapering with a polynomial of fifth order starting at $r \approx 4.5$Å and with a specially designed smoothing function (applied to attractively charged particles) that matches the Coulomb energy at a certain cutoff $r_{cut} \ll 0.01$ Å and tends to a constant value at $r = 0$, thereby representing deviations from point charge behaviour.

RESULTS

The reproduction of the training set data is displayed in Tables 1, 2 (boron nitride modifications) and 3, 4 (silicon nitride modifications). We have relaxed the structures with Newton-Raphson optimization procedures while constraining the respective space group symmetries. However, the absence of imaginary phonons at the gamma point demonstrates the correctness of the symmetry applied.

TABLE 1. Data for static relaxation of hexagonal boron nitride, a, b, c are in Å, cell volumes in Å^3, α, β, and γ in degrees; all remaining units are fractional

Observable	Initial	Final	Abs. Deviation
Volume	36.14	35.82	-0.31
a	2.50	2.49	-0.01
b	2.50	2.49	-0.01
c	6.65	6.66	0.00
α	90.00	90.00	0.00
β	90.00	90.00	0.00
γ	120.00	120.00	0.00
1 x	0.00	0.00	0.00
1 y	0.00	0.00	0.00
1 z	1/2	1/2	0.00
2 x	1/3	1/3	0.00
2 y	2/3	2/3	0.00
2 z	0.00	0.00	0.00
3 x	0.00	0.00	0.00
3 y	0.00	0.00	0.00
3 z	0.00	0.00	0.00
4 x	1/3	1/3	0.00
4 y	2/3	2/3	0.00
4 z	1/2	1/2	0.00

TABLE 2. Data for static relaxation of cubic boron nitride, units as above

Observable	Initial	Final	Abs. Deviation
Volume	47.24	47.68	0.44
a	3.61	3.62	0.01
b	3.61	3.62	0.01
c	3.61	3.62	0.01
α	90.00	90.00	0.00
β	90.00	90.00	0.00
γ	90.00	90.00	0.00
1 x	0.00	0.00	0.00
1 y	0.00	0.00	0.00
1 z	0.00	0.00	0.00
2 x	1/4	1/4	0.00
2 y	1/4	1/4	0.00
2 z	1/4	1/4	0.00

TABLE 3. Data for static relaxation of α-Si$_3$N$_4$, units as above

Observable	Initial	Final	Abs. Deviation
Volume	295.94	295.14	-0.79
a	7.81	7.78	-0.03
b	7.81	7.78	-0.03
c	5.59	5.62	0.03
α	90.00	90.00	0.00
β	90.00	90.00	0.00
γ	120.00	120.00	0.00
1 x	0.00	0.99	0.01
1 y	0.31	0.31	0.00
1 z	0.24	0.24	0.00
2 x	0.08	0.07	0.00
2 y	0.56	0.56	0.00
2 z	0.20	0.20	0.00
3 x	1/4	1/4	0.00
3 y	0.08	0.08	0.00
3 z	0.00	0.99	0.00
4 x	0.34	0.34	0.00
4 y	0.39	0.38	0.00
4 z	0.48	0.47	0.00
5 x	1/3	1/3	0.00
5 y	2/3	2/3	0.00
5 z	0.14	0.16	0.01
6 x	0.00	0.00	0.00
6 y	0.00	0.00	0.00
6 z	0.00	0.99	0.01

TABLE 4. Data for static relaxation of β-Si$_3$N$_4$, units as above

Observable	Initial	Final	Abs. Deviation
Volume	144.97	146.84	1.87
a	7.59	7.62	0.03
b	7.59	7.62	0.03
c	2.90	2.91	0.01
α	90.00	90.00	0.00
β	90.00	90.00	0.00
γ	120.00	120.00	0.00
1 x	0.23	0.22	0.00
1 y	0.40	0.40	0.00
1 z	0.01	0.01	0.00
2 x	0.32	0.33	0.00
2 y	0.02	0.03	0.00
2 z	0.02	0.01	0.01
3 x	1/3	1/3	0.00
3 y	2/3	2/3	0.00
3 z	0.00	0.01	0.01

Percentage errors for cell constants (not given above) were always well below one percent. As for the reproduction of second-order properties, we suffer from the scarceness of experimental data for comparison. Where such values are available, we reproduce them reasonably in nearly all cases to extremely well in others. In particular, the LO/TO splitting of the E_{2g} infra-red active mode in hexagonal boron nitride [7] near $\mathbf{k} = 0$ (1370/1610 cm^{-1}) is nicely reproduced with 1368 vs. 1612 cm^{-1} at an arbitrarily chosen small value of $\mathbf{k} = (0.001, 0.001, 0.001)$.

To further probe the predictive power of the potential, we applied the parameter set to (static, rational-function) optimizations of *ab initio*-structures that had been generated at the LDA-level of theory by Kroll and Hoffmann [8]. As may be seen from Table 5, for the most stable of these configurations, unit cell sizes agree to within about 4 pm, atomic positions vary at most 0.02 fractional units. The same is true for all the other *ab initio* structures at hand. Also, the *ab initio*

TABLE 5. Results for most stable proposal of crystalline $Si_3B_3N_7$, units as above

Observable	Initial	Final	Abs. Deviation
Volume	256.94	252.62	-4.31
a	7.48	7.44	-0.04
b	7.48	7.44	-0.04
c	5.29	5.26	-0.03
α	90.00	90.00	0.00
β	90.00	90.00	0.00
γ	120.00	120.00	0.00
1 x	0.77	0.78	0.01
1 y	0.17	0.18	0.00
1 z	0.36	0.35	0.00
2 x	0.19	0.19	0.00
2 y	0.87	0.87	0.00
2 z	0.12	0.11	0.00
3 x	0.98	0.99	0.00
3 y	0.70	0.70	0.00
3 z	0.11	0.10	0.00
4 x	0.03	0.05	0.01
4 y	0.32	0.33	0.00
4 z	0.36	0.35	0.00
5 x	2/3	2/3	0.00
5 y	1/3	1/3	0.00
5 z	0.34	0.32	0.02

energetic sequence of Kroll's and Hoffmann's potential crystalline polymorphs and one simple, fully ionic model based on radii and charges [9] are being confirmed by the potential as it is today.

Dynamical simulations have not yet been performed to such a stage that reporting on them would be worthwhile. Early tests of Monte-Carlo (MC) and molecular dynamics (MD) simulations though show promising results.

SUMMARY

We have fitted an empirical potential based on a balanced description of charged particles. The parameter set at its present stage nicely reproduces the training set of input observables (structural parameters of modifications of boron and silicon nitrides) and with slight imperfections on the corresponding curvatures, viz.: elastic constants, bulk moduli, phonons etc. The model is subject to further testing and appropriate refinement in dynamical investigations.

ACKNOWLEDGMENTS

This work was supported by the Deutsche Forschungsgemeinschaft in the framework of SFB 408. We are indebted to Christa Oligschleger and Alexander Hannemann for helpful discussions. We greatly appreciate having had access to Peter Kroll's and Roald Hoffmann's data prior to publication.

REFERENCES

1. Mühlhäuser, M., Gastreich, M., Marian, C. M., Jüngermann, H. and Jansen, M., *J. Phys. Chem.*, **100**, 16551 (1996).
2. Riedel, R., Bill, J. Kienzle, A., *Appl. Organomet. Chem.*, **10**, 241–256 (1996).
3. Srivastava, D., Duesler, E. N. and Paine, R. T., *Eur. J. Inorg. Chem.*, 855–859 (1998).
4. Baldus, H.-P. and Jansen, M., *Angew. Chem. Int. Ed. Engl.*, **36**, 328–343 (1997).
5. Weinmann, M, Haug, R, Bill, J., Aldinger, F, Schuhmacher, J., and K. Müller, *J. Organomet. Chem.*, **541**, 345–353 (1997).
6. Gale, J. D., *J. Chem. Soc. Faraday Trans.*, **93** 629 (1997).
7. Geick, R, Perry, C. H., and Ruprecht, G., *Phys. Rev.*, **146**, 543–547 (1966).
8. Kroll, P. and Hoffmann, R., *Angew. Chem. Int. Ed. Engl.*, **37**, 2527–2530, (1998).
9. Schön, J. C., and Jansen, M., *Angew. Chem.*, **108**, 1358-1377 (1996).

Schematic mode-coupling approach to experimental data

V. Krakoviack and C. Alba-Simionesco

CPMA, bâtiment 490, Université Paris-XI, F-91405 Orsay

Abstract. The schematic mode-coupling approach is presented as an alternative to the use of the asymptotic critical properties of the mode-coupling equations for tests of the theory on experimental data. Technical aspects concerning the design of an efficient schematic model are discussed and a short review of our investigation of light scattering data is given.

Most tests of the mode-coupling theory for the structural glass transition (MCT) are based on fits of asymptotic laws valid only in the vicinity of the ideal glass transition point introduced by the theory [1]. The reasons for this restriction are clear concerning experimental systems. Common glass-formers are molecular liquids with complicated interactions and often with internal degrees of freedom, a typical example of these systems being glycerol. For non-rigid molecules, no first-principle theory is available so far. In the simpler case where a molecule can reasonably be considered rigid, the structural information needed to evaluate the coupling coefficients of the recently developed molecular MCT [2] can not be determined with sufficient accuracy. Moreover, even if the theoretical predictions could be obtained, relating them to experimental data like light scattering spectra will remain a difficult problem. In the case of computer simulations, the situation is much more comfortable: the interaction potential is chosen from the beginning, simple systems can be considered for which static correlations are obtained with high statistics and the computed 'experimental' data can a priori be directly compared to the theoretical predictions. Nevertheless, one has still to rely on asymptotic results of the theory at some point. Indeed, as is common with first-principle theory in condensed matter physics, MCT is not able to predict accurately the location of the ideal glass transition. It has thus to be determined from the simulation results, typically with the asymptotic power law variation predicted for transport coefficients. This would obviously also be necessary if the first-principle calculation could be performed on a real glass-former.

The application of the asymptotic laws does not go without limitations and difficulties. First of all, critical predictions apply to specific aspects of the overall dynamics and thus only limited parts of the available experimental information are considered in such tests of the theory. Moreover, MCT deals with structural relaxation only. In order to investigate its predictions, one has thus to be sure that this relaxation is well separated from vibrational motion, a requirement for which no clear-cut criterion exists. More fundamentally, theoretical developments beyond the ideal version of MCT as well as the experimental tests of the theory clearly lead to the conclusion that the ideal glass transition is avoided. Recently, the corrections to the asymptotic laws have been investigated in great details for a hard sphere model, allowing a determination of the range of validity of these laws [3]. These are found to be observable only very close to the transition. For the critical prediction to survive in experiments or simulations, the transition has thus to be narrowly avoided, a point that deserves further investigation.

For these reasons, it is desirable to have a means to test the MCT on experimental data which is not restricted to the critical predictions and is able to give an account of the overall dynamics with a minimal discarding of experimental data. Schematic mode-coupling models provide such a possibility. These are simple sets of mode-coupling equations retaining the essential non-linear structure of the theory [4]. They played a major role in the development of the theory as supports for the investigation of the universal properties of the general mode-coupling equations and as illustrative tools to demonstrate the potentialities of the theory. Recently, they have been used as fitting formulas for data obtained from dynamical experiments over a wide time or frequency range [5–7]. In this context, they are considered as effective mode-coupling models for the dynamics. These models are numerically very flexible and can give an account for the dynamics over a wide temperature range, even at quite low temperatures where ideal MCT is not expected to be an adequate description. Thus one should not overestimate the physical meaning of the fitted parameters. Nevertheless, by critically looking at the results, one can address from a renewed point of view the problem of testing the predictions of the MCT on experimental data.

In ref. [6], we have analyzed depolarized light scattering susceptibility spectra for two glass-formers, salol and CKN ($0.4\,\mathrm{Ca(NO_3)_2}\,0.6\,\mathrm{KNO_3}$) [8], with the schematic model:

$$\ddot{\phi}_0(t) + \Gamma_0 \Omega_0 \dot{\phi}_0(t) + \Omega_0^2 \phi_0(t) + \Omega_0^2 \int_0^t m_0(t-\tau)\,\dot{\phi}_0(\tau)\,d\tau = 0,$$

$$m_0(t) = v_1\,\phi_0(t) + v_2\,\phi_0^2(t),$$

$$\ddot{\phi}_1(t) + \Gamma_1 \Omega_1 \dot{\phi}_1(t) + \Omega_1^2 \phi_1(t) + \Omega_1^2 \int_0^t m_1(t-\tau)\,\dot{\phi}_1(\tau)\,d\tau = 0,$$

$$m_1(t) = r\,m_0(t),$$

and the fitting function defined as

$$\chi''(\omega) \propto \omega\,\mathrm{Im}(\mathcal{FT}(\gamma\,\phi_0(t)^2 + (1-\gamma)\,\phi_1(t)^2)),$$

over a temperature range from above the melting point down to around commonly proposed values for the ideal glass transition point. The only temperature dependent parameters were chosen to be an amplitude factor and the effective mode-coupling vertices v_1 and v_2. This schematic model describes a simple transition scenario where only the first correlator encodes the evolution of the dynamics and drives the system towards its ideal glass transition, the second one being completely enslaved to it. The model is designed so that it is able to give an account of the main properties of an arbitrary correlation function close to the ideal glass transition point. Indeed, from the factorization theorem applied to an observable A

$$\phi_A(t) = f_A + h_A \sqrt{\sigma}\, g_\lambda(t/t_\sigma),$$

one sees that three parameters are of special importance, the exponent parameter λ which is a global characteristic of the system under study, and two A dependent parameters, the non-ergodicity parameter f_A and the critical amplitude h_A (to be more precise, one has in fact to consider the product $h_A \sqrt{C}$, where C is the temperature derivative of the separation parameter σ at the transition point; its value results from the matching of the microscopic, fast β-relaxation and α-relaxation time scales as they evolve with temperature and depends on the schematic model). With our model, a given λ completely determines v_1 and v_2 at the transition and thus imposes the values f_0 and h_0. These non-genericities are compensated by the introduction of the second correlator through the choice of the coupling constant r and of the weighting factor γ, which fix respectively the values f_1 and h_1 and their contribution to the fitting function, making virtually realizable any pair (f_A, h_A). The same arguments apply to the other schematic studies [7], where a closely similar two correlator model is used. In the latter calculations, the short time dynamics is not included in the fitted data so that the constraint on the f_A value is removed and thus a successful fit with the second correlator only is possible.

The results of our calculation are displayed on figures 1 and 2. On figure 1, two temperature regimes can be identified. At high temperatures, a linear evolution is found, whereas at lower temperature the behavior changes and the effective vertices evolve much faster with v_1 decaying and v_2 growing rapidly when the temperature is lowered. When considered in the phase diagram of the schematic model (figure 2), the high temperature regime manifests itself as a straight line and shows the evolution of the liquids toward the dynamical transition expected from MCT. As for the second regime, it is found that the vertices follow the transition line becoming closer and closer but never crossing it. The first regime can be associated with the domain of validity of ideal MCT, whereas the second one indicates the breakdown of the idealized approach. Indeed, first-principle MCT shows that the dynamics is governed by purely static quantities, which change smoothly with temperature. It is thus reasonable to expect smooth variations of the effective vertices with temperature if the model is adequate. From the linear regime can be determined by extrapolation ideal glass transition temperatures, $T_c = 257$ K for salol and $T_c = 388$ K for CKN, in good agreement with previously determined values. Our calculation shows thus that MCT can consistently describe the beginning of the slowing-down of the dynamics of supercooled liquids.

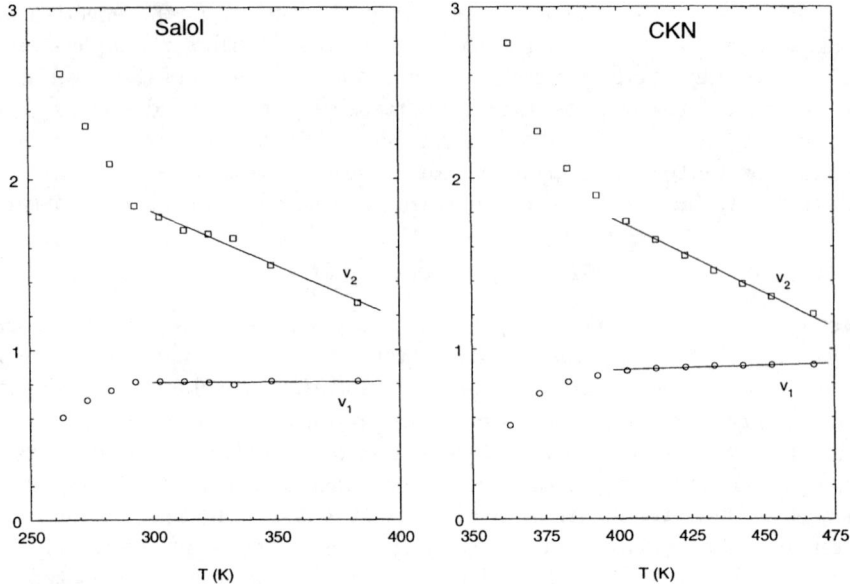

FIGURE 1. Effective vertices as functions of temperature for the light scattering spectra of salol and CKN (from ref. [6]).

Our results are especially suitable to investigate the relevance of the asymptotic laws as potential tests for the theory. It turns out that the situation is rather unfavorable. Indeed, the regime where a linear temperature variation of the effective vertices is found stops at a rather high temperature compared to the extrapolated critical temperature, in a domain where the asymptotic features have still not developed. In the low temperature domain, a wide range of the transition line is closely followed, so that the critical parameters like λ, which vary continuously along the transition line, are not well defined. It seems thus unclear from the present investigation that the asymptotic critical properties of the MCT equations can in general be observed unambiguously on experimental data and that the assumption of a narrowly avoided ideal glass transition is valid as regards such quantitative predictions of the theory.

The authors are indebted to Professor W. Götze and Dr. M. Fuchs for providing them the codes for some parts of the calculations. We are also grateful to Professor H.Z. Cummins and Dr. G. Li for the use of their experimental data. Dr. G. Tarjus is acknowledged for fruitful discussions.

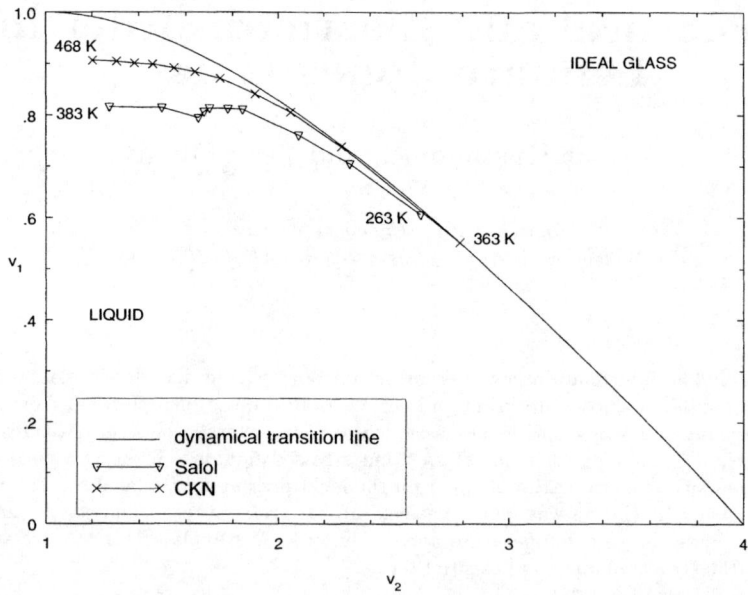

FIGURE 2. Phase diagram of the schematic model and effective vertices for the light scattering spectra of salol and CKN (from ref. [6]).

REFERENCES

1. For a recent review, see W. Götze, *J. Phys.: Condens. Matter* **11**, A1 (1999).
2. R. Schilling and T. Scheidsteger, *Phys. Rev. E* **56**, 2932 (1997).
3. T. Franosch, M. Fuchs, W. Götze, M.R. Mayr, and A.P. Singh, *Phys. Rev. E* **55**, 7153 (1997); M. Fuchs, W. Götze, and M.R. Mayr, *Phys. Rev. E* **58**, 3384 (1998)
4. W. Götze, *Z. Phys. B* **56**, 139 (1984); E. Leutheusser, *Phys. Rev. A* **29**, 2765 (1984).
5. C. Alba-Simionesco and M. Krauzman, *J. Chem. Phys.* **102**, 6574 (1995).
6. V. Krakoviack, C. Alba-Simionesco, and M. Krauzman, in *Non-equilibrium phenomena in supercooled fluids, glasses and amorphous materials*, Singapore: World Scientific, 1996, pp. 373-374; V. Krakoviack, C. Alba-Simionesco, and M. Krauzman, *J. Chem. Phys.* **107**, 3417 (1997).
7. T. Franosch, W. Götze, M.R. Mayr, and A.P. Singh, *Phys. Rev. E* **55**, 3183 (1997); A.P. Singh, G. Li, H.Z. Cummins, T. Franosch, M. Fuchs, and W. Götze, *J. Non-Cryst. Solids* **235-237**, 66 (1998).
8. G. Li, W.M. Du, A. Sakai and H.Z. Cummins, *Phys. Rev. A* **46**, 3343 (1992); G. Li, W.M. Du, X.K. Chen, H.Z. Cummins and N.J. Tao, *Phys. Rev. A* **45**, 3867 (1992).

Structural and Elastic disorder in Lennard Jones Glass

Tamar Kustanovich and Zeev Olami

Department of Chemical Physics,
The Weizmann Institute of Science, Rehovot 76100, Israel

Abstract. The interrelations between structural and elastic disorder in glasses and glass forming liquids pose important and yet unresolved questions. Here we show how the short range structure and local elastic features are related via local pressures for mono-atomic 6-12 Lennard Jones glasses and stressed liquids. There is a clear cut dependency of the coordination number on the local pressures (similar for liquids and glasses). A linear relation was found between pressure-related terms and elastic terms, both for glasses at zero temperature and for liquids. We explain this by a relation between the structure and the potential terms.

In monoatomic glasses one can distinguish between structural and elastic disorder. Structural disorder can be approached through the study of frustration effects [1] and variations in local environments [2]. The elastic disorder, on the other hand, is manifested by the variation of the local elastic constants. Complete information about it can be derived using the knowledge of the structure and the potential. Thus, correlations between the structural and elastic disorder will occur in any glass [3].

Qualitatively, structural and elastic disorders arise mainly due to the existence of internal pressures in glasses. In a liquid state, large variations in local environments induce a great variety in the local pressures felt by the atoms [4]. When a glass is created from a liquid, its atoms will reach one of the local minima of energy (defined by zero force), but not the global energy minima (corresponding to the crystal). Therefore, the local pressures in the compressed liquid will not be fully relaxed in the quench. Furthermore, the existence of different stressed/stretched local environments will result in varying local elastic constants, in compressed liquids and glasses. This elastic disorder (ED) determines the scale and noise of the interactions in the global dynamical matrix and thus it defines the spectra, the localization and the scattering effects. A major motivation for a study of ED and correlations in them are recent studies on ordered lattices with ED [5]. Those works have shown that effects on the spectra and the localization in ordered lattices and glasses are very similar. Thus it seems that ED is a main cause for scattering in glasses.

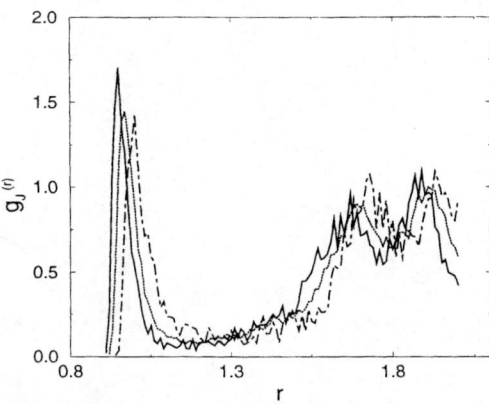

FIGURE 1. $g_J(r)$ vs. r for 3 averaged values of J - solid curve for compressed local environments, dotted curve for small values of $|J|$ and dashed-dotted line for stretched local environment (high values of J). r's in LJ units of $2^{1/6}\sigma$.

Consider a solid at equilibrium state at zero temperature. The average pressure on a surface of a cell of volume V_i around atom i can be defined as

$$P_i = \frac{1}{V_i} \sum_j U'(R_{ij}) R_{ij} \qquad (1)$$

where $'$ denotes derivative with respect to R_{ij}. To simplify the calculations and separate the elastic terms from structural, we consider the quantity

$$J_i = \sum_j U'(R_{ij}) \qquad (2)$$

For 6-12 monoatomic Lennard-Jones potential (LJP) (and alike short range repulsive-attractive potentials) $J_i = const1 \cdot P_i V_i + const2$ with a very good accuracy. Even in a fully relaxed glass, with zero global pressure, there are large variations of local pressures. (Details about the systems considered and further discussion of the distribution of J' appear in [6], [7]).

In glasses and liquids, the usual description of the radial structure is given through the radial pair distribution function $g(r)$. However, g(r) is not sensitive to variations in the local quantities. To observe the structural effects of the local pressures directly, we consider the conditional radial pair distribution $g_J(r)$, which is defined as the average number of neighbors at distance r for a specific given J.

As expected from the short range nature of the potential, J is dominated by contributions from the first peak of $g_J(r)$, while there is an almost constant contribution from further neighbors. A reasonable approximation to J is: $J = J_{fp} + J_{bg}$ where we denote $_{fp}$ and $_{bg}$ as the first peak and further neighbors contributions, respectively. Though the main effect on J's arises from the first peak in $g_J(r)$, there are also large structural effects in the second peak.

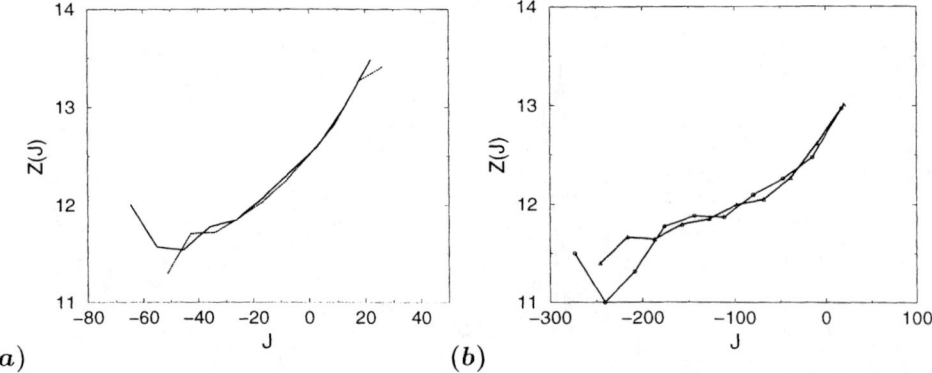

FIGURE 2. (a) $z(J)$ vs. J for two different samples of glass at 0K. The plot has a good quadratic fit for $J > 25$ (except for the upper point which is statistically meaningless as it represents $\sim 0.006\%$ of atoms). (b) $z(J)$ vs. J for two samples of liquid at 120K, from which glasses at (a) where created.

The first peak is described by the coordination number and the average peak position, given by the following definitions:

$$z(J) = \int_{fp} g_J(r) dr \qquad (3)$$

$$R(J)_{av} = \frac{\int_{fp} g_J(r) r dr}{z(J)} \qquad (4)$$

where the cutoffs of those integrals are defined by the end of the first peak. Two major effects arise from the analysis of the dependence of $z(J)$ and $R(J)_{av}$ on J: a shift of the average position of the peak $R(J)_{av}$ and a change in the coordination number $z(J)$.

We have found a square dependence of $z(J)$ on J for medium and high values of J's (both in 0K glass and in 120K liquid). For lower J's the environments become denser and one observes stronger fluctuations and lower coordination numbers. Notably the coordination numbers are all close to 12, which is the so called 'icosahedral' packing. However there are very strong variations. The number of neighbors increases with J until it reaches a value of 13.5 and then one observes no more environments. The system cannot sustain larger $z(J)$ and $R(J)_{av}$'s because such an environment will collapse.

The existence and the form of correlations between the J's and the local elastic constants is another related and important issue. It was generally shown by S. Alexander [3] that such correlations should exist in glasses. The form of those correlations varies for different types of potentials and thus yields an important tool for studying both structural and dynamical features of various glasses. For the LJP, those relations are simple and clear cut.

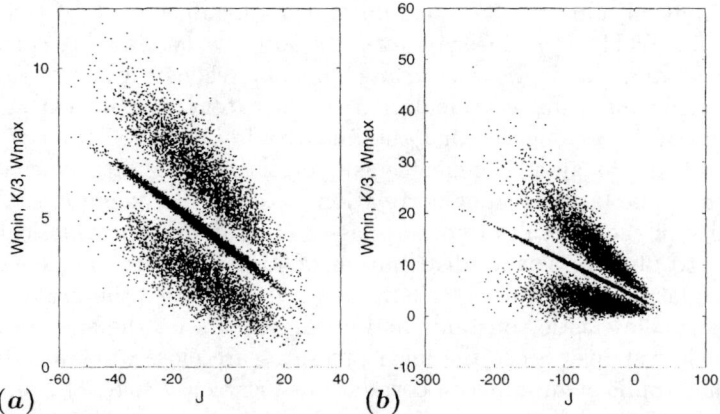

FIGURE 3. (a) Upper, middle and lower plots are, respectively, the largest ($Wmax_i$), the average ($K_i/3$) and the smallest ($Wmin_i$) eigenvalues of A_i vs. J_i for glass at 0K. (b) $Wmax_i$, $K_i/3$ and $Wmin_i$ vs. J_i for liquid at 120K.

Defining the local dynamics of an atom i (with other atoms' positions being fixed) by the matrix $A_i = \sum_j \partial_\alpha \partial_\beta U_{ij}$ (where α and β denote components of a vector), the eigenvalues of A_i vs. J_i (see Figure 3) were calculated for LJ glasses and liquids. A linear relation between trA and J was found both in liquids and glasses![1] Next, we suggest an explanation for those results.

For a radial pair potential the trace of A_i is:

$$K_i = \sum_j [U''_{ij} + \frac{2U'_{ij}}{R_{ij}}] \tag{5}$$

As for J's, K can be reasonably separated into K_{fp} and K_{bg} which is almost constant. For LJP, $U'_{ij}/R_{ij} \ll U''_{ij}$ with R_{ij} in the first peak of $g(r)$. Thus, the linear contribution of the second terms in eq. (5) is very small. One can use first order estimates for J_{fp} and K_{fp} to estimate the $K - J$ curve to be [7]:

$$K_{fp} = (J - J_{bg})F2(U(R_0)) + z(J)F1(U(R_0)) + K_{bg} \tag{6}$$

with $F1(U(R_0)) = \frac{[U'''(R_0)]^2 - U'(R_0)U''''(R_0)}{U'''(R_0)}$ and $F2(U(R_0)) = U''''(R_0)/U'''(R_0)$.
Estimates of K from eq. (6) using $R_0 = R(J)_{av}$ reveal that, although the corrections due to $z(J)$, $F1(U(R(J)_{av}))$ and $F2(U(R(J)_{av}))$ are quite large, those contributions cancel each other to create a linear curve (with error of fit $< 1\%$)!

For the LJP, there is a wide distribution of local J's and K's, which is strongly coupled to structural features, such as the position of the first peak of $g(r)$, and

---

[1] The same results where obtained with another short-range attractive-repulsive pair potential.

to the coordination number. For medium and large values of J, $z(J)$ has a good quadratic fit for J. Thus the linearity in the $K-J$ curve is related to constraints on both structural and elastic disorder. Note that our reconstruction of the linearity does not provide quantitative calculation of its extent and its extremely small noise. Qualitatively, we suggest that the linearity is a manifestation of the radial homogeneity of the local environments (as compared to the large variations in U' and U'') and of the lack of strong correlations with farther neighbors.

The validity of those results to other glasses is an important practical question. One expects to observe a wide distribution of local pressures in densely packed monoatomic glasses. The wide distributions in J imply also the existence of wide distributions in the elastic constants and strong effects on the spectra and localization in such systems. Since the local pressures are closely related to the local density of the atomic environments, correlations between J's, $R(J)_{av}$ and $z(J)$ will exist in such systems. Whether the linear and almost noiseless $K-J$ relation is a general feature, is a delicate question, which is under current study. This relation seems to be a general feature for short range monoatomic repulsive-attractive potentials. For other monoatomic glasses (e.g. Dzugutov potential [8]), an average linear $K-J$ relation is observed, with large noise. Since the disorder increases with the temperature, the relative noise decreases in the liquid state.

The obtained results for short range effects imply also that there will be longer range effects. This rises very interesting questions: determining the range where the interrelations between structural and elastic properties exist and characterizing the form they take at farther distances. An understanding of the local structural-elastic relations and knowledge of their farther correlations might also be constructive steps toward understanding the global spectra, which are mainly determined by the wide distribution of local elastic disorder.

Acknowledgments

We thank the late S. Alexander for many illuminating discussions about the subject. T. K. thanks S. Simdyankin for his remarks during the proceeding.

REFERENCES

1. Sadoc, J.F., *J. Non Cryst. Solids.* **44**, 1 (1981).
2. Jund, P., Caprion, D., Jullien, R., *Euro. Phys. Lett.* **37**, 547 (1997).
3. Alexander, S., *Phys. Rep.* **296**, 65 (1998).
4. Egami, T., Madea, K., Vitek, V., *Phil. Mag. A.* **6**, 883 (1980).
   Egami, T., Vitek, V., *Phys. Stat. Sol.* **144**, 145 (1987).
5. Galanti, B., Olami Z., cond-mat/9903392 (1999).
6. Kustanovich, T., Alexander, S., Olami. Z., *Physica A* **266**, 434 (1999).
7. Kustanovich, T., Olami, Z., cond-mat/9906445 (1999).
8. Dzugutov, M., *Phys. Rev. A.* **46**, R2984 (1992).

Solvable models of glass transition

Matthieu Micoulaut

Laboratoire de Gravitation et Cosmologie Relativistes
ESA 7065, Université Pierre et Marie Curie
Tour 22, Boite 142, 4, Place Jussieu, 75252 Paris Cedex 05, France

Abstract. Simple statistical agglomeration models can provide a universal link between the local structure and the glass transition temperature in network glasses. We first stress the physical features of the models and the relevancy of the hypothesis made and then show how to define the glass transition temperature. The models are applied to various types of binary, ternary and multicomponent chalcogenide glass networks and the predictions compared to experimental data.

I INTRODUCTION

Although much attention has been devoted to the understanding of the glass transition problem [1], a general relationship between the temperature of this transition (when measured under standard conditions, at e.g. constant heating rate), and some easily reliable quantities is still lacking [2]. In this paper, we show that there are some aspects of structure or connectivity that apparently play an important part in determining the absolute magnitude of T_g. The construction and the prediction of this temperature from solvable agglomeration models is parameter-free and can be easily extended from binary to ternary, etc. glassy systems.

II AGGLOMERATION MODEL

Let us imagine a liquid that is slowly cooled and atomic motions are progressively arrested. In network glasses, the Arrhenius-like increase of viscosity upon cooling to the glass transition is intimaly related to a decrease of dangling bonds as the starting network is polymerized. One typical physical process taking place in the supercooled liquid should therefore be a kind of sticking process in which clusters (or macromolecules) agglomerate together. Also, one should remark that the most important determinant of chemical and physical properties of a glass is the concentration of different types of atoms involved. Thus the simplest level of description of such agglomeration processes should use local structural configurations (LSC) defined by the concentration, corresponding to short-range order (SRO), and consequently to a random network description of the glass. We should stress here that the next level of description, using intermediate range order, is very similar to the SRO construction [3]. The LSC's in Ge_xSe_{1-x} binary can for instance be the germanium tetrahedrally coordinated to selenium atoms, or in silica based glasses the silicon tetrahedra $Q^{(k)}$ (the subscript k refers to the number of bonding oxygens on each $SiO_{4/2}$ tetrahedron).

Consider a typical cluster with a certain LSC distribution $\{p_i^0\}_{i=1..N}$. As long as the viscosity is not too high, other LSC can stick on this typical cluster, creating new covalent bonds i-j with probability:

$$p_{ij}^L(T) = \frac{W_{ij}}{\mathcal{Z}} p_i^0 p_j^0 e^{-E_{ij}/k_B T} \tag{1}$$

where W_{ij} is a statistical factor corresponding to the number of equivalent ways to stick a LSC i on a LSC j being part of a cluster (W_{ij} is thus related to the coordination numbers m_i and m_j of the LSC) and E_{ij} is the i-j LSC bond energy. $\mathcal{Z}$ normalizes the bond distribution. The creation of these new bonds produces a local variation in the probability (or concentration) distribution of the cluster and it can be encoded in the following master equation:

$$\frac{dp_i}{dt} = \frac{1}{\tau}\left[\frac{1}{2}\sum_{j=1}^N (1+\delta_{ij})p_{ij}^L(T) - p_i^0\right] \tag{2}$$

where τ represents the mean agglomeration time and (2) represents a system of $(N-1)$ non linear differential equations. At solidification temperature T_s ($T_s = T_m$ for a crystal or $T_s = T_g$ for a glass) one should reach a stationary state and $dp_i/dt = 0$, i.e. the variation of local probability distribution should be minimized.

How can we distinguish a glass from a crystal ? Imagine that a local fluctuation ϵ_i appears in the vicinity of a stationary solution, satisfying a linearized version of equ. (2). As usual, we can distinguish three types of singular points by means of the linearization. If all the roots of the characteristic equation of the linearized system have a negative real part, the solution is a stable attractor, i.e. it will show the preferential agglomeration process, which corresponds to nucleation of the crystal. Thus there will be no possibility for a fluctuation to grow, and we identify the stable stationary solution with a crystal, and $T_s = T_m$. If all the real parts are positive, one gets an unstable stationary solution. If both are present, a saddle point solution is obtained (fig. 1). The glass correponds to the latter characteristic, because it is neither a stable nor an unstable system, it has metastable character, and $T_s = T_g$. There is indeed still a chance for the system to escape (in other words for a fluctuation to grow) from the stationary saddle point and to fall on the stable crystalline attractor, which always happen experimentally when a glass is annealed.

One can extend such a description to binary and ternary glasses as well. First, we consider the case when $N = 2$, i.e. when there are only two different types of

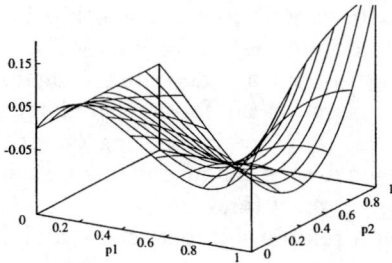

FIGURE 1. The right-hand side of one of the equation (2) for $N = 3$ in a simple polygon model [4]. The system out of equilibrium can fall on the attractive $p_3 = 1$ stationary solution (crystallization, and $T_s = T_m$) and never on $p_2 = 1$ or $p_1 = 1$. In some situations, the liquid can stay in the metastable state characterized by the saddle point solution at ($p_1 = 0.6$, $p_2 = 0.2$, $p_3 = 0.2$) and $T_s = T_g$. Note that the plot has to be truncated in order to have $p_1 + p_2 + p_3 = 1$.

LSC. We denote them by A and B with their respective coordination numbers m_A and m_B. The system (2) reduces then to a single equation with only one variable, $p^0 = x$, e.g. the probability of occurence of the LSC B (and set equal to the concentration of B species). The solution of (2) yields:

$$T_g = \frac{\Delta_B}{k_B \ln\left[\frac{m_B(2x-1)}{m_A(x-1)}\right]} = \frac{T_0 \ln\left[\frac{m_B}{m_A}\right]}{\ln\left[\frac{m_B(2x-1)}{m_A(x-1)}\right]} \quad (3)$$

where $\Delta_B = E_{AB} - E_{AA}$ is introduced when computing the probabilities p_{AA} and p_{AB} from equ. (1) (we have neglected the possibility of BB bonds because we are dealing in the following with low modified glasses only, and below the stoichiometric composition). One can see in the second part of equation (3) that the relationship can be made parameter-free, by considering the limit $x \simeq 0$, when $T_g \simeq T_0$. The initial network (with glass transition temperature T_0) is made of A LSC only (e.g. the selenium network in Ge_xSe_{1-x} systems) and one gets from the first part of equ. (3): $\Delta_B = k_B T_0 \ln[m_B/m_A]$. In order to predict the glass transition temperature in a binary glass $A_{1-x}B_x$, there is just need of the coordination number of the involved LSC's and the initial glass transition temperature T_0 of the A network. Finally, we can obtain a linear equation at the very beginning of structural modification:

$$\left[\frac{dT_g}{dx}\right]_{x=0,T_g=T_0} = \frac{T_0}{\ln\left[\frac{m_B}{m_A}\right]} \quad (4)$$

The application to ternary glass networks $A_{1-x-y}B_xC_y$ is slightly different, because when $N = 3$, there are two non-linear equations to solve in terms of two probabilities $p_B^0 = x$ and $p_C^0 = y$. However, one obtains a saddle point solution from (2), yielding again a parameter-free relationship between x, y and T_g, because the new bond energy differences $\Delta_C = E_{AC} - E_{AA} = k_B T_0 \ln[m_C/m_A]$ and $E_{BC} - E_{AA} = k_B T_0 \ln[m_C m_B/m_A^2]$ are determined again from boundary conditions (from the binary AC glass for the former, similarly to Δ_B, from the binary slope equation for the latter) [5]. One interesting quantity in such systems (and in multicomponent chalcogenides) is the average coordination number, defined by $\bar{r} = m_A(1 - x - y) + m_B x + m_C y$ (and $m_A = 2$). From the saddle point solution of equ. (2), we obtain a relationship between $\bar{r}$ and T_g, to be compared with experiment:

$$\bar{r} = \frac{2 m_B m_C \left[m_B m_C \alpha \gamma (\gamma \alpha - \gamma - \alpha) + 2 r_C \alpha^2 (1 - \gamma) + 2 r_B \gamma^2 (1 - \alpha) \right]}{\left(2 r_C \alpha - 2 r_B \gamma - r_B r_C \alpha \gamma \right)^2 + 8 r_B r_C \alpha \gamma} \quad (5)$$

where $\alpha = (2/m_C)^{(T_0/T_g)}$ and $\gamma = (2/m_B)^{(T_0/T_g)}$. The slope in the limit $\bar{r} = 2$ (i.e. x=y=0) has also a simple expression:

$$\left[\frac{dT_g}{d\bar{r}} \right]_{\bar{r}=2, T_g=T_0} = \frac{T_0}{(m_B - 2) \ln\left[\frac{m_B}{2}\right] + (m_C - 2) \ln\left[\frac{m_C}{2}\right]} \quad (6)$$

III COMPARISON WITH EXPERIMENTAL DATA

The obtained relationships (4) and (5) can be compared to the experimentally measured glass transition temperatures in binary, ternary and multicomponent chalcogenide glass systems.

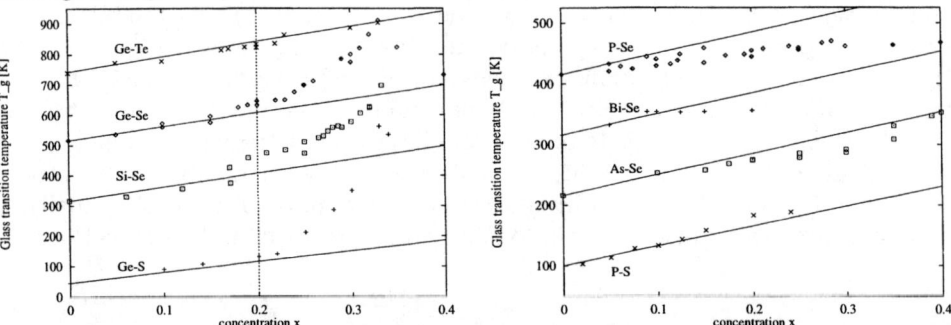

FIGURE 2. Binary IV-VI (left side) and V-VI (right side) chalcogenide glasses (e.g. $Ge_x Se_{1-x}$). The lines represent the slope equation (4) with $m_A = 2$ and $m_B = 4$ for the IV-VI glasses, and $m_B = 5$ for V-VI glasses. Data have been displaced by 200 K and 100 K for a clearer presentation [8]. The vertical shaded line corresponds to the critical average coordination number $\bar{r}_c = 2.4$, predicted by Phillips and Thorpe [9].

Given the initial glass transition temperature T_0 of vitreous sulphur (245 K [6]), selenium (316 K [6]) and tellurium (343 K extrapolated from the data in [7]), we have plotted the equations (4) and (5) for chalcogenides including elements of Group IV and V. We can see that equation (4) predicts the $T_g(x)$ trend at low concentration for all the binary systems IV-VI and V-VI systems displayed (fig.2). From obvious structural considerations, we can insert in equ. (4) the value $m_B = 4$ (Group IV) or $m_B = 5$ (Group V), $r_A = 2$ and T_0, to be compared with the plotted experimental measurements on glass transition temperatures. The prediction gives also an indirect evidence of the stiffness transition (occuring at $\bar{r} = 2.4$ following the theory of Phillips and Thorpe [9]). For $\bar{r} > 2.4$ ($x > 0.2$ in IV-VI glasses), the network looses its random character and chemical ordering occurs, due to the chemical stability composition at x=0.333. Thus the description in terms of a random network of $A - A$ and $A - B$ bonds should fail at this concentration. This is clearly seen for the Ge_xSe_{1-x} data, which start to deviate from the equation (4) at $x = 0.18$, and even more at $x = 0.24$, consistently with Mössbauer spectroscopy [10]. The sulfide system behaves very similarly, as seen on fig. 2 (and still $r_B = 4$), although the structure of the initial glass (x=0) is rather different (chains and S_8 rings). The addition of germanium leads to a random network composed of $GeS_{4/2}$ linkages between S chains and rings. Note that there is no deviation for the Ge-Te compound at x=0.2. This can be related to the fact that $c - GeTe_2$ does not exist (in contrast with the existence of $c - SiSe_2$, $c - GeSe_2$, etc.) and probably that chemical ordering probably does not occur at the same concentration. As a consequence, the network of Ge_xTe_{1-x} can be thought as random.

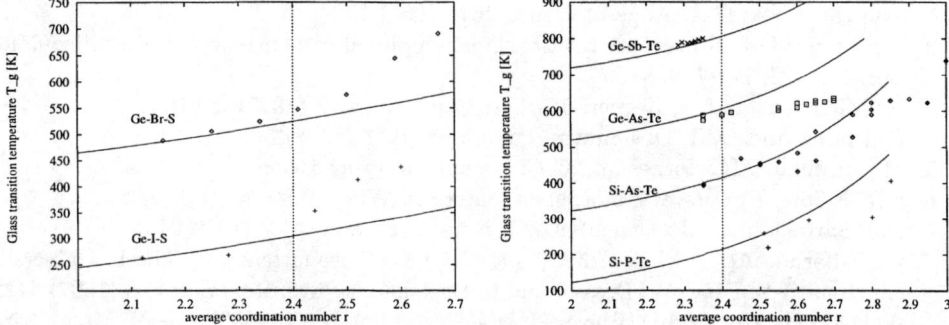

FIGURE 3. Glass transition temperature in ternary chalcogenides as a function of average coordination number $\bar{r}$. Left: chalcohalide glasses. Right: ternary tellurides. The curves represent equ. (5) with $m_A = 2$ and m_B and m_C inserted following the Group of the Periodic Table. Data have been displaced for simplicity and are taken from [11]

Agreement of the prediction with experimental measurements is also obtained in V-VI network glasses. For all the systems displayed in figure 2, the slope equation (4) gives the correct trend in the variation of the glass transition temperature with network modification. Deviation of the stochastic prediction of (4) is here also supposed to occur at $\bar{r} = 2.4$, corresponding to $x = 0.4$ in the As-Se system. There is some evidence that the Bismuth atom could be five-fold coordinated in binary Bi-Se glasses [8]. If this is the case, the stiffness transition may occur $x \simeq 0.13$ (corresponding to $\bar{r} = 2.4$ and may be observable in the T_g data, as well as the deviation of the stochastic prediction of equ. (4)). We can observe the same kind of agreement in ternary chalcogenides, where the glass transition temperature is given as a function of the average coordination number by equ. (5) or (6). The figure 3 shows that in very different systems (germanium chalcohalide glasses, thus $m_B = 4$, $m_C = 1$, and telluride glasses) the network is random for $\bar{r} < 2.4$ and the description in terms of A-A, A-B, A-C and B-C bonds only is accurate. The same deviation is observable at $\bar{r} = 2.4$ which can again be interpreted by the occurence of chemical ordering due to the stiffness transition (e.g. occurence of a Sb_2Te_3 phase in the Ge-Sb-Te system). However, we should point out that experimental measurements in the low modified region of telluride systems (plot of the right side of fig. 3) should be realized in order to definitely confirm the prediction. Other systems exhibit the same universal trends in the glass transition temperature variation and the construction can be extended to quaternary and multicomponent chalcogenide systems in a simple fashion [5].

Acknowledgements: The author would like to thank P. Boolchand, R. Kerner, G.G. Naumis, J.C. Phillips and M.F. Thorpe for very interesting comments and discussions on this subject.

REFERENCES

1. read this book; C.A. Angell, Science **267** (1995) 1924;
2. For a review of all previous relationships (empirical or heuristic), see M. Micoulaut, Eur. Phys. J. B **1** (1998) 277
3. M. Micoulaut and R. Kerner, J. Phys: Cond. Matt. **9** (1997) 2551
4. R. Kerner and D.M. Dos Santos, Phys. Rev. B**37** (1988) 3881
5. M. Micoulaut, R. Kerner and G.G. Naumis, in preparation
6. S.R. Elliott, Physics of Amorphous Materials, Wiley 1989
7. D.J. Sarrach and J.P. Deneufville, J. Non-Cryst. Solids **22** (1976 245
8. G. Saffarini, Appl. Phys. A**59** (1994) 385; D. Selvanathan MS Thesis University Cincinnati; X. Feng, W. Bresser and P. Boolchand, Phys. Rev. Lett. **78** (1997) 4422; A. Feltz, H. Aust and D. Blayer, J. Non-Cryst. Solids **55** (1983) 179; Y. Monteil and H. Vincent, Z. Anorg. Allg. Chem. **428** (1977) 259; D. Lathrop, M. Tullius, T. Tepe and H. Eckert, J. Non-Cryst. Solids **128** (1991) 208; M.B. Myers, J.C. Schotmiller and W.J. Hillegas, Anal. Calor. **2** (1970) 309
9. J.C. Phillips, J. Non-Cryst. Solids **34** (1979) 153; M.F. Thorpe, J. Non-Cryst. Solids **57** (1983) 355
10. W. Bresser, P. Boolchand and P. Suranyi, Phys. Rev. Lett. **56** (1986) 2493
11. J. Heo and J.D. Mackenzie, J. Non-Cryst. Solids **111** (1989) 29; A.B. Seddon and M.A. Hemingway, J. Non-Cryst. Solids **161** (1993) 323; A. Srinivasan and E.S.R. Gopal, J. mater. Sci. **27** (1992) 4208; P. Lebaudy, J.M. Saiter, J. Grenet, M. Belhadji and C. Vautier, J. Mater. Sci. Eng. **132A** (1991) 273

Crystallization in hard sphere systems: a structural analysis

Patrick RICHARD[1], Annie GERVOIS[2], Luc OGER[1] and Jean-Paul TROADEC[1]

[1] *Groupe Matière Condensée et Matériaux, UMR CNRS 6626, Université de Rennes I, 35042 Rennes Cedex, France*
[2] *Service de Physique Théorique, Direction des Sciences de la Matière, CEA/Saclay, 91191 Gif-sur-Yvette Cedex, France*

Abstract. We present numerical simulation results of crystallization of hard sphere systems. The study of the distribution of a local bond order parameter (Q_6) gives precise informations on the structure and allows to detect without ambiguity the presence of fcc or hcp crystalline zones. We have found, as expected, that the fcc symmetry is more stable than the hcp one in hard sphere crystallized systems. The fraction of hcp clusters in the system is found to decrease as the propensity to crystallize increases. The method developed here to detect fcc and hcp order can also be applied to bcc order.

One of the most striking properties of a hard sphere system is the existence of an entropy-driven ordering transition when the packing fraction C increases [1–3]. This transition can be explained by the fact that the system cannot maximize its entropy by adopting an average configuration which is disordered. Disordered systems at packing fractions higher than the packing fraction of freezing can exist, but they are metastable. The maximum packing fraction (corresponding to the so-called random close packing, RCP) that such packings can reach is approximately $C_{RCP} \simeq 0.64$. [4–7]. The phase diagram for hard sphere systems is made of four branches (see Figure 1). A fluid branch starts at $C = 0$ (perfect gas) and continues up to the freezing-point volume fraction ($C \approx 0.494$). Then, the phase diagram splits into two parts. The upper one corresponds to a branch which is the metastable extension of the fluid branch; it diverges at $C = C_{RCP}$. The lower one corresponds to the stable branch. It begins with a plateau where fluid and solid coexist until the melting-point ($C \approx 0.545$). At this point the system belongs to the solid branch that diverges at $C = 0.7405$ (fcc or hcp packing fraction).

Steinhardt et al. [8] have proposed a quantitative measure of the local orientational order. The method consists in assigning the quantity

$$Q_{lm} = Y_{lm}(\theta(\vec{r}), \phi(\vec{r})) \tag{1}$$

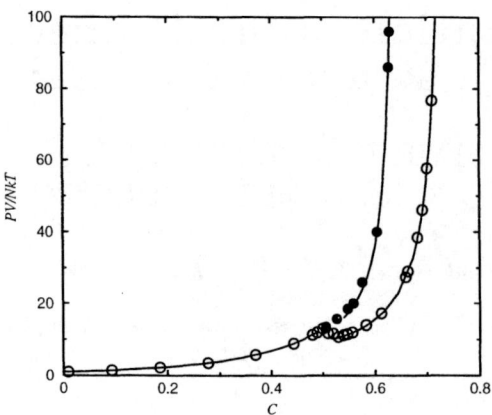

FIGURE 1. Evolution of PV/NkT for thermodynamically stable (o) and metastable (•) packings

to every 'bond' joining a sphere to its neighbors (defined in the present work as the Voronoï neighbors), where $Y_{lm}(\theta(\vec{r}),\phi(\vec{r}))$ are spherical harmonics, each bond being identified by its midpoint $\vec{r}$ and its polar angles $\theta(\vec{r})$ and $\phi(\vec{r})$. As Q_{lm} depends on the reference coordinate system, one must consider rotationally invariant combinations, such as

$$Q_l = \left(\frac{4\pi}{2l+1}\sum_{m=-l}^{l}|\langle Q_{lm}\rangle|^2\right)^{1/2} \quad (2)$$

where $\langle Q_{lm}\rangle$ is the average of Q_{lm} over the bonds.

Since the lowest nonzero Q_l in common to the cubic, hexagonal and icosahedral symmetries corresponds to $l = 6$ [8], Q_6 can be taken as an order parameter. Relation (2) can be used in two ways. One can either average Q_{lm} over all the bonds of the packing (Q_6^{global}) [9] or calculate Q_6 for each sphere (Q_6^{local}) [10]. In the rest of this paper, we focus on the Q_6^{local} and its distribution which allows to describe the local configurations of our systems.

The first step is the determination of the Q_6^{local} distribution, $p(Q_6^{local})$ for slightly disordered fcc and hcp lattices of atoms, i.e. ideal structures with random infinitesimal displacements of the atoms in order to remove the degeneracy of unstable vertices of the tessellation by adding a small face in place of each of them. This can be done theoretically following the work of Troadec et al. [11]. A Voronoï cell can then have a number f of faces between 12 and 18. For each configuration reported in [11], we have analytically determined the Q_{lm} of the bonds, then, by Eq. (2), we have calculated the value of Q_6^{local}. Assuming that there are no correlation between neighboring cells, we can also calculate the probabilities to have a f-faceted cell (see [11]). In the cases where there are several values of Q_6^{local} for one value of f we have also determined the theoretical fraction of each configuration. All these results are reported in [10], where the theoretical values have been compared with numerical simulations. We have plotted in Figure 2 the numerical distribution of Q_6^{local} for slightly disordered fcc (a) and hcp (b) packings of almost 15000 spheres.

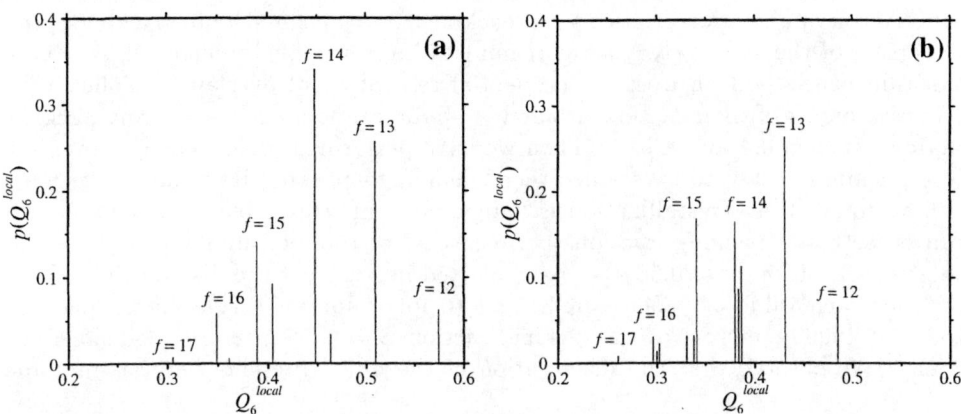

FIGURE 2. Distribution of Q_6^{local} for the slightly disordered fcc (a) and hcp (b) lattices. The values for $f = 18$ at $Q_6^{local} = 0.2651$ (fcc) and at $Q_6^{local} = 0.2237$ (hcp) are too rare to be visible.

Then we have studied the same distribution on packings built in the following manner. We start from ordered fcc or hcp packings diluted to a given packing fraction between 0.545 (melting point) and 0.7405. Then we perform molecular dynamics simulations [9] on them. We have reported in Figure 3 the distributions of Q_6 for each value of f between 12 and 16 for respectively (a) fcc and (b) hcp initial packings of almost 15000 spheres diluted to $C = 0.56$. We observe a widening

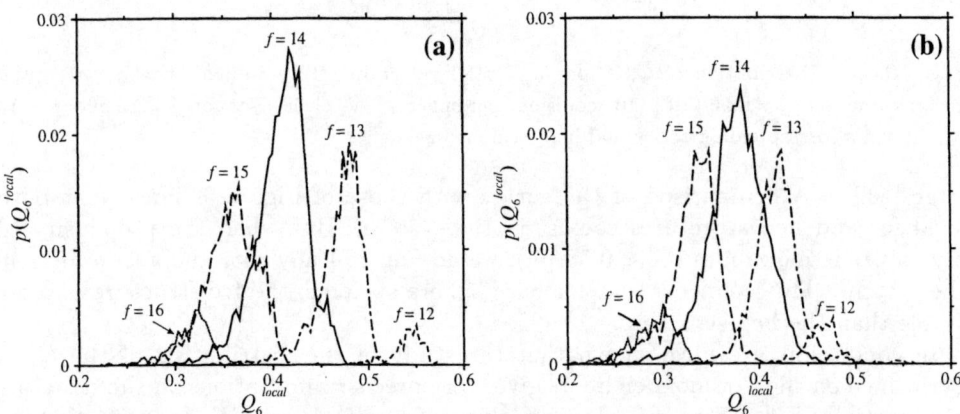

FIGURE 3. Distribution of Q_6^{local} after equilibration of an initial fcc (a) and an initial hcp (b) diluted to $C = 0.56$.

of the peaks with regard to the slightly disordered cases (on the left for the fcc and more symmetrically for the hcp). Due to this widening some details (for example the 3 peaks corresponding to $f = 14$ for the hcp) are no more separated.

The next step is to compare with the previous ones the distribution of Q_6^{local} during the crystallization of disordered packings. We have first built packings with the version of the Jodrey-Tory's algorithm [12] improved by Jullien et al. [7]. This algorithm consists in an iterative sequential resorption of overlaps of spheres by successive moves of their centers. It allows to build random packings at any packing fraction between 0.4 and C_{RCP}. Then we have performed molecular dynamics on these packings to try to crystallize them. For high packing fractions ($C > 0.6$), we have found that crystallization is not possible in reasonable times [9]. So we limit ourselves to packing fractions between that of the melting point and that of the glass transition $C \approx 0.58$. We have plotted in Figure 4 the distribution of the Q_6^{local} for each value of f between 12 and 16 for a Jodrey-Tory-Jullien's packing of almost 15000 spheres with a packing fraction $C = 0.56$ after 10^5 collisions per sphere. We observe that the distribution of the Q_6^{local} presents here again some

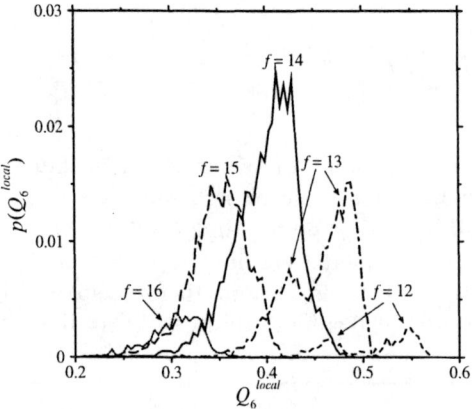

FIGURE 4. Distribution of Q_6^{local} for a crystallized Jodrey-Tory-Jullien's packing of initial packing fraction $C = 0.56$ (after 10^5 collisions per sphere). We clearly see for $f = 12$ and $f = 13$ two maxima corresponding to fcc and hcp local arrangements.

large peaks. A comparison of the curves with those of Figure 3 shows that both local fcc and hcp structures coexist in the system. But when the propensity to crystallize is maximum ($C \approx 0.58$ [9]), we do not find any more hcp structures in the system. This confirms that for hard sphere systems, the fcc structure is more stable than the hcp structure.

In conclusion, we have shown that the study of the distribution of the Q_6^{local} performed on the Voronoï neighbors gives very precise informations on the structure of a partially ordered or disordered packing of hard spheres. We have studied the crystallization of disordered packings and observed that the crystallized structure is a mixture of fcc and hcp clusters. The fraction of hcp clusters in the packing decreases to zero as the propensity of the packing to crystallize increases. This shows, as expected, that for hard sphere systems, the fcc structure is more stable than the hcp one. This method can also be used in the case of soft potentials. The next step of these studies is to understand how order appears and propagates in an initially disordered packing of hard spheres.

ACKNOWLEDGMENTS

We thank P. Jund for fruitful discussions about order and disorder in sphere packings.

REFERENCES

1. Alder B. J., and Wainwright T. E., *Jour. Chem. Phys.* **27**, 1208 (1957).
2. Wood W. W. and Jacobson J. D., *Jour. Chem. Phys.* **27**, 1207 (1957).
3. Hoover W. G. and Ree F. R., *Jour. Chem. Phys.* **49**, 3609 (1968).
4. Bernal J. D., *Nature* **183**, 141 (1959).
5. Finney J. L., *Proc. Roy. Soc. Lond. A.* **419**, 479 (1970).
6. Berryman J. G., *Phys. Rev. A* **27**, 1053 (1983).
7. Jullien R., Jund P., Caprion D. and Quitmann D. *Phys. Rev. E* **54**, 6035 (1996).
8. Steinhardt P. J., Nelson D. R. and Ronchetti M. *Phys. Rev. B* **28**, 784 (1983).
9. Richard P., Oger L., Troadec J. P. and Gervois A. *Submitted to Phys. Rev. E* (1999).
10. Richard P., Gervois A., Oger L. and Troadec J. P. *Submitted to Euro. Phys. Lett.* (1999).
11. Troadec J.P., Gervois A. and Oger L. *Europhys. Lett.* **42**, 167 (1998).
12. Jodrey W. S. and Tory E. M. *Phys. Rev. A.* **32**, 2347 (1985).

EPR, NMR and Molecular Dynamics in fluoride glasses

G.Silly*, J.Y. Buzaré*, B. Bureau*, C. Legein[†], B. Boulard[†]

*Laboratoire de Physique de l'Etat Condensé, UPRES-A CNRS n° 6087
[†]Laboratoire des Fluorures, UPRES-A CNRS n° 6010
Université du Maine, 72085 LE MANS Cedex 9

Abstract. This paper deals with the method of investigation of local order in Transition Metal Fluoride Glasses that has been developed in the past few years using magnetic resonance techniques associated to molecular dynamics calculations. The results presented here will concern lead fluoride PZG (PbF_2-ZnF_2-GaF_3) glasses of the PbF_2-$Mt^{II}F_2$-$Mt^{III}F_3$ family, but the method remains valid for any other disordered compound. The only limitations are that the sample may be doped with transition metal ions ($S \geq 1$) for EPR measurements or contain quadrupolar nuclei ($I \geq 1$) to perform NMR investigations.

INTRODUCTION

From previous structural investigations by EXAFS, Raman and diffraction techniques (1-5) the TMFG networks are usually described as mixed packing of F$^-$ ions and Pb^{2+} cations (due to the closeness of their ionic radii) with transition metal in octahedral holes. It is also commonly assumed that the network of these glasses is built up of chains of fluorine corner sharing octahedra $(ZnF_6)^{4-}$ and $(GaF_6)^{3-}$ with Pb^{2+} ions at the interstitial sites. But at this stage short range order remained not precisely known. NMR of $I=1/2$ nuclei ^{19}F and ^{207}Pb, have been performed and corroborate this model. On the one hand, ^{19}F studies (6,7) permitted to evidence the so called "shared fluorines" belonging to two octahedra, the "unshared fluorines" belonging to only one octahedron and the "free fluorines" which do not belong to any octahedron. As the proportion of each fluorine ions have been measured it was also possible to determine the degree of connectivity of the octahedron chains and the dimensionality of the network ranging between 1D and 3D according to the composition of the glass. On the other hand, ^{207}Pb NMR (8) permitted to investigate the interstitial sites and was able to give information on medium range order around them. In this paper we give details on the method used to obtain quantitative estimation of the fluorine octahedra distortion as seen from their center either by EPR of the probes Cr^{3+} and Fe^{3+} or by NMR of the quadrupolar nuclei $^{69,71}Ga$. Actually, the related experimental spectra give evidence for fine structure quadrupolar parameter distributions which mirror octahedron distortion distributions.

QUADRUPOLAR FINE STRUCTURE PARAMETER DISTRIBUTION

To determine the energy levels of an electronic or nuclear spin I≥1 submitted to a magnetic field, the hamiltonian that have to taken into account is the sum of the Zeeman contribution and of the quadrupolar contribution responsible of the zero field splitting. The general form of the quadrupolar fine structure spin hamiltonian, limited to second order, written in an arbitrary axis system and developed on an irreducible tensor spin operator basis (T_n^m) is :

$$H_Q = \left(A_2^0 T_2^0 + A_2^{-1} T_2^{-1} + A_2^1 T_2^1 + A_2^{-2} T_2^{-2} + A_2^2 T_2^2\right)$$

the A_2^m are the 5 symmetric fine structure tensor components which can be represented by a 3x3 traceless matrix. In its eigenframe the fine structure spin hamiltonian can be expressed with only two parameters :

$$H_Q = A_2^0 T_2^0 + A_2^2 T_2^2$$

In practice, each spectroscopist community has its own notations :
– In EPR the pertinent parameters are b_2^0 and b_2^2 (or b_2^0 and λ) ;

$$H_Q = \tfrac{1}{3} b_2^0 \left(O_2^0 + \lambda O_2^2\right)$$

with $\lambda = b_2^2 / b_2^0$ and O_n^m are the Steven's operators [9]
– In NMR the pertinent parameters are ν_Q and η_Q;

$$H_Q = \frac{h\nu_Q}{6\hbar^2}\left[(3I_z^2 - I^2) + \eta_Q(I_x^2 - I_y^2)\right]$$

with $\nu_Q = (1-\gamma_\infty)\dfrac{3eQV_{ZZ}}{2I(2I-1)h}$ and $\eta_Q = \dfrac{V_{XX} - V_{YY}}{V_{ZZ}}$; V_{XX}, V_{YY} and V_{ZZ} are the Electrical Field Gradient components, the Sternheimer's constant γ_∞ is a characteristic of the considered ion ($\gamma_\infty = -9.5$ for Ga^{3+}) and Q is the quadrupolar moment of the nucleus (Q = 0.178 10^{-28} m^2 for ^{71}Ga) [10].

QUANTIFICATION OF THE SHORT-RANGE ORDER IN TMFG USING MOLECULAR DYNAMICS

The TMFG experimental spectrum were very similar to the amorphous GaF$_3$ ones, either in NMR or in EPR, so this material has been used as a structural model to perform molecular dynamic calculations (9) in order to relate experimental results with atomic position distributions.

The initial set of atomic positions corresponded to crystalline rhombohedral GaF$_3$ and the box size was 4a x 4b x 2c with 768 atoms (Z = 192). The "temperature T" is the main parameter of the calculation. It fixes the total energy of the system. The higher it is, the faster the ions move and the larger the octahedron distortion. Atomic positions were calculated for several "temperature" values. As shown in figure 1, from the atomic positions, angular and radial distributions can be checked.

Electrical Field Gradient calculations have been undertaken in order to build the fine structure parameter distributions and to calculate the corresponding spectra.

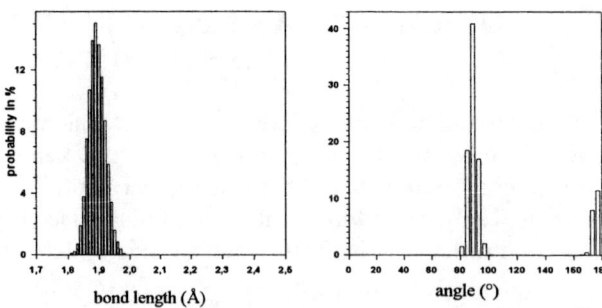

Figure 1 : radial Ga^{3+}-F^- bond length and angular F^--Ga^{3+}-F^- angle distributions in GaF_6 octahedra at "120 K".

The last step consisted to compare these spectra with the experimental one to estimate the degree of disorder in the material.

EPR study

EPR of Cr^{3+} and Fe^{3+} have been extensively studied (9, 11). The great sensitivity of the EPR spectra to octahedron distortions and calculation of the zero field splitting spin hamiltonian parameters using the superposition model (12) permitted to give quantitative results on GaF_6 octahedra distortions.

Previous studies (9, 11) have shown that the atomic positions generated at "120 K" give a good description of the system. In figure 2 are presented the corresponding fine structure parameter distribution and the experimental and calculated spectra for Cr^{3+}.

This led to standard deviations :
- Ga-F bond length :   0.021Å
- angle for adjacent Ga-F bonds : 2.4°
- angle for opposite Ga-F bonds : 4.3°

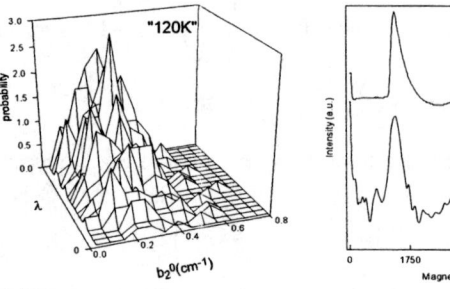

FIGURE 2 : fine structure parameter distribution and comparison of calculated and experimental EPR spectra of Cr^{3+} at "120K".

NMR study

NMR of $^{69,71}Ga$ indicates that the gallium atoms remain at the fluorine octahedron center since the isotropic chemical shift value is the same as in the crystalline phase

$(\delta_{iso} = -60\ \text{ppm})$. Quadrupolar parameter distribution is obtained through EFG calculation with a deformable point charge model. The F⁻ polarisability is an adjustable parameter.

Figure 3 shows the quadrupolar parameter distribution obtained at "120K" with 1.75 Å³ for F⁻ polarisability. The corresponding calculated spectrum fitted perfectly the experimental one. In fact due the large dispersion of the polarisability values (0.8 - 1.8 Å³) any "temperature" lying between "120K" and "500K" could be acceptable. Nevertheless, whatever the "temperature" the octahedra remain slightly distorted.

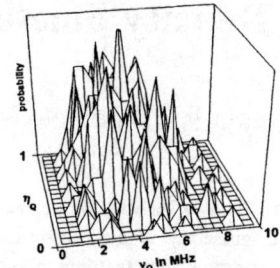 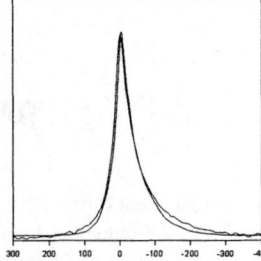

FIGURE 3: quadrupolar parameter distribution and comparison of calculated and experimental NMR spectra at "120 K".

CONCLUSION

NMR studies indicate that the network is built of $(GaF_6)^{3-}$ octahedra. The fine structure parameter distributions we obtained to reconstruct the NMR spectra of ^{69,71}Ga and the EPR spectra of Cr^{3+} and Fe^{3+} ions substituted in the TMFG networks, are similar to CZJZEK like distributions. In addition molecular dynamics and Electric Field Gradient calculations lead to small radial and angular distortions of the octahedra. In NMR, incertitude on adjustable parameters prevents us from accurate quantification of disorder on the contrary to EPR measurements.

REFERENCES

1. Dupas C, Le Dang K, Renard J P, Veillet P, Miranday J P, Jacoboni C, *J. Physique* **42**, 1345 (1981)
2. Lebail A, Jacoboni C, De Pape R, *Mater. Sci. Forum* **6**, 441 (1985)
3. Boulard B, Jacoboni C, Rousseau M, *J. Solid State Chem.* **80**, 17 (1989)
4. Lebail A, Jacoboni C, De Pape R, *J. Solid State Chem.* **48**, 168 (1983)
5. Lebail A, Jacoboni C, De Pape R, *J. Physique* **C8**, 163 (1985)
6. Bureau B, Silly G, Buzaré J.Y., Emery J, Legein C, Jacoboni C, *J.Phys.:Condens. Matter* **9**, 6719 (1997)
7. Bureau B, Silly G, Buzaré J.Y., Jacoboni C, accepted in *Journal of Non-Crystalline Solids*.
8. Bureau B, Silly G, Buzaré J.Y., accepted in *Solid State NMR*.
9. Legein C, Buzaré J.Y., Boulard B, Jacoboni C, *J. Phys : Condens. Matter* **7**, 4829 (1995)
10. Bureau B, Silly G, Buzaré J.Y., Legein C, Massiot D, *Solid State NMR*, in press.
11. Legein C, Buzaré J.Y., Emery J, Jacoboni C, *J. Phys. : Condens. Matter* **7**, 3853 (1995)
12. Newman D, *Adv. Phys.* **20**, 197 (1971).

Slow Dynamics of Supercooled Colloidal Fluids: Spatial Heterogeneities and Nonequilibrium Density Fluctuations

M. Tokuyama,[1] Y. Enomoto,[2] and I. Oppenheim[3]

[1] Statistical Physics Division, Tohwa Institute for Science, Tohwa University, Fukuoka 815, Japan.
[2] Graduate School of Engineering, Nagoya Institute of Technology, Nagoya 466, Japan.
[3] Department of Chemistry, Massachusetts Institute of Technology, Cambridge, MA 02139, U.S.A.

Abstract. The importance of the dynamic anomaly of the self-diffusion coefficient on the slow dynamics of density fluctuations near a colloidal glass transition ϕ_g is emphasized. That is, the self-diffusion coefficient $D_S(\Phi)$ becomes zero at ϕ_g as $D_S(\Phi) \sim D_0 |1 - \Phi(x,t)/\phi_g|^\gamma$ with an exponent γ, where $\Phi(x,t)$ is the local volume fraction of colloids and D_0 the single-particle diffusion constant. This dynamic anomaly results from the many-body correlations due to the long-range hydrodynamic interactions and can enhance even small initial disturbances near ϕ_g, leading to the long-lived, cluster-like glassy domains for intermediate times. Those spatial heterogeneities are responsible for the slow relaxation of nonequilibrium density fluctuations. Thus, the self-intermediate scattering function is shown to obey a two-step relaxation around the β-relaxation time $t_\beta \sim |1 - \phi/\phi_g|^{-1}$, and also to be well approximated by a Kohlrausch-Williams-Watts function with an exponent $\beta(z_0)$ around the α-relaxation time $t_\alpha \sim |1 - \phi/\phi_g|^{-\eta}$ with a power-law exponent $\eta(z_0) = \gamma/\beta(z_0)$, where ϕ is the particle volume fraction, and z_0 indicates the initial inhomogeneities. Finally, it is shown to obey an exponential decay around the time $t_L \sim |1 - \phi/\phi_g|^{-\gamma}$ with $\gamma = 2$ where $t_\beta \ll t_\alpha \ll t_L$. Those nonequilibrium effects would be observable as a waiting time effect.

INTRODUCTION

In recent years there has been growing interests in the study of the dynamics of concentrated hard-sphere colloidal suspensions[1,2]. This is mainly due to the experimental discovery[2] that those systems can exhibit a transition from a liquid phase to a glass phase, similar to that in glass-forming materials.

There are two types of theoretical approaches, depending on which initial state the system starts from, an equilibrium state or a nonequilibrium state. As long as the system is away from a critical point, the relative magnitude of the density fluctuations $\delta\phi(x,t)$ to the mean density $\Phi(x,t)$ is small even near the glass transition point in both initial states; $|\delta\phi/\Phi| \ll 1$. In the nonequilibrium system, therefore, the causal motion dominates the system and describes the spatial heterogeneities, obeying the nonlinear deterministic equation. Recently, we have shown that the long-lived spatial heterogeneities play an important role in the slow dynamics of hard-sphere suspensions near the glass transition, leading to the von Schweidler law and the Kohlrausch-Williams-Watts (KWW) formula (or a stretched exponential) of the self-intermediate scattering function[3]. Thus, those heterogeneities turn out to influence the dynamics of density fluctuations, and hence they might be observable as the waiting time effect in experiments and simulations. On the other hand, the dynamical properties of monodisperse hard-sphere suspensions in equilibrium systems near the colloidal glass transition have been investigated thoroughly experimentally[2,4], recovering long-known phenomena in glass-forming materials, such as the stretching of the α process and the von Schweidler law. On contrast to the nonequilibrium case, however, it would be difficult to observe the spatial heterogeneities experimentally in the equilibrium state. This is mainly because the equilibrium fluctuations can easily destroy those heterogeneities even though those are generated by nonlinear density fluctuations.

In order to discuss the spatial heterogeneities, therefore, Tokuyama has recently proposed a linear response-like theory based on the coupled diffusion equations for the average number density $\Phi(x,t)$ and the nonequilibrium density fluctuations $\delta\phi(x,t)$ around $\Phi(x,t)$[5]. This theory is applicable to a suspension in a nonequilibrium state before equilibration after the quench, where the initial state of the system is spatially nonuniform. Since $|\delta\phi/\Phi| \ll 1$, $\Phi(x,t)$ then describes the nonlinear deterministic behavior of heterogeneous structure, while $\delta\phi(x,t)$ describes a linear relaxation of the nonequilibrium density fluctuations on such nonuniform structure. In fact, the nonequilibrium effect has been observed as the waiting time dependence of the intermediate scattering function[6]. The most important feature of this theory is that the self-diffusion coefficient $D_S(\Phi)$ contained in the coupled diffusion equations becomes zero at the glass transition volume fraction ϕ_g as $D_S \sim D_0 |1 - \Phi(x,t)/\phi_g|^\gamma$ with the exponent $\gamma = 2$, where $\phi_g = (4/3)^3/(7\ln3 - 8\ln2 + 2) \simeq 0.57184 \cdots$.[7] As time goes on, therefore, $D_S(\Phi)$ becomes smaller and smaller near ϕ_g. When the system is initially in a nonequilibrium state, this theory can describe how even the small disturbances are enhanced by the dynamic anomaly near ϕ_g, leading to long-lived, cluster-like glassy domains with $\Phi(x,t) \geq \phi_g$, and influence the dynamics of the density fluctuations, leading to a two-step relaxation and a nonexponential decay.

MODEL

The dynamics of spatial heterogeneities of colloidal suspensions is described by the local volume fraction $\Phi(x,t)$. On the other hand, the dynamics of density fluctuations $\delta\phi(x,t)$ can be measured by dynamic light scattering through the intermediate scattering function[8] which is given by the Fourier transform, $F(k,t)$, of the autocorrelation function of the density fluctuations. For scattering vectors much larger than the maximum position k_m of the structure factor $S(k) = F(k,0)$, the scattering function $F(k,t)$ reduces to the self-intermediate scattering function $F_S(k,t)$, where $F_S(k,0) = 1$. Hence we start with the following coupled diffusion equations already described elsewhere:[5]

$$\frac{\partial}{\partial t}\Phi(x,t) = \nabla \cdot [D_S(\Phi(x,t))\nabla \Phi(x,t)], \quad (1)$$

$$\frac{\partial}{\partial t}F_S(k,t) = -k^2 \sum_q D_S(k-q,t)F_S(q,t) \quad (2)$$

with the Fourier transform, $D_S(k,t)$, of the self-diffusion coefficient

$$D_S(\Phi(x,t)) = D_S^S(\Phi)\frac{(1-9\Phi(x,t)/32)}{[1+(\Phi(x,t)D_S^S(\Phi)/\phi_g D_0)(1-\Phi(x,t)/\phi_g)^{-2}]}, \quad (3)$$

and the conservation law $(1/V)\int dx\, \Phi(x,t) = \phi$, where $D_S^S(\Phi)$ is the short-time self-diffusion coefficient (see Ref. 7 for details). Here the second singular term in the denominator of Eq.(3) results from the many-body correlations between particles due to the long-range hydrodynamic interactions, while the numerator of Eq.(3) results from the coupling between the direct and the short-range hydrodynamic interactions between particles. For short times $t \ll t_\gamma$, $D_S(\Phi)$ reduces to $D_S^S(\phi) \sim |\sigma|^0$ since the direct interactions and the correlations are negligible, while for long times $t \geq t_L$, it reduces to the long-time self-diffusion coefficient $D_S^L(\phi) = D_S(\phi) \sim |\sigma|^2$, where $t_\gamma \sim a^2/D_S^S$ and $t_L \sim a^2/D_S^L$ are the characteristic times of the short- and long-time self-diffusion processes, respectively, and $\sigma = \phi/\phi_g - 1$. Those coefficients, D_S^S and D_S^L, agree with the experimental data[4,9,10] well. Thus, there exists a crossover from the short-time process described by D_S^S to the long-time process described by D_S^L for intermediate times, where the dynamic anomaly of $D_S(\Phi)$ plays an important role.

RESULTS

We now discuss the numerical solutions of Eqs. (1) and (2), starting in a completely random configurations toward an equilibrium uniform configuration. In order to solve

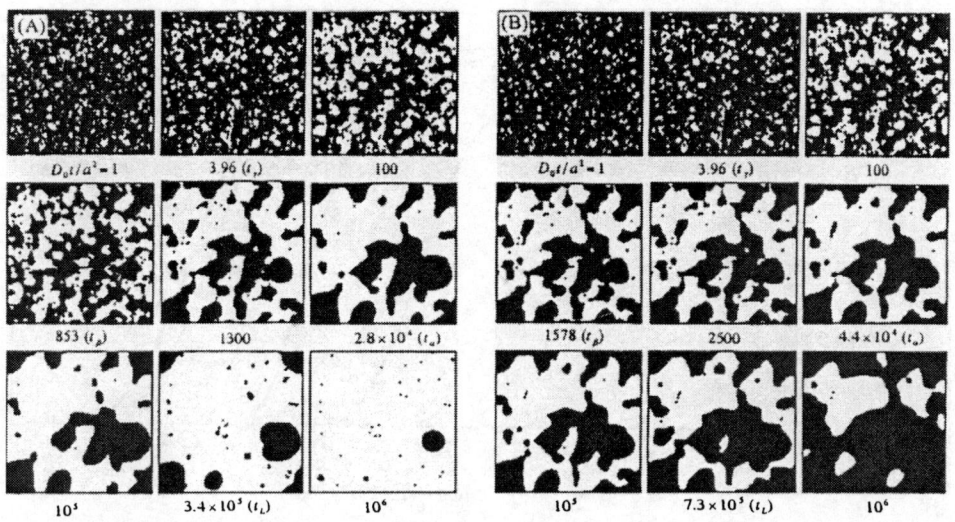

FIGURE 1. Typical configurations of glassy domains for (A) $\phi=0.571$ and (B) 0.573 at $z_0=0.8$.

them self-consistently, we fix the values of the two parameters, ϕ and z_0, as the initial conditions, where $z_0 = 1 - (1/V)\int dx\,|1 - \Phi(x,0)/\phi|$. Figure 1 shows a sequence of snapshots projected onto a plane of a typical configuration of the glassy phase at $z_0 = 0.8$ just below and just above the colloidal glass transition, where the glassy phase with $\Phi(x,t) \geq \phi_g$ is colored black. In Fig. 2 we show the time evolution of $F_S(k,t)$ at $z_0 = 0.8$ and 0.95 and $ka = 3$ for different volume fractions. There are four characteristic stages. The first is the early stage with $t \leq t_\gamma$, where the spatial inhomogeneities are described by $\Phi(x,t) = \exp[-tD_S^S(\phi)\nabla^2]\Phi(x,0)$, and the density fluctuations obey the short-time exponential decay, $F_S(k,t) = \exp[-k^2 D_S^S(\phi)t]$. After this stage, the dynamical behavior becomes complicated because of the anomalous property of $D_S(\Phi)$. For intermediate times $t_\gamma < t < t_\beta$, the glassy domains seem to be freezed near ϕ_g. After t_β, those domains start to be dissolved, disappearing very slowly in the supercooled region for $\phi_\beta \leq \phi < \phi_g$, where ϕ_β is a crossover volume fraction[11], over which the two-step relaxation appears. This is the so-called β stage, where $F_S(k,t)$ obeys two kinds of power-law decays with exponents $b_0(\phi,z_0,ka)$ and $b(\phi,z_0,ka)$ around t_β, where $b_0(0.571, 0.95, 3) = 0.3$ and $b(0.571, 0.95, 3) = 0.68$. These power-law decays continue up to t_α. The third is the α-relaxation stage with $t_\alpha \leq t \ll t_L$, where $\Phi(x,t)$ reaches nearly to the equilibrium value ϕ and only the rearrangement of domains occurs slowly. Hence one can assume in this stage that $\Phi(x,t)$ is scaled, near

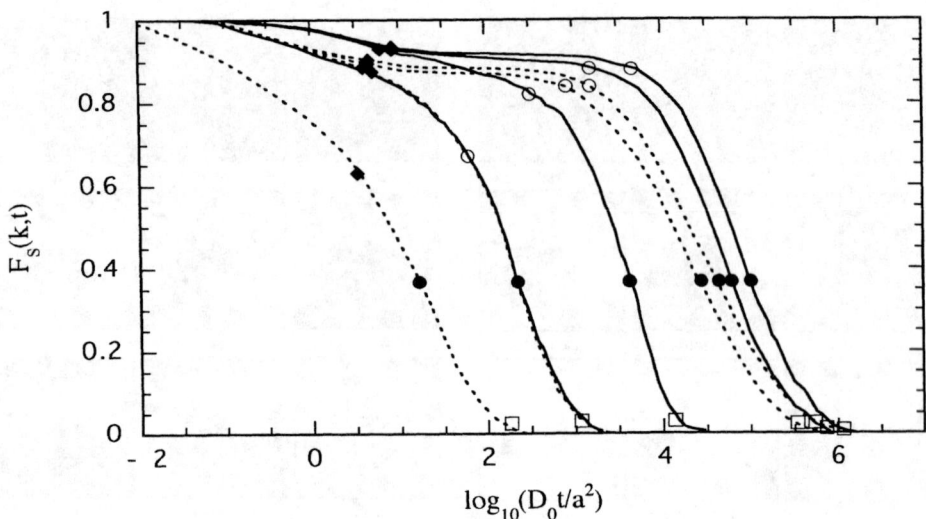

FIGURE 2. Self-intermediate scattering function $F_s(k,t)$ for $\phi = 0.543$, 0.559 (ϕ_β), 0.571, and 0.573 (dotted lines from left to right) at $z_0 = 0.8$, and $\phi = 0.559$, 0.565 (ϕ_β), 0.571, and 0.573 (solid lines from left to right) at $z_0 = 0.95$. The symbols indicate the time scales: t_γ (◆), t_β (○), t_α (●), and t_L (□).

ϕ_g, as

$$\Phi(x,t) = \phi[1 + Z(t^\mu x)], \tag{4}$$

where $\mu(z_0)$ is an exponent to be determined. Then, $F_s(k,t)$ is shown to obey the KWW function[3]

$$F_s(k,t) \propto \exp[-(t/t_\alpha)^\beta] \tag{5}$$

with the α relaxation time

$$t_\alpha \propto (k|\sigma|)^{-\eta}, \qquad \eta(z_0) = \gamma/\beta(z_0), \tag{6}$$

where $\eta(0.95) = 2.69$ ($\beta = 0.744$, $\mu = 0.086$) and $\eta(0.8) = 3.65$ ($\beta = 0.548$, $\mu = 0.15$). Thus, the KWW formula can be explained by the existence of long-lived, glassy domains. This stretched behavior continues up to t_L. The last is the late stage with $t \geq t_L$, where we have $\Phi(x,t) = \phi$, and the fluctuations obey the long-time exponential decay $F_s(k,t) = \exp[-k^2 D_s^L(\phi)t]$. As is seen from Fig. 2, below ϕ_β, $F_s(k,t)$ decays quickly to zero, obeying the simple exponential decay. On the other hand, above ϕ_β, the shape of $F_s(k,t)$ becomes very sensitive to the value of ϕ, forming a shoulder with the height $F_s(k,t\to\infty;\sigma=0)=f_k^c(z_0)$. With increasing volume fraction at a fixed z_0, we thus observe a progression from normal colloidal fluid ($0<\phi<\phi_\beta$), to supercooled colloidal fluid ($\phi_\beta \le \phi < \phi_g$), and to glass ($\phi \ge \phi_g$).

CONCLUSIONS

The main results reported here are as follows. (i) The existence of long-lived, irregularly shaped glassy domains with $\Phi(x,t) \ge \phi_g$ in the supercooled region $\phi \ge \phi_\beta$. (ii) Four different types of relaxations of nonequilibrium density fluctuations in the supercooled region $\phi \ge \phi_\beta$. Thus, we have shown that the long-lived, heterogeneous structure, which results from the dynamic anomaly of the self-diffusion coefficient, does cause the two-step relaxation and the stretched behavior. Finally, we mention that even above ϕ_g $F_s(k,t)$ decays to zero near ϕ_g, obeying the four relaxation processes (see Fig. 2). On the other hand, in the normal region $0<\phi<\phi_\beta$ there is no anomalous behavior.

ACKNOWLEDGEMENT

This work was supported by the Tohwa Institute for Science, Tohwa University.

REFERENCES

1. Pusey, P. N., "Colloidal suspensions", in *Liquids, Freezing and the Glass Transition*, edited by D. Levesque, J. P. Hansen, and J. Zinn-Justin (Elsevier, Amsterdam, 1991).
2. Pusey, P. N., and van Megen, W., Nature **320**, 340 (1986).
3. Tokuyama, M., Enomoto, Y., and Oppenheim, I., Physica A **270**, 165 (1999).
4. van Megen, W., and Underwood, S. M., J. Chem. Phys. **91**, 552 (1989); Phys. Rev. E **49**, 4206 (1994).
5. Tokuyama, M., Physica A **229**, 36 (1996); Phys. Rev. E **54**, R1062 (1996).
6. van Megen, W., Mortensen, T. C., Williams, S. R., and Müller, J., Phys. Rev. E**58**, 73 (1998). Mortensen, T. C. and van Megen, W., in *Slow Dynamics in Complex Systems*, eds. M. Tokuyama and I. Oppenheim (AIP, New York, 1999).
7. Tokuyama, M., and Oppenheim, I., Physica A **216**, 85-119 (1995).
8. Berne, B. J., and Pecora, R., *Dynamic Light Scattering*, New York: Wiley, 1976.
9. Segrè, P. N., Behrend, O. P., and Pusey, P. N., Phys. Rev. E **52**, 5070 (1995).
10. van Megen, W., Underwood, S. M., Müller, J., Mortensen, T. C., Henderson, S. I., Harland, J. L., and Francis, P., Prog. Theor. Phys. Suppl. 126, 171 (1997).
11. Tokuyama, M., Prog. Theor. Phys. Suppl., No. 126, 43 (1997).

Author Index

A

Affouard, F., 217
Alba-Simionesco, C., 243

B

Barrat, J.-L., 47
Benoit, M., 222
Binder, K., 68
Bordat, P., 227
Bouchaud, E., 231
Bouchaud, J.-P., 63
Boulard, B., 264
Brown, R., 227
Buchenau, U., 3
Bureau, B., 264
Buzaré, J. Y., 264

C

Caprion, D., 129

D

Descamps, M., 217

E

Elliott, S. R., 171
Enomoto, Y., 268

G

Gale, J., 237
Gaskell, P. H., 38
Gastreich, M., 237
Gervois, A., 259
Gleim, T., 68

I

Ispas, S., 222

J

Jullien, R., 129, 222
Jund, P., 129, 222

K

Kalia, R. K., 149
Kivelson, D., 83
Kob, W., 68
Krakoviack, V., 243
Kustanovich, T., 248

L

Legein, C., 264
Levelut, C., 24

M

Marian, C. M., 237
Micoulaut, M., 253

N

Nakano, A., 149

O

Oger, L., 259
Olami, Z., 248
Oppenheim, I., 268

P

Pelous, J., 24

R

Richard, P., 259
Rivier, N., 119

S

Sadoc, J.-F., 105, 129
Schober, H. R., 191
Silly, G., 264

T

Taraskin, S. N., 171
Tarjus, G., 83

Teichler, H., 203
Terki, F., 24
Tokuyama, M., 268
Troadec, J.-P., 259

V

van Brutzel, L., 231
Vashishta, P., 149